Techniques and System Design of Radar Active Jamming

Guangfu Tang · Yifeng Cai · Rongbing Gan · Yaodong Zhao

Techniques and System Design of Radar Active Jamming

国防工業出版社
National Defense Industry Press

Springer

Guangfu Tang
Xihua University
Chengdu, Sichuan, China

Rongbing Gan
Southwest China Research Institute
of Electronic Equipment
Chengdu, Sichuan, China

Yifeng Cai
Southwest China Research Institute
of Electronic Equipment
Chengdu, Sichuan, China

Yaodong Zhao
Southwest China Research Institute
of Electronic Equipment
Chengdu, Sichuan, China

ISBN 978-981-19-9943-7 ISBN 978-981-19-9944-4 (eBook)
https://doi.org/10.1007/978-981-19-9944-4

Jointly published with National Defense Industry Press
The print edition is not for sale in China (Mainland). Customers from China (Mainland) please order the print book from: National Defense Industry Press.
ISBN of the Co-Publisher's edition: 978-711-81-1949-7

This Springer imprint is published by the registered company Springer Nature Singapore Pte Ltd.
The registered company address is: 152 Beach Road, #21-01/04 Gateway East, Singapore 189721, Singapore

Preface

In 2014, Dr. Laizhao Hu proclaimed good news to everyone that academician Xiaoniu Yang and researcher Jian Meng were leading the writing of a series of books on electronic warfare. Upon hearing this news, we were all much excited. For a long time, electronic warfare-related books were rather fragmented, and there was a lack of systematic introduction of electronic warfare-related technologies. We felt honored to be involved in the writing of this series. But at the same time, we also felt pressured because it is not easy to write a good book and much of the content was previously lacking in refinement, lest there be something inappropriate to mislead readers and bring negative impact to the industry. Eventually, with the encouragement of our colleagues, we started, and on top of that, we had a large intellectual collective that provided suggestions for the book to be better completed.

Before writing this book, we summarized the existing electronic warfare-related books, especially those related to radar countermeasures. The previous books mainly introduced the basics of electronic warfare, such as *EW 101: A First Course in Electronic Warfare* by David Adamy, translated by Wang Yan; *Principles of Radar Countermeasures* by Prof. Zhao Guoqing, and *Introduction to Modern Electronic Warfare Systems* by Andrea De Martino, translated by Jiang Daoan, etc., which provided a good explanation of the basics of electronic warfare but lacked guidance for practitioners on how to design jamming systems. Therefore, considering that the content of the book should be as rich as possible, and adding the author's actual research field, we chose the research content of the book in the category of "radar active jamming". Finally, we decided the title of the book as *Techniques and System Design of Radar Active Jamming*.

In the rewriting process, we have referred to several previously published books on electronic countermeasures, consulted experts in specialized fields such as antennas, receivers, jamming sources, transmitters, and system control, and combined our understanding from practical work. From their perspective, designers or ongoing ones, this book seeks to know about radar active jamming techniques and system design. Positioned as a guide for designers of radar active jamming systems, the book is divided into ten chapters starting from basic concepts and focusing on the description of requirements analysis, key indicators, and design methods for radar jamming system design and each subsystem design.

Chapter 1 introduces the concepts related to radar jamming. Based on the basic concepts, combined with the author's understanding, the author analyzes the related concepts, introduces the typical applications of radar active jamming systems, as well as briefly reviews the development of radar jamming.

Chapter 2 introduces the principle of radar active jamming technology. The author introduces the principle of radar detection and radar jamming, gives a mathematical model description of multi-false target, and summarizes the latest radar anti-jamming and jamming technologies.

Chapter 3 mainly focuses on the overall design method of radar active jamming system, which includes the basic concept of radar jamming system, design process, design requirement analysis, overall scheme design, design of key technical indicators and engineering design principles, etc. Finally, it introduces the typical radar active jamming systems and its development trends.

Chapters 4–8 respectively introduce the design of antennas, receivers, jamming sources, transmitters and system control extensions, and introduce their requirements analysis, design, index calculations, development trends, etc.

Chapter 9 introduces the relevant theories and methods of radar jamming effect evaluation which mainly includes the role of jamming effect evaluation, evaluation criteria, evaluation indicators along with evaluation methods.

Chapter 10 introduces the frontier technology of radar and radar active jamming. The author introduces the development of radar technology and the development of radar jamming technology, respectively.

Researcher Gan Rongbing planned and arranged the content of the book and wrote Chaps. 1, 5, and 8. Senior engineer Zhao Yaodong is responsible for the draft of this book and wrote Chaps. 3 and 6. Dr. Tang Guangfu from Xihua University wrote Chaps. 2, 9, and 10. Senior engineer Lan Zhu wrote Chap. 4. Professor Guan Zhaohui wrote Chap. 7.

The writing of this book was carried out under the unified deployment of academician Yang Xiaoniu and researcher Meng Jian. During the writing, we have received guidance from senior experts in electronic warfare, academician Zhang Xixiang, Dr. Hu Laizhao, and researcher Yi Zhenghong, receiver technology expert He Weiguo, antenna technology expert Tang Yimin, senior engineer Hou Lingli, signal processing expert Wang Dazhao as well as other experts, and we have received the enthusiastic help from leaders and peers such as editor Wang Xiaoguang of National Defense Science and Technology Press and Hua Yun, Gu Jie, He Juncen, Liu Jiang, Xiao Xia of Key Laboratory of Electronic Information Control. The content of the book involves some of the research achievements of colleagues such as Zheng Kun, Chang Jindan, An Hong, Yang Li, Gao Youbing, Shi Xiaowei, Wu Guangjie, Xu Wang as well as others. Master Yuan Rufang collected and processed a large amount of information during the writing of this book. Thank them for their help in writing and publishing this book.

Due to the limited level and experience of the authors, there are inevitably some improprieties or errors in the book, and relevant experts and readers are welcome to criticize and correct.

Chengdu, China
May 2019

Guangfu Tang
Yifeng Cai
Rongbing Gan
Yaodong Zhao

Translators' Note

In January 2020, this book was first published in China and was given a high evaluation by peers in the field of radar active jamming. To make this book a reference for more peers around the world, the writing group of this book decided to translate this book into English for publication.

Dr. Guangfu Tang from Xihua University, as one of the authors of the original book, translated Chaps. 2, 3, 4, 6, 7, 9, and 10, as well as drafted the translation of the entire book. Dr. Yi feng Cai translated Chaps. 1, 5, and 8. The authors of the original book, Dr. Rongbing Gan and Dr. Yaodong Zhao, have provided many valuable suggestions and guidance on the translation of this book. Two graduate students of Xihua University, Zhu Shu and Zhou Yunchun, have also provided a lot of help in the translation of the book.

Due to the limited level of translators, there are inevitably some improprieties in this book. We urge peers to criticize and correct.

Translators
September 2022

Contents

Abbreviations

AD, A/D, ADC	Analog-to-digital converter
AESA	Active electronically scanned array
AGC	Automatic gain control
AM	Amplitude modulation
AOA	Angle of arrival
B, BW	Bandwidth
BIT	Build-in test
c	Speed of light = 3×10^8 m/s
CFAR	Constant false alarm rate
CMA	Covariance matrix adaptation
CMOS	Complex metal oxide semiconductor
CPU	Central processing unit
CR	Cognitive radar
CVR	Crystal video receiver
DA, D/A, DAC	Digital-to-analog converter
DAR	Digital array radar
DARPA	Defense Advanced Research Projects Agency
dB	Decibel
DBF	Digital beam forming
dBm	Decibel referenced to the power of one milliwatt
DDC	Digital down convert
DDS	Direct digital synthesizers
DOA	Direction of arrival
DPCA	Displaced phase center antenna
DRFM	Digital radio frequency memory
DSP	Digital signal processor
DTR	Digital transmit receive
E	Electric field strength, energy
EA	Electronic attack
ECCM	Electronic counter countermeasures
ECM	Electronic countermeasures

ELINT	Electronic intelligence
EMC	Electromagnetic spectrum control
EMI	Electromagnetic jamming
EP	Electronic protection
ERP	Effective radiated power
ES	Electronic support
ESPRIT	Estimating signal parameters via rotational invariance techniques
EW	Electronic warfare
FFT	Fast Fourier transform
FIR	Finite impulse response
FM	Frequency modulation
FPGA	Field-programmable gate array
FSK	Frequency shift keying
GaAs	Gallium arsenide
GaN	Gallium nitride
GHz	Giga hertz
GMTI	Ground moving target indicator
GO	Geometrical optics
GPS	Global positioning system
HF	High frequency (3–30 MHz)
HOJ	Home-on-Jam
Hz	Hertz (cycles per second)
IBW	Information band width
IEEE	Institute of Electrical and Electronic Engineers
IEWS	Integrated electronic warfare system
ISAR	In-verse synthetic aperture radar
JSR	Jamming-to-signal ratio
K	Kelvin
k	Kilo (10^3) or Boltzmann constant
Km	Kilometer
LDMOS	Laterally diffused metal oxide semiconductor
LFM	Linear frequency modulation
LMS	Least mean square
LPF	Low-pass filter
LPI	Low probability of intercept
LVDS	Low-voltage differential signaling
m	Milli (10^{-3}), meter, or electron mass
MEMS	Micro-electronic mechanical system
MHz	Mega hertz (10^6 Hz)
MIMO	Multiple input multiple output
MLFMM	Multilevel fast multipole method
MMCM	Microwave multi-chip modules
MoM	Method of moment
MOP	Modulation on pulse
MTD	Moving target detection

MTI	Moving target indicator
MUSIC	Multiple signal classification
MW	Mega watt
NCO	Numerically controled oscillator
NGJ	Next-generation jammer
NLFM	Nonlinear frequency modulation
NLMS	Normalized least mean square
ns	Nanosecond
OODA	Observe Orient Decide Act
PA	Pulse amplitude
PC	Pulse compression
PD	Pulse Doppler
PDW	Pulse descriptor word
PESA	Passive electronically scanned array
PLL	Phase locking loop
PO	Physical optics
PRF	Pulse repetition frequency
PRI	Pulse repetition interval
PW	Pulse width
R	Range
Radar	Radio detection and ranging
RAT	Ram-air turbo-generator
RCS	Radar cross section
RF	Radio frequency
RLS	Recursive least square
ROM	Read only memory
RWR	Radar warning receiver
s, S, or sec	Seconds
SA	Helical antenna
SAR	Synthetic aperture radar
SDLVA	Successive detection log video amplifier
SiP	System in package
SLB	Sidelobe blanking
SLC	Sidelobe cancelation
SMA	Sub-miniature A connector
SoC	System-on-Chip
STAP	Space–time adaptive processing
T	Time (also t), temperature
T/R	Transmit/receive
TG	Technical generator
THz	Tera hertz
TOA	Time of arrival
TSV	Through silicon via
TTNT	Tactical targeting network technology
TWT	Traveling-wave tube

UHF	Ultra-high frequency (300 MHz–3 GHz)
UTD	Uniform geometrical theory of diffraction
UWB	Ultra-wide band
VCO	Voltage controled oscillator
VSWR, SWR	Voltage standing wave ratio
W	Watt
WBDF	Wide band Dicke fix
YIG	Yttrium iron garnet

Chapter 1
Summary of Radar Jamming

It has been over 70 years since radar jamming emerged as a countermeasure to the radar used for military purposes. During this period, radar jamming from birth to development has become an indispensable force in modern warfare.

As the fundamental, essential electronic warfare (EW) concepts will be introduced in this chapter. In this chapter, we introduce the basic concepts of EW and radar jamming, including the most recent proposals; in Chap. 2, the effect of radar jamming in military battle is given; in Chap. 3, we briefly introduce the development history of radar jamming.

1.1 Basic Concept

1.1.1 Basic Concept of Electronic Warfare

Radar jamming falls into the category of EW. We first briefly introduce the basic concepts of EW before discussing radar jamming.

The general definition of EW is the tactics and technologies used to ensure friendly unimpeded access and to deny the opponent's taking the advantage of electromagnetic (EM) spectrum [1]. This concept keeps EW within the bound of electromagnetic and defines from the aspect of the achieved effect (to ensure friendly unimpeded access, and to deny the opponent's taking the advantage of electromagnetic spectrum) but does not restrict the types of strategy and technology.

In Joint Publication 3–13.1: Electronic Warfare, the term EW refers to military action involving the use of EM energy and directed energy (DE) to control the electromagnetic spectrum (EMS) or to attack the enemy. The definition above originated from operational principles and emphasizes military action. The methods that can be used are restricted to EM energy and DE. The purpose of EW is to control EMS or to attack the enemy, where attacking the enemy is taken into consideration besides controlling the EMS. For example, it can destroy hostile facilities or kill enemy

G. Tang et al., *Techniques and System Design of Radar Active Jamming*,
https://doi.org/10.1007/978-981-19-9944-4_1

personnel by employing high-power microwave (HPM), high-energy laser weapons, and so on.

In Integrated Electronic Warfare [2], EW is defined as all kinds of tactics, technologies, and actions employed in order to determine, disturb, weaken, break, and destroy the enemy's electronic information systems and electronic devices, as well as to protect the electronic information systems and electronic devices of our own side, by EW energy, directed energy, and underwater acoustic energy.

The definitions above differ in description methods and points of view. Now, we summarize the definitions from the perspectives of scope, means, and purposes.

Scope: The maximum scope consists of EM energy, directed energy, and underwater acoustic energy, while the minimum scope only contains EW energy. Since the utilization of underwater acoustic and underwater acoustic warfare both utilize electronic systems to process data, it is reasonable to put underwater acoustic energy into the scope of EW. Directed energy such as high-power microwave (HPM) and high-energy laser is within the range of EM energy, and therefore, EW contains both EM energy and underwater acoustic energy from the aspect of scope.

Means: There is no description of particular EW means in the definitions. Regular EW means contain reconnaissance, jamming, anti-radiation attack, and electronic protection.

Purposes: The widest extension of the definitions is to control the EWS or to attack the enemy, in which the actions in order to control the EWS and the actions of attacking the enemy by the EWS are both included.

EW contains two aspects that struggle with each other, that is, electronic countermeasure (ECM, e.g., electronic reconnaissance, electronic jamming, electronic stealth, and electronic destruction) and electronic counter countermeasure (ECCM, e.g., anti-reconnaissance, anti-jamming, anti-stealth, and anti-destruction) [3].

In the traditional concept, EW is divided into three parts, i.e., electronic support (ES), electronic attack (EA), and electronic protection (EP). In 2010, US Lieutenant General Robert J. Elder proposed the definition of electromagnetic spectrum control (EMC) and extended the concept of EW [4], as shown in Fig. 1.1.

Electronic warfare support (ES) is the actions tasked by, or under the direct control of, an operational commander to search for, intercept, identify, and locate or localize sources of intentional and unintentional radiated electromagnetic energy.

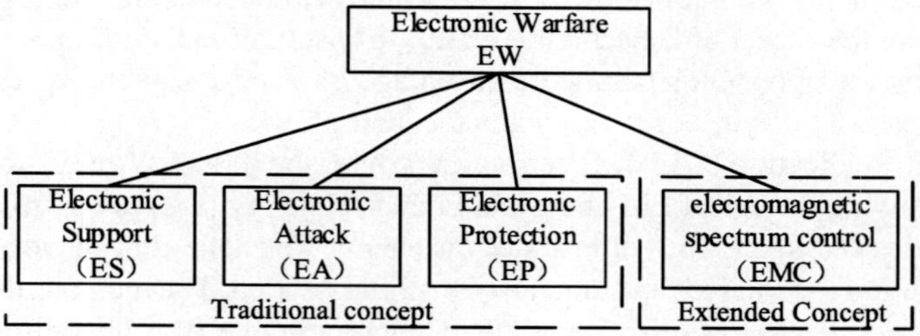

Fig. 1.1 Components of EW

The purpose of ES is to identify the threat, aim at the target, plan, and conduct future operations. ES data can produce signal intelligence, which provides target indication for the immediate electronic attack or destructive attack on one hand, and enrich the target database used for target feature studies that provide priori information for the upcoming EW battles.

Electronic attack (EA) utilizes EM energy, directed energy, or anti-radiation weapons to attack personnel, facilities, or devices in order to degrade, neutralize, or destroy the operational capability of the enemy. EA contains the actions in order to prevent or degrade the effective utilization of EMS by the enemy, such as jamming, electronic deception, and the weapons that employ EM energy or directed energy as their main destructive equipment (e.g., laser, radio frequency weapon, particle beams).

Electronic protection (EP) is the actions taken to protect the personnel, facilities, and devices of our own side from any effects of friendly or enemy use of the EMS. The potential effects contain degrading, neutralizing, or destroying friendly combat capability. The measures of EP include spectrum management, electromagnetic reinforcement, radiation control, and wartime preparation mode.

Electromagnetic spectrum control (EMC) is realized by effective management and coordination of friend systems to counter enemy systems. In fact, EMC integrates ES, EA, and EP to achieve its goal of controlling the EMS, that is, to ensure friendly use of EMS while effectively stopping the use of EMS by the enemy.

Electronic countermeasure consists of radio frequency (RF) countermeasure, electro-optical (EO) countermeasure, and underwater acoustic countermeasure based on the spectrum used. RF countermeasure includes radar countermeasure, communication countermeasure, navigation countermeasure, and IFF countermeasure. We do not explore it in this book, since detailed information can be found in plenty of books related to electronic countermeasure.

1.1.2 Basic Concept of Radar Jamming

Radar jamming disturbs, weakens, and breaks the normal work of the radar systems of the enemy by all means, so that the enemy radars cannot detect and track real targets accurately.

Radar jamming can be divided into two types based on the effect which are suppressing jamming and deception jamming.

(1) Suppressing jamming

Suppressing jamming is also called cover jamming [5]. The predicted effect is to make the enemy radar fail to detect the target or to degrade the capability of detection of the enemy radar. Suppressing jamming consists of active suppressing jamming and passive suppressing jamming. Passive suppressing jamming affects the target detection and measurement of the radar by the clutters produced by the chaff ejected in the air. Active suppressing jamming generates high-power jamming signals by active devices. The jamming signal is received by the radar and affects its target detection and measurement. Common active suppressing jamming contains noise suppressing jamming and dense false target jamming. Noise suppressing jamming lowers the signal-to-noise ratio (SNR) of the radar receiver by noise jamming signal, thereby degrading the target detection capability of radar. Dense false target jamming produces a large number of dense false targets; hence, the radar detects more targets than it can process, or the operator cannot find the real target.

(2) Deception jamming

Deception jamming makes the radar system get the wrong target information after processing through emitting, retransmitting, or reflecting electromagnetic waves to generate signals similar to the target echo. Based on the content of the generated false information, deception jamming can be divided into four parts which are range deception, velocity deception, angle deception, and false target deception. Range deception makes radar generate incorrect range information of the target, which is usually realized by the range gate pull-off technique. Velocity deception makes radar generate incorrect velocity information of the target, which is usually realized by the velocity gate pull-off technique. Angle deception makes radar generate incorrect angle information of the target, which is usually realized by inverse gain jamming for conical scan radar and cross-eye jamming or cross-polarization jamming for single pulse radar. False target deception adds fake targets into the radar detection results to affect troop deployment and employment. False target deception is realized by sophisticated modulation in range, angle, velocity, and energy dimension for the jamming signal that is generated and similar to the target echo.

Based on the signal generation method, radar jamming can be divided into two types, which are active jamming and passive jamming.

(1) Active Jamming

Active jamming breaks and weakens the detection and track capabilities of enemy radar by actively emitting or transmitting electromagnetic waves. Typical active jammer includes noise jammer, repeater jammer, etc.

(2) Passive Jamming

Passive jamming utilizes devices that do not emit electromagnetic waves themselves to reflect or absorb electromagnetic waves to break and weaken the detection and track capabilities of enemy radar. Typical passive jammer includes angle reflector, passive decoy, chaff, etc.

1.2 Role of Radar Jamming in the Military Struggle

Radar jamming achieves the goal of breaking the capture of air intelligence, the indication of targets, and the guidance of missile through disturbing, weakening, and breaking the radar systems of the enemy. As a result, radar jamming can protect our operational systems and improve strike capability. In this subsection, we introduce the effect of radar jamming in military struggles from the aspects of air force penetration, ship defense, and key position air defense.

1.2.1 *Role in the Air Force Penetration*

The typical usage of radar jamming in air force penetration includes stand-off jamming, escort jamming, and self-protection jamming [6].

Stand-off Jamming: The electronic jamming aircraft jams the early warning radars, thereby weakening and breaking the target detection capability of the enemy's early warning radars in the area that our troop is going to penetrate so that the enemy cannot organize effective air or ground defense actions in advance. The scenario of stand-off jamming is shown in Fig. 1.2. The figure only shows the situation of jamming ground early warning radar, while the early warning aircraft Radar and reconnaissance satellite radar also can be the targets of stand-off jamming. In the past, stand-off jamming is realized by dispensing chaff over a large region, and now most stand-off jamming aircrafts have the capability of high-power active jamming. A typical example of stand-off jamming aircraft is the EA-6B Prowler aircraft of the US navy.

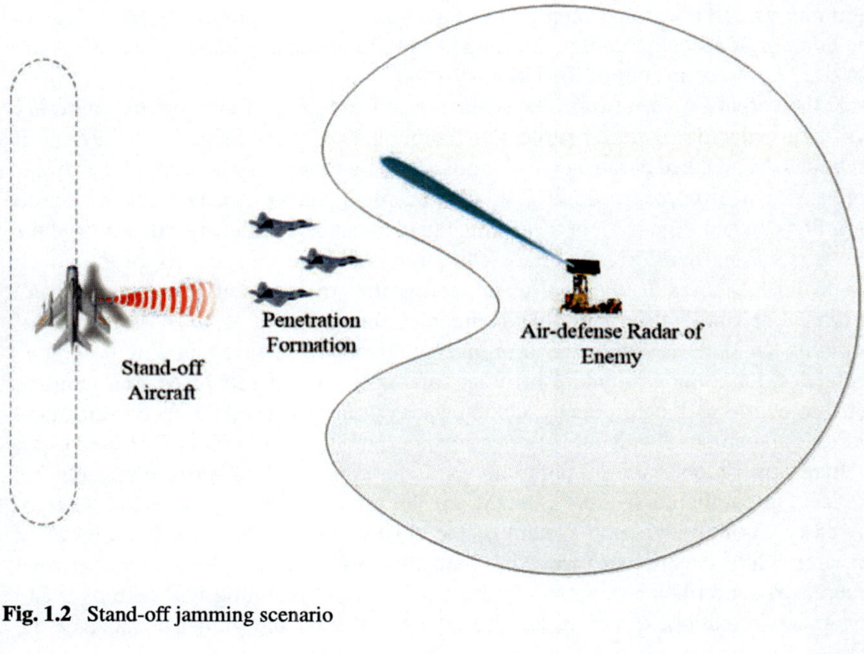

Fig. 1.2 Stand-off jamming scenario

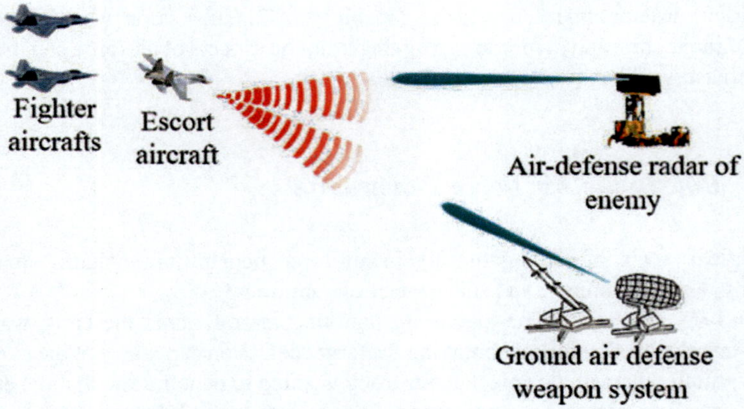

Fig. 1.3 Escort jamming scenario

Escort Jamming: The electronic jamming aircraft and the strike aircraft enter the combat airspace in formation. The strike aircraft takes the charge of the attack, and the electronic jamming aircraft jams the enemy's early warning and fire control radars so that our penetration formation would not be precisely located, tracked, or attacked. A typical escort jamming scenario is shown in Fig. 1.3. The scenario of jamming ground early warning radar and fire control radar is given in Fig. 1.3. The jamming of early warning aircraft radar and airborne fire control radar is also in the domain of escort jamming. A typical example of escort jamming aircraft is the EA-18G Growler aircraft of the US navy.

Self-protection Jamming: The strike aircraft is equipped with a radar jamming system in order to realize self-protection jamming. During the fight, if a strike aircraft is about to be attacked by surface-to-air missile or air-to-air missile, it could jam the air defense radar, airborne fire control radar, or terminal guide radar. The self-protection jamming can make the radar of the enemy fail to track the jammer aircraft or introduce track error, and therefore, the attack condition could not be satisfied, or the attack would fail. As a result, the goal of protecting the strike aircraft itself is achieved. A typical self-protection jamming scenario is shown in Fig. 1.4. The scenarios of jamming air defense radar and terminal guide radar are given in Fig. 1.4. Self-protection jamming equipment includes internally carried self-protection jammer, self-protection pod, chaff, decoy, etc. Nowadays, the advanced strike aircraft in the world is all equipped with a self-protection jamming system, such as F-16, and F-35.

In real operation, stand-off jamming, escort jamming, and self-protection jamming are used in combination. That is, stand-off jamming is used to reduce the range of the early warning detection system of the enemy and covers the strike aircraft or missiles while they move forward. When the strike aircraft fleet enters the area where the stand-off jamming cannot cover, the escort jamming will take over and cover the formation. If our strike aircraft is locked by the guidance radar of the

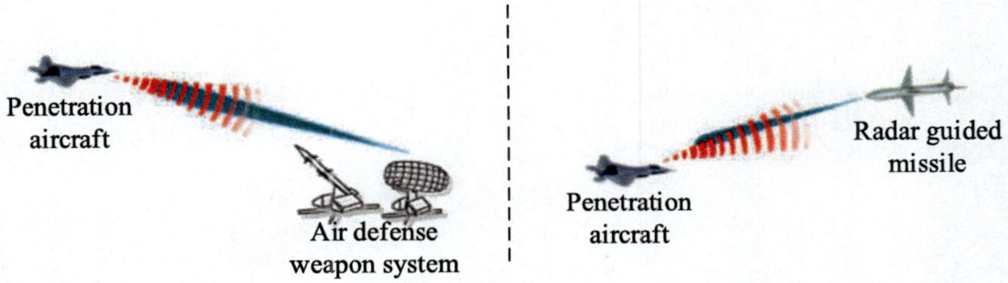

Fig. 1.4 Self-protection jamming scenario

enemy, the self-protection system would make a response immediately, using the self-protection jammer, chaff or decoy to protect the aircraft.

1.2.2 Role in the Warship Defense

In ship defense, the defense measures for radar-guided anti-ship weapons contain active jamming, angle reflector, and chaff. Since this book focuses on active radar jamming, only the effect of active radar jamming in ship defense will be discussed. There are mainly two types of active radar jamming in ship defense, which are off-board active jamming and shipborne active jamming.

(1) Off-board Active Jamming

Off-board active jamming locates in some special area near the ship and actively emits jamming signals through receiving the signal emitted by the terminal guide radar. Therefore, it makes the anti-ship missile cannot hit the ship and the goal of protecting the ship is achieved. The exact operation application methods of off-board active jamming include hovering, towed, and floating. Figure 1.5 gives the operation application scenario of off-board active jamming verse anti-ship missiles. The effect of off-board active jamming is to inveigle the anti-ship missile into deviating from the ship and attacking another area, the centroid of the jammer and the ship, the jammer, etc. Hence, the ship is protected from attack. The active radar jamming system Nulka, which is jointly developed by the US and Australia, exploits hover rocket and autonomous control techniques and is equipped in a great number of ships of the US, Canada, and Australia. The off-board active decoy such as the C-Gem system developed by Israel, the Eager and FLYRT systems developed by the US Naval Research Laboratory, make use of parachute, tethered, and unman aircraft respectively to make the jammer stay in the air shortly.

(2) Shipborne Active Jamming

Shipborne active jamming identifies and warns the threats through receiving the signal of the terminal guide radar in the coming anti-ship missile and emits jamming

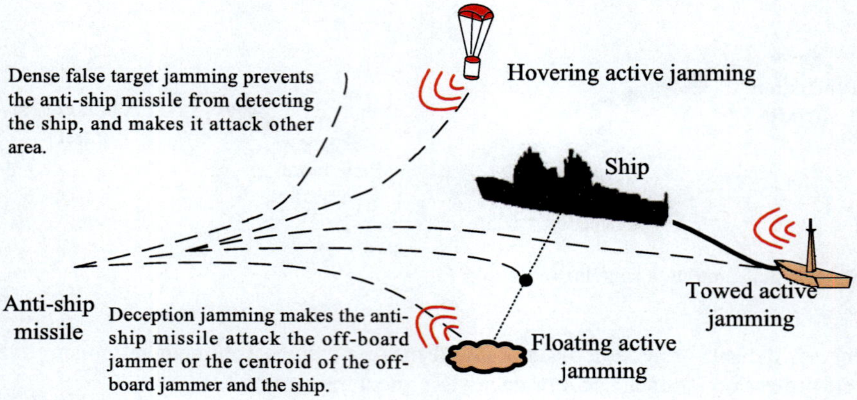

Fig. 1.5 Off-board active jamming scenario

signals via the jamming subsystem to affect the target interception and track of the terminal guide radar. The ship is protected from the attack since the missile is misled by the jamming. Shipborne active jamming contains noise jamming, deception jamming, and combination jamming. The operational effect is shown in Fig. 1.6. The objective of shipborne noise jamming is to make the anti-ship missile cannot hit the target since its radar cannot find the target or lost the target because of the jamming. Shipborne deception jamming includes range deception, velocity deception, and angle deception. The objective of shipborne deception jamming is to make the anti-ship missile radar track a false target, and the real target is therefore lost. Hence, the goal of self-protection is achieved. The *Syke3* shipborne EW system of the Russian navy can find the direction with high precision and jam multiple targets simultaneously. The AN/SLQ-32 Sidekick EW system has multiple different models and has been equipped in a number of warships. The Halifax-class Frigate of the Canadian navy is equipped with AN/SLQ-503 EW system, which is also called the RAMSES system. It is reported that the RAMESES system has multiple working modes and reprogramming capability.

1.2.3 Role in the Air Defense for Strategic Points

The local wars launched by the US after the Cold War indicate that important military facilities, such as command and control centers, airports, important radar stations, and ground missile positions, would be attacked during the initial period of the war. Hence, the defense for strategic points becomes the most important part of the defense fight. In key position defense, the main effects of radar jamming include: (1) jamming the battlefield reconnaissance radar to prevent them from finding and locating our important military facilities; (2) jamming the radar-guided weapons to

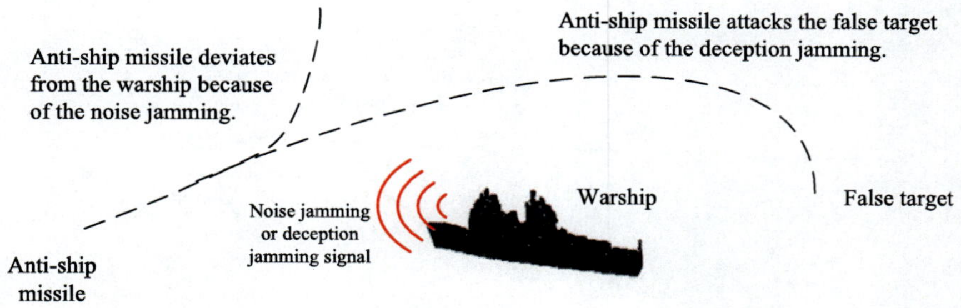

Fig. 1.6 Shipborne active jamming scenario

protect our important military facilities from enemy attack; (3) jamming the anti-radiation weapon of the enemy to cover our ground to air radars.

(1) Battlefield Reconnaissance Radar Jamming

The battlefield reconnaissance duty is mainly fulfilled by synthetic aperture radar (SAR) and its derivative radar systems [7, 8]. SAR has the advantages of all weather, all time, long operation distance, and high resolution, and has become an important method of battlefield reconnaissance. Most advanced countries have spaceborne SAR that is able to get images of the surface and the sea. The images can be processed to obtain the important targets and their locations in the intelligence center. SAR also can be equipped with reconnaissance aircraft, strike aircraft, unmanned aircraft, and missiles to identify the targets on the battlefield and assist the attack.

It can protect the disposition and transfer of our forces, especially the ships at the port, aircraft at the airport, and assembled vehicles, from being detected by the enemy by jamming the battlefield reconnaissance radar. Figure 1.7 illustrates the operational application method of jamming the battlefield reconnaissance radar. In Fig. 1.7, a radar jamming vehicle is jamming the spaceborne radar and airborne to cover the disposition of the vehicle fleet [9, 10].

(2) Radar-Guided Weapon Jamming

As the development of imaging techniques, SAR already can be used as the sensor for weapon systems to guide the attack toward ground targets. Hence, the main target of radar-guided weapon jamming is still SAR in key position air defense.

(3) Anti-radiation Weapon Jamming

Technically, anti-radiation weapon jamming is not against radar but the receiver of an EW system. It is referred to as an important part of key position air defense but will not be described in detail in this book. Anti-radiation weapon jamming includes two types which are deception jamming and disruption jamming. Deception jamming sets radar decoys to mislead anti-radiation missiles by transmitting radar-like signals. Disruption jamming emits jamming signals affecting the opponent receiver to protect the radiation source since the anti-radiation cannot intercept or measure the direction of the radiation source.

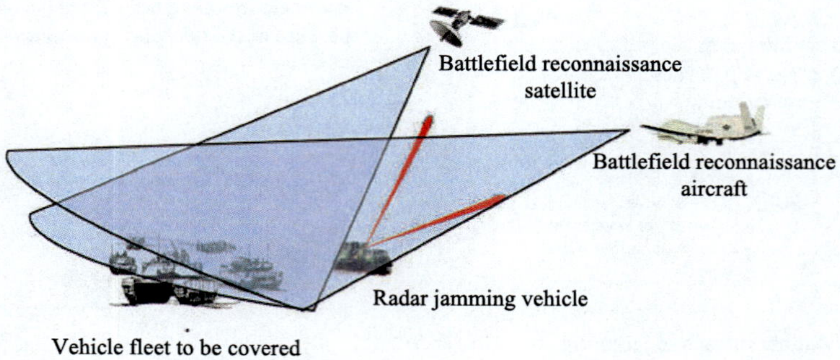

Fig. 1.7 Jamming the battlefield reconnaissance radar

1.2.4 The Development History of Radar Jamming

Radar jamming was born after the invention and utilization of radar in war. The development of radar jamming usually goes with the development of radar. In this section, using historical examples, we will introduce the features of the radar jamming technique in each era from the birth of radar till the present.

There are mainly three stages in the development of EW. The first stage, from 1861 to 1918, was the initial period of EW development. The second stage, from 1919 to 1945, was the mature period of EW development. The third stage, from 1946 to the present, was the rapid developing period of EW when EW played an important role in local wars. Since radar and radar countermeasure, which were first used in World War II, appeared after communication and communication countermeasure, the development stage of radar countermeasure differs from that of EW.

According to the application scale and technique feature, the history of radar jamming can be divided into three stages. The first stage, from 1943 to 1945 when World War II was going on, was the initial period of radar jamming. The second stage, from 1946 to 1991 when the Cold War ended, was the rapid developing period of radar jamming. In this stage, radar jamming significantly contributed to the war. The third stage, from 1991 to the present, was the mature period of radar jamming. In this stage, radar jamming plays a leading role in local wars.

1.2.5 Radar Jamming from 1941 to 1945

Radar was widely used in war during World War II, and radar jamming appeared at the same time. Radar jamming, especially passive jamming and noise jamming, was applied in war.

From the 1930s, the powerful countries in the world started to develop radar systems and gradually became fighting forces. In 1934, the first experimental radar of the US navy research lab detected a close-range aircraft. In 1941 when the US joined the war, there were already 19 radars serving on the US navy ships. Germany made the first successful experiment of radar in 1933, and developed *Freya* early warning radar, which worked in 120–130 MHz. The UK demonstrated the essential principle of radar in February 1935. In 1937, the first ground radar *Chain Home* was equipped by Royal Air Force (RAF), and then 20 *Chain Home* radar stations were deployed across the coast from Southampton to River Tyne shortly. The first airborne radar, whose working frequency was 240 MHz, finished its test flight in an aircraft *the Anson* of RAF in July 1937. Japan developed a double base, continuous wave coherent system named Type A in 1936. RUS-1 and RUS-2 radars of the Soviet Union stated to serve in the army from September 1939.

Because of the use of radar, EW had shifted its emphasis from wireless guidance countermeasure to radar countermeasure since May 1941. The US established National Defense Research Council. In 1941, the Radiation Laboratory of Massachusetts Institute of Technology (MIT) signed a contract with General Radio Corporation to develop airborne wideband electronic intelligence (ELINT) receiver. On February 12, 1942, under the cover of a great amount of radar jamming, German battle cruiser *Scharnhorst* and *Gneisenau* arrived in Germany by quickly crossing the English Channel from Brest, a town locating in northern France, since the radars of Britain failed to detect them because of the jamming. Between July 24 and 25, 1943, the UK first dispersed chaff in large numbers in operation and severely affected German *Wurzburg* and *Wurzburg-Riese* radar and *FUG202 Liechtenstein* airborne interception radar that was equipped in the night fighter of Luftwaffe and worked in 490 MHz. With the jamming of chaff, the loss rate of British fighters decreased by 2%. In the Normandy landing in June 1944, multiple deception methods, including fake intelligence of wireless radio, boats with reflection balloons and transponder jammers, etc., were used by the Soviet Union, the US, and the UK. The utilization of deception measures made Germany misjudge the landing place and helped the successful landing of the Allies. The US started using active jamming to deal with fire control radar in April 1945. EW aircraft B-29 had been used in the bomber troops to strengthen the jamming since June 1945.

In general, radar jamming has been used to assist the operation since World War II. Radar jamming included two types which were active jamming and passive jamming. Radar jamming measures contained suppressing jamming and deception jamming. Suppressing jamming mainly utilized noise jamming and chaff, and deception jamming exploited reflection equipment.

1.2.6 Radar Jamming from 1946 to 1990

In the years after the end of World War II, all the countries reduced their interest in EW, and plenty of EW systems used in World War II were abandoned. However, during the period of the Cold War, EW continued its development, triggered by the local wars that followed. As the progress of electronic technology, EW started its rapid development during this period. During this period, radio frequency memory technique and wideband travel wave tube were invented and could be used to realize deception jamming for gun control radar and missile guidance radar. At the same time, professional electronic jamming aircraft and external ECM pods were developed, which were able to increase the ECM capability of operational aircraft.

In the Korean War which took place from 1950 to 1953, the US did not use chaff to jam the channels used by the fighter aircraft. As a result, they lost several B-29 fighters. In the last seven months, the US relaxed its restriction on EW. Because of the utilization of jamming, the casualties were significantly decreased through jamming, and only 3 out of 4000 B-29 fighters were shot down.

In the Vietnam War which started in 1955, the US invented anti-radiation missiles against Vietnam's radar. The anti-radiation missile had a special guidance way, which guided the missile to fly toward the direction of the electromagnetic radiation source and destroy the radar position. Anti-radiation missiles opened the area of hard kill as well as a new direction of EW anti-radiation countermeasure. The emergence of anti-radiation missiles caused trembling among radar operators and put great pressure on them.

In the war of Bekka Valley between Syria and Israel on June 9, 1982, EW aircraft, early warning aircraft, and anti-radiation missiles operated with each other and created a classic new EW pattern. The Israel army used RC-707 EW aircraft to perform strong electronic jamming, E-2 the *Eagle Eye* early warning aircraft to command from the air, and destroyed 19 anti-air missile basements, which were painstakingly built up by the Syria army for ten years, with anti-radiation missiles equipped in F-15 and F-16.

In the Falklands War between Argentina and the UK in 1982, the guided missile destroyer *Sheffield* was sunk by the AM-39 air-to-ship missiles *Flying fish* since it did not take a suitable countermeasure. The British started to use jamming, and hence the air-to-ship missiles of the Argentina Army failed to hit the target in their following attacks.

During this period, the main progress of radar countermeasure techniques was radio frequency memory, wideband traveling wave tube (WTW), and anti-radiation missiles. Among them, digital radio frequency memory (DRFM) solved the problem of coherent jamming signal production and could better jam pulse Doppler (PD) radar. The emergence and widespread use of WTW raised the radiation power of the jammer and enhanced the jamming performance. The invention of the anti-radiation missiles broke new ground in the area of EW.

1.2.7 Radar Jamming from 1991 to the Present

With the wide employment of microelectronics, computers, and digital technologies since 1991, the EW techniques have made great progress and shown strong power in many local wars.

The utilization of military command, control, communication, and high-tech weapons has been more dependent on electronics technology since the 1990s. EW technology entered a brand new era in the aspects of adaption of complicated EM signal environment, enlargement of the spectrum, enhancement of signal sorting capability, adding new jamming patterns, increase in jamming power, decrease in system response time, integrative architecture, artificial intelligence (AI), self-adaption, jamming of multiple targets and the jamming capability of new electronic devices, etc. EW made further progress in the wars at the end of last century and the beginning of this century, such as the Gulf War, Kosovo War, and Iraq War. The control of EM became the high ground for the controls of air, sea, and land, and the attainment of EM supremacy became the premise and guarantee of winning the war.

In the Gulf War in 1991, US EW aircraft such as EF-111A entered Iraq and jammed the early warning radar, height-finding radar, and tracking radar. The US EW aircraft emitted deception jamming to make the Iraq army launch their radars to guide fire and therefore, the locations of the radars were known and destroyed by the US. During the 42-days battle, there were 250 radars of Iraq destroyed, and the aircraft loss rate of the multi-national force was only 0.425%.

During the Kosovo war from March to June 1999, the US sent more than 30 EW aircraft, such as EA-6B, to jam the radars of the Federal Republic of Yugoslavia (FRY). However, in one assault in which the F-117A stealth bomber failed to get the cover from EA-6B, the bomber was shot down by the SA-3 of FRY.

The US paid more attention to the EW in the Iraq war and Libya war and effectively suppressed the air defense system of the enemy. The EW showed its great power in the war and became the necessary element in taking control of the air in modern wars. After 2000, the US developed the new EW aircraft EA-18G which had a jamming capability of a new level. EA-18G can execute SOJ and EJ and has the capability of air combat with AIM-120 middle-range air-to-air missile and striking the ground with anti-radiation weapons with AGM-88 high-speed anti-radiation missile.

Nowadays, radar jamming technology has been widely used in ground, sea, and air-based equipment. On the ground, the jamming equipment is usually to support jamming and strategic point cover. All the advanced ships are equipped with self-jamming systems in the sea. In the air, except support jamming aircraft, almost all the advanced fighter aircraft, transport aircraft, and special aircraft are equipped with a self-jamming system. From the aspect of jamming technique, phased array jammers, digital receivers, modern signal process techniques, digital radio frequency memory, digital jammers, high-performance transmitters, and so on rapidly developed and were applied in the advanced jamming systems.

References

1. Adamy DL (2001) EW101: a first course in electronic warfare [M]. Artech House Inc., Norwood
2. Key Laboratory of electronic information control. Comprehensive electronic warfare technology entry [M]. Chengdu: the 29th research institute of electronics of the ministry of information industry (2001)
3. Guoqing Z (2003) Principle of radar countermeasures [M]. Xidian University Press, Xi'an

4. Robert JH (2010) The 21st century's electronic warfare [J]. Int Electron Warfare 6:11–15
5. Yiyu Z, Wei A, Fucheng G (2014) Electronic countermeasure principle and technology [M]. Electronic Industry Press, Beijing
6. Neri F (2001) Introduction to electronic defense system [M]. Artech House Inc., Norwood
7. John CC (1991) Synthetic aperture radar: systems and signal processing [M]. California Institute of Technology, California
8. Musman S, Kerr D, Bachmann C (1996) Automatic recognition of ISAR ship images [J]. IEEE Trans Aerosp Electron Syst 32(4):1392–1404
9. Goj WW (1993) Synthetic aperture radar and electronic warfare [M]. Artech House, Boston London
10. Hong L, Yingke Y, Baomin X (2010) Introduction to synthetic aperture radar countermeasure [M]. National Defense Industry Press, Beijing

Chapter 2
Principle of Radar Active Jamming Technology

Radar active jamming is to disrupt or block the target detection and tracking of enemy radar by generating radio signals from electronic equipment. It has the advantages of flexible and controllable jamming power, jamming mode and jamming effect, and is an important radar countermeasure. The principle of radar active jamming technology is the theoretical basis for the design of radar active jamming system.

This chapter introduces the basic principles related to radar and radar jamming. Section 2.1 introduces the basic principles of radar detection, including the physical model of radar detection, radar equations, typical radar technologies, and typical radar anti-jamming technologies; Sect. 2.2 introduces radar jamming models; Sect. 2.3 introduces the radar jamming equation; Sect. 2.4 introduces the principle of radar blanket jamming, which not only analyzes noise blanket jamming, but also introduces the newer dense false targets suppression jamming; Sect. 2.5 introduces the principle of radar deception jamming.

2.1 Basic Principles of Radar Detection

Radar stands for radio detection and ranging, its original meaning is "radio detection and ranging"; that is, the target is found by radio and its position in space is determined [1]. With the continuous development of radar technology, radar can not only measure the range and angle of the target but also the speed, polarization, size, and shape of the target.

G. Tang et al., *Techniques and System Design of Radar Active Jamming*,
https://doi.org/10.1007/978-981-19-9944-4_2

2.1.1 The Physical Model of Radar Detection

Figure 2.1 is a schematic diagram of a typical radar. Its fundamental working process is as follows: the electromagnetic wave generated by the radar transmitter is transmitted to the antenna through the transceiver switch, and the antenna radiates this electromagnetic wave into a predetermined airspace. The electromagnetic wave propagates in the space at the speed of light. If the target is within the beam range of the antenna, which will reflect part of the wave. The radar antenna receives the electromagnetic wave reflected by the target and transmits it to the receiver through the transceiver switch. The receiver amplifies and processes the weak signal to obtain the required information, sending the result to the terminal to display [2].

Although the term radar is derived from radio detection and ranging, with the unceasing development in radar technology, the information that radar can provide is more than just range. In general, the speed of the target can be measured by doppler frequency shift detection; the radar can also measure the pitch and azimuth angle of the target through monopulse angle measurement technology; one-dimensional or two-dimensional high-resolution imaging technology can measure the size and shape of the target; what is more, polarization and other technologies have the capacity to gauge the symmetry, surface roughness, and dielectric properties of the target [3].

The radar emits signals, which are reflected by the target in space. The radar receives the reflected signals and resolves the attributes of the target. This is the process of solving a system given its inputs and outputs [4]. As shown in Fig. 2.2.

Assuming that the radar transmitted signal is $s(t)$, the radar received signal is $s_r(t)$, provided that the system transfer function from the transmitted signal to the received signal is $h(t)$, ignoring the receiver noise, then:

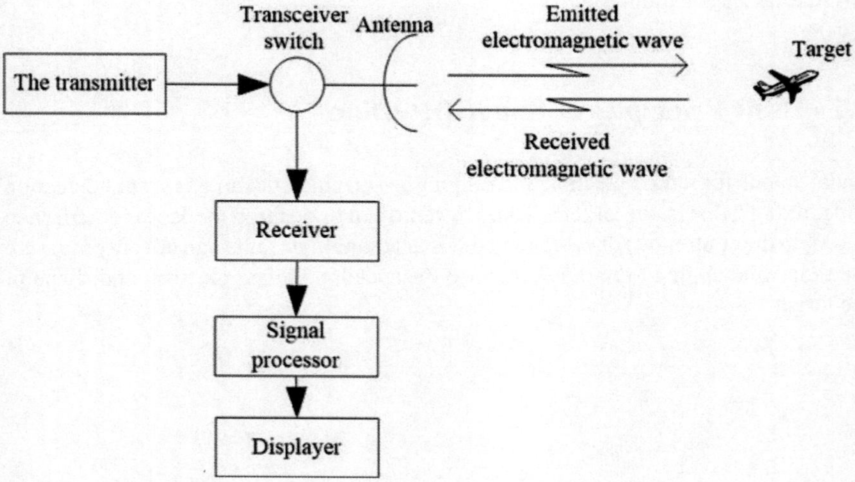

Fig. 2.1 Radar principle and its basic composition

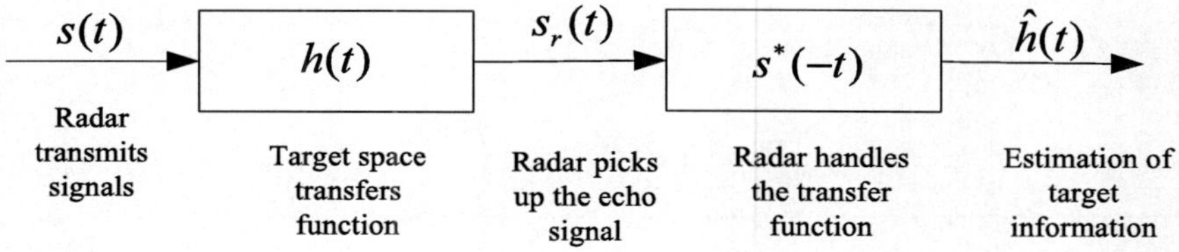

Fig. 2.2 Physical model of radar detection

$$s_r(t) = s(t) \otimes h(t) \tag{2.1}$$

$\otimes$ represents convolution. The radar performs matched filtering on the received signal and convolves the echo with the conjugate flip signal of the transmitted signal, and then the attribute estimation of the space and the target $\hat{h}(t)$ can be obtained.

$$\hat{h}(t) = s_r(t) \otimes s^*(-t) = h(t) \otimes p(t) \tag{2.2}$$

$p(t) = s(t) \otimes s * (-t)$ is the point spread function of the radar processing the target. $\hat{h}(t)$ includes the target's spatial position (range and azimuth), scattering intensity, speed, polarization characteristics, etc.

The following introduces the basic principles of radar ranging, speed, angle, size, and shape measurement.

2.1.1.1 Range Measurement

By measuring the time between the radar signal and the target, the radar can calculate the range of the target. As shown in Fig. 2.3, if the time that the received signal lags the transmitted signal is Δt, the range between the target and the radar is:

$$R = \frac{c\Delta t}{2} \tag{2.3}$$

In the formula, R is the one-way range in meters from the radar to the target; Δt stands for the time interval between the electromagnetic wave and the target; c represents the speed of light $c = 3 \times 10^8$ in m/s. Since electromagnetic waves travel at the speed of light, the time unit commonly used in practice is μs. For example, when the echo lags the transmitted signal by 1 μs, the corresponding target range is 150 m.

The narrower the pulse, the higher the ranging accuracy, so narrow pulse is the common waveform of ranging. A wide pulse can be compressed into a narrow pulse through pulse compression and other technologies, which not only achieves the equivalent narrow pulse ranging performance but also features a long range of action in the wide pulse. In addition, frequency-modulated and phase-modulated continuous

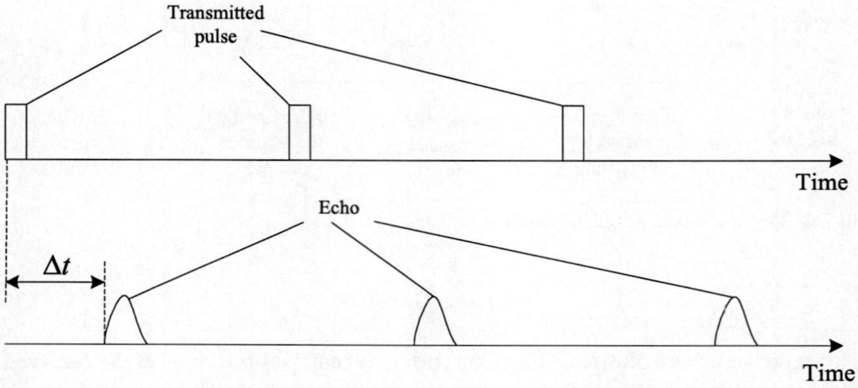

Fig. 2.3 Principle of radar ranging

waves can also be able to measure the range to a target. By comparing the phases of two or more continuous wave frequencies, the range of a single target can be measured. Continuous wave ranging is widely used in airborne radar altimeter and surveying instrument.

2.1.1.2 Velocity Measurement

The rate of range change, namely radial velocity, can be obtained by continuous measurement of the target range, but the accuracy of this method is not high. In practice, the common method of radar velocity measurement is Doppler measurement technology.

When there is a relative velocity between the target and the radar, the carrier frequency of the received echo signal will produce a Doppler shift relative to the carrier frequency of the transmitted signal:

$$f_d = \frac{2v_r}{\lambda} \tag{2.4}$$

In the formula, f_d is the Doppler frequency shift in Hz; v_r serves as the radial velocity in m/s between the radar and the target, and λ is the carrier wavelength in meters. When the target moves toward the radar, $v_r > 0$, the echo carrier frequency increases and visa-versa. If the Doppler frequency shift of the echo signal f_d is measured, the relative radial velocity between the target and the radar can be determined with a high measurement accuracy.

In addition to speed measurement, Doppler frequency shift is more widely used in moving target display, pulse Doppler radar, to distinguish fixed clutter and moving target echo.

2.1.1.3 The Angle Measurement

In order to determine the spatial position of the target, after measuring the distance of the target, it is necessary to measure the angle of the target, including azimuth angle and pitch angle. The theoretical basis of radar angle measurement is the linear propagation characteristic of electromagnetic wave and the directivity of radar antenna.

Radar angle measurement generally has two methods: phase method and amplitude method.

(1) Angle measurement by phase method

The phase method utilizes the phase difference between the echo signals received by multiple antennas to measure the angle. As shown in Fig. 2.4, there is a far-region target in the θ direction, and the electromagnetic wave at the receiving antenna is approximately a plane wave. Hence, there is a wave path difference ΔR between the signals received by the two receiving antennas, then:

$$\varphi = \frac{2\pi}{\lambda}\Delta R = \frac{2\pi}{\lambda} d \sin\theta \tag{2.5}$$

In the formula, φ is the phase difference, d stands for the range between the two receiving antennas, λ represents the radar wavelength, and the angle θ of the target will come out with measuring the phase difference φ with the phase method.

(2) Angle measurement by amplitude method

The amplitude method angle measurement uses the amplitude of the echo signal received by the antenna to measure the angle, which can be divided into the maximum signal method and the equal signal method.

The basic principle of maximum signal method is as follows: when the antenna is scanning in space with a certain rule, for single-base radar with common antenna, the amplitude of the signal received is modulated by the antenna two-way direction diagram, and when the amplitude of the signal is at its maximum, the beam is pointing in the direction of the target.

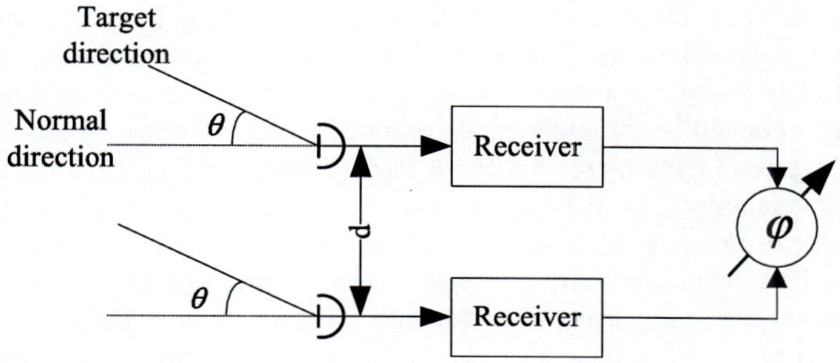

Fig. 2.4 Principle of angle measurement by phase method

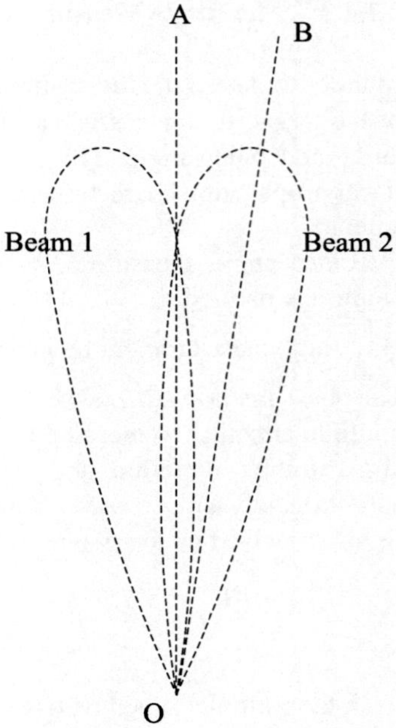

Fig. 2.5 Angle measurement principle of isosignal method

The equal signal method uses two identical and overlapping beams to measure angles, as shown in Fig. 2.5. OA direction is the overlapping axis of two beams, also known as equal signal axis. When the target is in OA direction, the signals received by beam 1 and beam 2 are equal. When the target is in OB direction, the signal received by beam 2 is stronger than that received by beam 1. Since the shape of the beam is known, the angle of the target can be calculated from the amplitude of the signals received by the two beams.

2.1.1.4 Measurement in Size and Shape

Radar can obtain high resolution in the radial range dimension by transmitting a large bandwidth signal. At the same time, it can also get a one-dimensional range image of the target to finely describe its size and structure. Figure 2.6 is the one-dimensional range image obtained by the radar irradiation of the aircraft target, from which the size and structure of the target in the one-dimensional domain are depicted in detail.

As radar technology is growing, synthetic aperture radar (SAR) and inverse synthetic aperture radar (ISAR) utilize the relative movement of the target and the radar to synthesize the smaller real antenna aperture through data processing into a large equivalent antenna aperture to obtain high lateral resolution, so that target's size and shape can be described in detail in the two-dimensional domain [5–7]. Figure 2.7

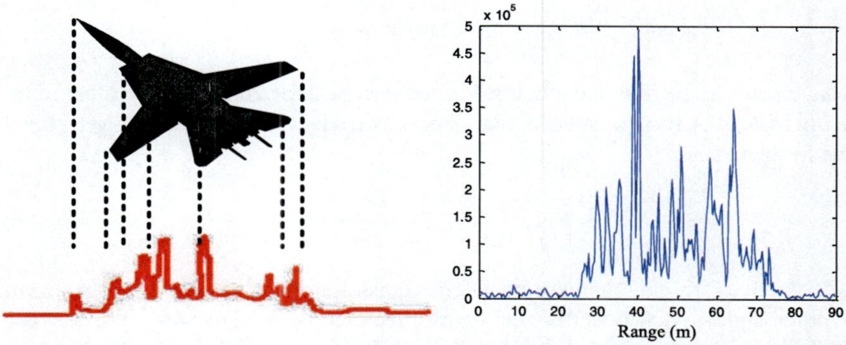

a) One dimensional range image of a fighter b) One-dimensional image of a civil aviation aircraft with 1 meter resolution

Fig. 2.6 One-dimensional range image of aircraft target

is a SAR image with a resolution of 4 inches, which has reflected them in a very detailed two-dimensional domain.

In addition, symmetry, surface roughness, and dielectric properties of the target can be theoretically obtained from the polarization information of the target radar echo.

Fig. 2.7 SAR image

2.1.2 Radar Equation

Radar equation describes the relationship between radar performance and other influencing factors. A form of radar equation performed by the power of received signal is as follows:

$$P_r = \frac{P_t G_t}{4\pi R^2} \times \frac{\sigma}{4\pi R^2} \times A_e \tag{2.6}$$

The above formula ignores atmospheric transmission losses, and the right side of the equal sign is written as the product of three factors, each of which represents a physical process when the radar detects a target. The first factor $P_t G_t / 4\pi R^2$ represents the power density at a range R, with a radiation power of P_t, and an antenna gain of G_t. The second factor $\sigma / 4\pi R^2$ represents the degree of attenuation of the power density of the scattering echo from a target with a cross-sectional area σ at range R. The product of the first two factors signifies the power density of the radar wave returning to the radar. The third factor A_e is the effective receiving area of the receiving antenna.

If the maximum radar operating range is defined as the radar operating range when the received power P_r is equal to the receiver's minimum detection signal $S_{\min}$, then the radar equation can be written as:

$$R_{\max}^4 = \frac{P_t G_t A_e \sigma}{(4\pi)^2 S_{\min}} \tag{2.7}$$

When the same antenna is used for transmitting and receiving, the relationship between the gain of the transmitting antenna G_t and the effective receiving area A_e in the receiving antenna can be expressed as:

$$G_t = \frac{4\pi A_e}{\lambda^2} \tag{2.8}$$

where λ represents the radar wavelength. Incorporating Eq. (2.8) into Eq. (2.7), the following conclusions can be drawn:

$$R_{\max}^4 = \frac{P_t G_t^2 \lambda^2 \sigma}{(4\pi)^3 S_{\min}} \tag{2.9}$$

$$R_{\max}^4 = \frac{P_t A_e^2 \sigma}{4\pi \lambda^2 S_{\min}} \tag{2.10}$$

The radar Eqs. (2.9) and (2.10) can be used to roughly estimate the radar capability and do not include the various losses introduced by the radar and the environment. Additionally, the target radar cross section (RCS) and the minimum detectable signal are statistics, so the radar operating range obtained is also a statistic.

In order to ensure that the signal can be reliably detected, $S_{\min}$ represents the product of SNR and receiver noise, where SNR is the SNR required for reliable detection, and receiver noise is represented by thermal noise generated relative to the ideal receiver. Thermal noise is equal to kTB, where k is Boltzmann constant, T is thermodynamic temperature, and B is receiver bandwidth. Receiver noise is the product of thermal noise and receiver noise, where receiver noise is measured relative to the reference temperature ($T_0 = 290$ K). Then, $S_{\min}$ can be written as:

$$S_{\min} = kT_0BF_n\frac{S}{N} \tag{2.11}$$

In general, the ratio of signal energy to noise spectral density (denoted as E/N_0) is a more basic parameter than the ratio of signal power to noise power. For rectangular pulse with width τ, signal power E/τ, noise power N_0B, signal energy E, noise energy N_0 (assuming noise is evenly distributed in the frequency domain), and receiver bandwidth B, then:

$$S_{\min} = kT_0F_n\frac{E}{N_0\tau} \tag{2.12}$$

Substitute it into Eq. (2.7) to get:

$$R_{\max}^4 = \frac{E_tG_tA_e\sigma}{(4\pi)^2kT_0F_n(E/N_0)} \tag{2.13}$$

In the formula, $E_t = P_t\tau$ is the energy of a rectangular pulse with width τ. Although the transmitting waveform of Eq. (2.13) is assumed to be a rectangular pulse, this formula is correct as long as E_t is the energy contained in the transmitting waveform and the receiver with noise figure of F_n is designed to match the signal.

2.1.3 Radar Technology

There are many technologies related to radar. This chapter only briefly introduces a few typical radar technologies closely connected to radar jamming, including moving target indication, pulse Doppler, constant false alarm detection, frequency agility, pulse compression, etc. For other radar technologies, please refer to pertinent literatures.

2.1.3.1 Moving Target Indication

Moving objects are usually the targets to be detected by radar, such as airborne aircraft, missiles, ships, and vehicles. However, there are often various settings around the target, such as a variety of ground objects, clouds and rain, sea waves, and metal wire interference from the enemy. Moving target indication (MTI) radar is to filter out the fixed clutter and retain the echo of the moving target through a filter, thereby greatly improving the detection ability of moving targets under the clutter background.

When a radar pulse of frequency f meets with an aircraft moving at a certain radial velocity V_R, the frequency of the reflected echo received by the radar is:

$$f_\gamma = f + f_d \tag{2.14}$$

In the formula, $f_d = V_R / \lambda$, f_d is the Doppler frequency, and $\lambda = c/f$ is the wavelength of the radar carrier, PRF of the pulse acts as $F_R = 1/T$, where T represents the pulse repetition period.

When the Doppler effect is used to distinguish the echo of a moving target from a fixed target, the difference from common pulse radar is that a reference voltage must be added to the input of the phase detector. This voltage should be coherent to the frequency of the transmitted signal and saved the initial phase, including persistent existence during the entire received signal period. The phases of various echo signals are compared with the reference voltage. The video pulse signal output from the phase detector includes the constant amplitude pulse signal of a fixed target and the amplitude-modulated pulse signal of a moving target. Before the signal is sent to the terminal (display or data processing system), the fixed clutter needs to be filtered out by clutter filter, and only the moving target signal is retained.

A simple MTI delay line canceller requires two radar echoes to reach a steady state. This filter has the advantage of being simple and easy to implement. The disadvantage is that clutter of ground objects may not be completely suppressed. For example, clutter of swaying trees is not fixed, and the spectrum will have a certain width, resulting in poor clutter filtering effect. MTI filters with multiple delay lines are generally used. The response of filters with different delays is matched with clutter spectrum by weighting each unit. Figure 2.8 shows the implementation block diagram of MTI dual cancelators (two single delay line cancelators). In order to obtain the effective output of the filter, at least three signal pulses are required. As can be seen from the response curve in Fig. 2.9, clutter is well suppressed.

It can also be seen from the figure that when $f_d = 1/T$, the output of the filter is zero again, and the corresponding speed is called "blind speed". In general, when the

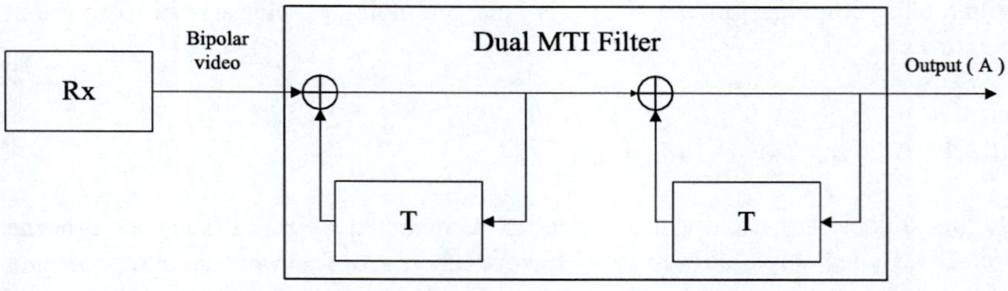

Fig. 2.8 Block diagram of double cancelation MTI

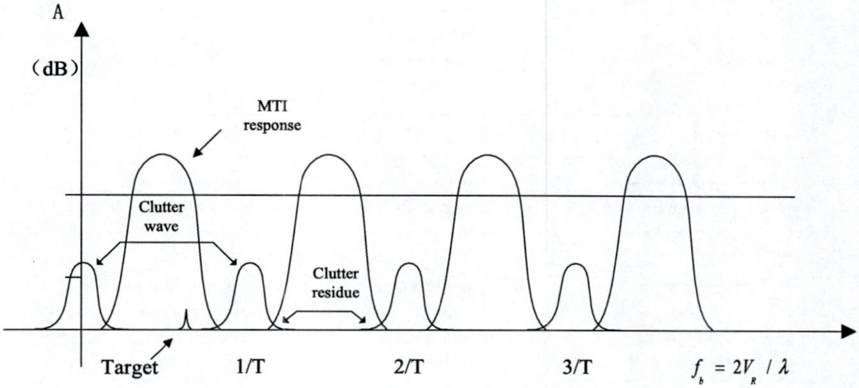

Fig. 2.9 Schematic diagram of dual cancelation MTI performance

radar has a low PRF, the velocity ambiguity tends to occur, and when the radar has a high PRF, the range ambiguity tends to occur. In order to get rid of the blind speed problem, the PRF staggering technology can be adapted to make the PRF of the radar change according to certain rules, so that the Doppler frequencies corresponding to the integer multiples of the PRF in the MTI filter can be offset. Please refer to related references for specific principles.

2.1.3.2 Pulse Doppler

Pulse Doppler (PD) radar is a kind of pulse radar which uses Doppler effect to detect target information. It is a new radar system developed based on MTI radar. This kind of radar has the range resolution of pulse radar and the velocity resolution of continuous wave radar. It can distinguish the moving target echo in the strong clutter background and has stronger clutter suppression and anti-jamming ability. It has been widely used in the aircraft platform.

The classification of PD radar is shown in Fig. 2.10, which can be divided into three types: high PRF, medium PRF and low PRF, among which high PRF and medium PRF are fully coherent systems, while low PRF systems can be fully coherent or receive coherent systems. The high PRF system is ambiguous in range measurement and unambiguous in velocity measurement. The low PRF system is ambiguous in velocity measurement and unambiguous in range measurement. The medium PRF system is ambiguous both in velocity and range measurement.

Like other radars, the main components of PD radar also include antenna, transmitter, receiver, T/R components, signal processing system and indication system. Signal processing system is an important part of PD radar, and its function is to filter noise, fixed clutter and jamming signal to the maximum extent, so as to retain the moving target signal. PD radar has a complex signal processing system, which

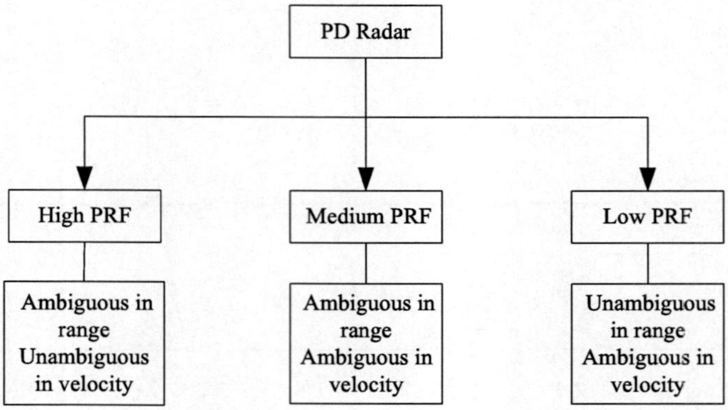

Fig. 2.10 Classification of PD radar

Fig. 2.11 PD radar signal processing flow

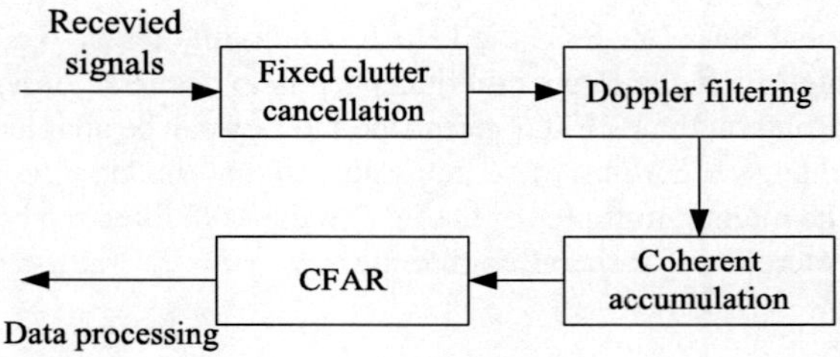

mainly includes clutter cancelation, Doppler filtering, phase coherent accumulation, Constant false alarm detection, etc., as shown in Fig. 2.11.

Among them, the fixed clutter cancelation is to use the difference in velocity between the fixed clutter and the moving target and use technologies such as moving target indication to suppress clutter and improve the signal-to-clutter ratio. Doppler filtering is a key part of PD radar, which will be further introduced later. Coherent accumulation is to perform phase coherent accumulation on the filtered signal to further improve the signal-to-clutter ratio (signal-to-noise ratio); finally, the target is detected by constant false alarm rate (CFAR).

Doppler filtering is a key component of PD radar, and it is a set of contiguous narrowband filters covering the Doppler frequency shift of the expected target. The narrowband filter bank can not only distinguish and measure the velocity but also improve the ability to detect moving targets under clutter. When the target has different radial velocities relative to the radar, the corresponding Doppler frequency shift is also different, which will fall into different narrowband filters. There are two ways to implement a Doppler filter bank: one is the use of Fourier transform in the frequency domain to achieve a Doppler filter bank; the other is to implement a finite impulse response (FIR) filter in the time domain. The former one is inferior in flexibility and has a higher sidelobe level, and there are no pits near the zero frequency;

the latter one has good flexibility, and different weights can be selected according to the actual situation to obtain a filter that meets the requirements. So far, pulse Doppler filter banks are mostly implemented by FIR filters, and the relationship between input and output signals is:

$$y(n) = \sum_{i=0}^{N-1} w_i x(n - iT_r) \tag{2.15}$$

where N is the number of filter coefficients; w_i is the weighting coefficient of the filter; T_r is the pulse repetition period. The structure of the filter bank is shown in Fig. 2.12. The characteristics of the Doppler filter bank are shown in Fig. 2.13.

Since the advent of PD radar, it has attracted extensive attention in radar field, mainly because of its strong anti-jamming ability. PD radar has the following characteristics:

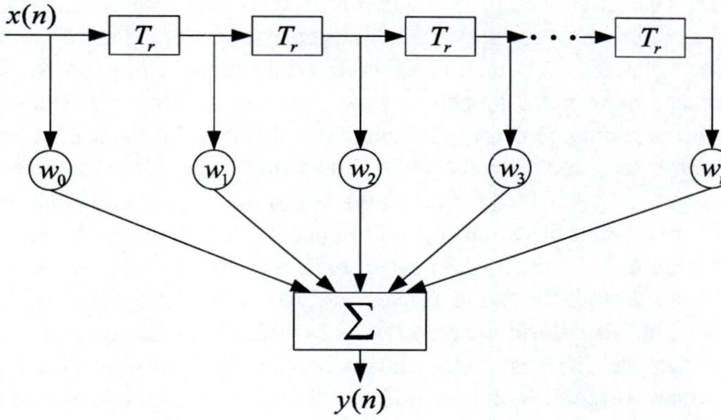

Fig. 2.12 Doppler filter bank structure

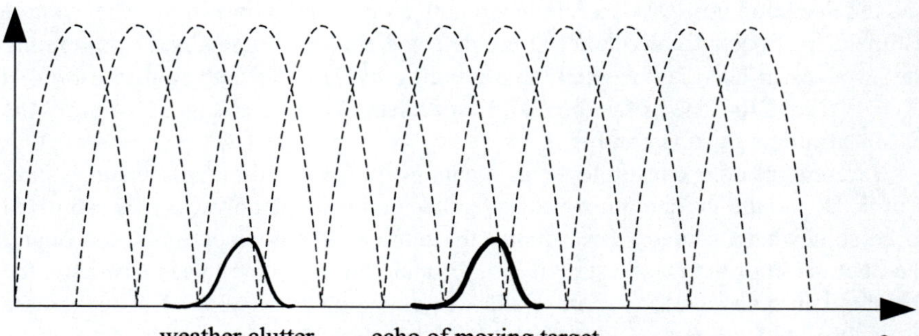

Fig. 2.13 Doppler filter bank characteristics

(1) Good clutter suppression ability. The most basic feature of PD radar is that it has good clutter suppression capabilities. It can detect moving target signals in a strong clutter background, so it is widely used in airborne radars.
(2) High coherent processing gain. PD radar is an advanced fully coherent system radar, through coherent accumulation, higher gains can be obtained.
(3) Strong anti-jamming ability. PD radar has good anti-airborne chaff jamming performance. Besides, it has a strong discrimination of range and velocity, which it can not only anti-range pull-off jamming through the comparison of Doppler frequency and range change rate but also anti-velocity pull-off jamming through the correlation between the range change rate and the Doppler frequency.

2.1.3.3 Constant False Alarm Detection

When the radar is detecting the target, the target echo enters the receiver together with internal thermal noise, clutter (ground objects, rain, snow, sea waves, etc.) and jamming (active jamming and passive jamming). How to detect useful target echoes under complex signal background and get target parameters is the key and difficult point of radar target signal detection. If the fixed threshold detection is used in the automatic detection of radar targets, a small increase in clutter power will make the false alarm rate change drastically, which will cause the radar data processing equipment overloaded and the radar will fail to work. Then it is impossible to make a correct judgment even if the signal-to-noise ratio is large. Therefore, the detector needs to have a constant false alarm performance while extracting the echo signal.

Constant false alarm rate (CFAR) detection is an echo signal processing method that provides automatic detection thresholds. It can achieve radar signal detection with constant false alarm characteristics by using various constant false alarm methods. CFARs in different fields have different performance or adaptability; however, the basic principles and methods are the same. Let us take the cell average CFAR as an example to introduce the basic principle and implementation method of the constant false alarm detector.

The input signal x_i is sent to the delay line composed of $(2L + 1)$ delay units, and the detection unit D takes L units on both sides as reference units. The average estimate $\hat{\mu}$ of background clutter at the detected cell can be obtained by summing the x values in all reference cells and dividing by $2L$. The detection threshold is $U_0 = K\hat{\mu}$, and the size of the threshold coefficient K can be changed to control the false alarm rate, as shown in Fig. 2.14.

$2L$ reference units constitute the data window for calculating mean value estimation $\hat{\mu}$. Due to the limited number of reference units, the mean value $\hat{\mu}$ is estimated to be somewhat fluctuate. The smaller the number of reference units, the bigger the fluctuation of the mean estimate $\hat{\mu}$. . To maintain the same false alarm rate, the threshold must be increased appropriately (by adjusting the K value). But an increase in the threshold will reduce the probability of detection, it is necessary to increase the signal-to-noise ratio to maintain the specified probability of detection. Figure 2.15 is a schematic diagram of the threshold of constant false alarm detection. As we can

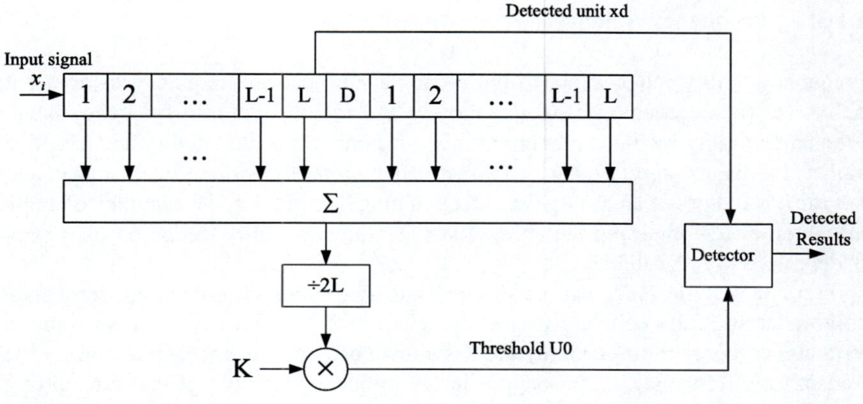

Fig. 2.14 Schematic diagram of unit average constant false alarm detection

see from Fig. 2.15a, when the clutter fluctuates in a fixed range, the constant false alarm threshold remains basically unchanged. In Fig. 2.15b, when the fluctuation range of the clutter becomes larger and larger, the constant false alarm threshold also gradually increases with the clutter, which always maintains effective detection of the target.

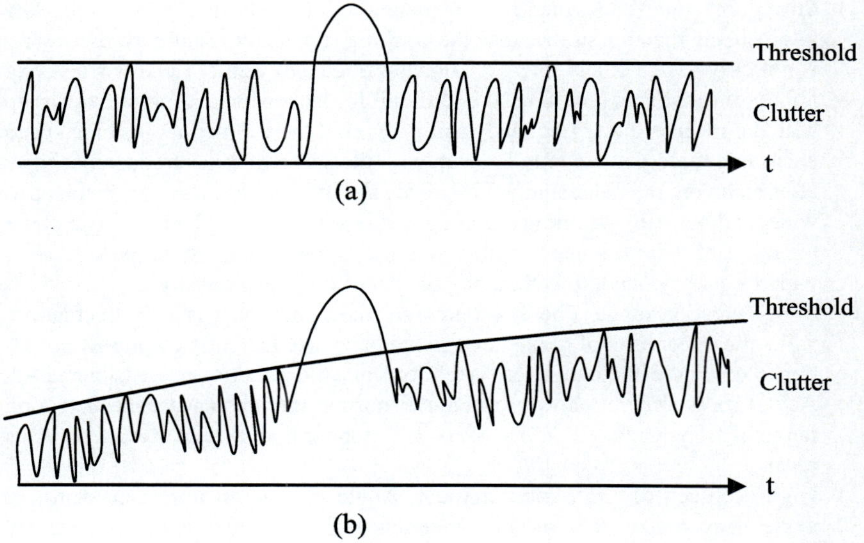

Fig. 2.15 Schematic diagram of constant false alarm threshold

2.1.3.4 Frequency Agile

Frequency agility radar refers to the radar that the carrier frequency of adjacent pulses (or pulse groups) changes randomly and rapidly within a frequency band. Frequency agility includes random agility, program-controlled agility and adaptive agility. Frequency agility radar can effectively counter the aiming interference, and has the advantages of enlarging the detection range, improving the accuracy of angle measurement, suppressing the wave clutter. Frequency agility technology has been widely used in radar field.

Frequency agile radar can be divided into two types: non-coherent frequency agile radar and fully coherent frequency agile radar. The former one was simple in structure and easy to implement, and has a low cost, but it is not easy to control the transmitting frequency, and it is poor in the frequency stability of the transmitting signal, as well as it cannot be compatible with the moving target indication system. The latter one fully coherent frequency agile radar, whose core is the agile frequency synthesizer which can generate fast and agile transmit signal and local oscillator signal. It is high in the frequency stability; it is easy to realize controllable agility and can be compatible with pulse compression, moving target indication along with other systems, but it is expensive and sophisticated in technology.

Frequency agility radar has the advantages of strong anti-interference ability, long detection range, high angle measurement accuracy and good wave clutter suppression.

(1) Strong anti-interference ability. Frequency agility radar has good signal low intercept characteristics, because the working frequency is agile, so the enemy is not easy to detect and intercept signals; frequency agility radar has a strong ability to resist spot jamming. It is difficult for narrowband jammer to keep up with the speed of radar frequency agility. Even if the jammer adopts high speed electronic tuning, it can only keep up with the speed of radar frequency agility after receiving the radar signal. Frequency agility radar has a good resistance to wideband blocking jamming. Although wideband blocking jamming can cover the range of radar frequency agility, jammers spread the jamming power over a wide frequency band, thus reducing the jamming power density.
(2) Long detection range. The target echo of traditional radar is slow fluctuation, while the target echo of frequency agile radar is fast fluctuation without correlation, thereby the fluctuation loss is reduced, and the detection range is increased. Actual measurement shows that, under normal circumstances, the detection range of frequency agile radar is 20–30% longer than that of fixed frequency radar.
(3) High accuracy of angle measurement. Angle glint is an important cause of angle measurement error of radar. Frequency agility technology can effectively suppress angle glint and improve angle measurement accuracy.
(4) Good effect of suppressing wave clutter. Wave clutter at the same range unit usually has a longer time correlation, thereby it is not the ideal method of relying on signal accumulation to suppress sea clutter. However, the correlation of sea

clutter can be removed by frequency agility and achieve suppressing sea clutter effectively.

2.1.3.5 Pulse Compression

Pulse compression (PC) radar transmits large time-width and large bandwidth signals and uses matched filtering technology to achieve pulse compression during reception to obtain narrow pulse signals. It has the advantages of long operating range and high range resolution. Pulse compression has been widely used in radar field.

To obtain better range accuracy and range resolution, the radar is required to transmit large bandwidth signal; to obtain better velocity measurement accuracy and velocity resolution, the radar is required to transmit signals with large time-width; to increase the radar's operating range, the radar is required to transmit high signal energy. Therefore, if the radar system needs to have the characteristics of long range, high measurement accuracy, and good resolution accuracy, the radar signal needs to have large time-width, bandwidth product, and high signal energy. The range measurement performance, velocity measurement performance, and radar observation range of conventional pulse signal are contradictory. Adopting a radar signal with a large time-width bandwidth product is an effective method to solve the contradictions mentioned above. This kind of radar system is pulse signal compression radar, or pulse compression radar for short.

If the pulse compression radar transmits a wide pulse signal with a width of τ and the transmitted signal bandwidth is B, the range resolution is:

$$D_R = \frac{c}{2B} \tag{2.16}$$

In the formula, c is the speed of light.

The pulse compression radar performs matched filtering on the received signal and outputs a narrow pulse signal with an effective width of $\tau_0 = 1/B$, and the range resolution D_R is:

$$D_R = \frac{c\tau_0}{2} \tag{2.17}$$

That is, by using a wide pulse with a width of τ, the pulse compression radar can obtain the range resolution which equivalent to a conventional radar with a transmitted pulse effective width of τ_0. The ratio of the width τ of the transmitted signal to the effective width τ_0 of the received signal after being compressed becomes the pulse compression ratio, which is recorded as:

$$D = \frac{\tau}{\tau_0} \tag{2.18}$$

Because $\tau_0 = 1/B$, therefore

$$D = \tau B \tag{2.19}$$

That is, the pulse compression ratio is equal to the product of the signal's time-width and bandwidth, and the time bandwidth product represents the improvement of the signal-to-noise ratio after and before pulse compression, τB is one of the important parameters of pulse compression radar systems.

Pulse compression radar has many advantages:

(1) Under the condition that the peak power of the transmitter is limited, pulse compression radar makes full use of its allowable average power to increase the energy of the transmitted signal and increase the power range of the radar.
(2) Pulse compression radar performs matched filter processing on the received signal, compressing wide pulses into narrow pulses. It has good range accuracy and range resolution.
(3) Pulse compression radar improves the anti-jamming capability of the radar system. For noise blanket jamming, the jamming is also required to have a wide bandwidth because the radar signal has a wide bandwidth, which is equivalent to reducing the power spectral density of the jamming signal. For response jamming, due to the modulation characteristics of the transmitted signal pulse, the pulse compression is matched to the transmitted signal, but not to response jamming, at least not completely matched, so as to better suppress the jamming. For passive jamming, the signal-to-noise ratio is improved obviously after pulse compression, and the resolution of radar is improved, and the capability of resisting passive jamming is also improved.

According to the modulation law of transmitted signal, pulse compression can be divided into linear frequency modulation pulse compression, nonlinear frequency modulation pulse compression, phase coding pulse compression, time frequency coding pulse compression, and so on. Linear frequency modulation pulse compression is the most common one, and its pulse compression principle is briefly introduced below. For pulse compression of other signal modes, please refer to relevant literature. For other signal modes, please refer to relevant literatures.

Linear frequency modulation (LFM) is a pulse compression signal which was researched the earliest and used the most widely. It modulates the radar's carrier frequency to increase the radar's transmission bandwidth and realizes pulse compression when receiving. The principle of linear frequency-modulated pulse compression is shown in Fig. 2.16.

The linear frequency-modulated waveform is composed of a rectangular transmission pulse with a width of T, as shown in Fig. 2.16a. The carrier frequency f increases linearly according to $\Delta f = f_2 - f_1$ within the pulse width, with modulation slope $\mu = 2\pi \Delta f / T$, as shown in Fig. 2.16b. Figure 2.16c shows the frequency-delay characteristics of the compression network. It changes in a linearly decreasing manner, which is opposite to the linear frequency modulation slope of signal. The low-frequency component f_1 of the LFM signal enters the filter first, and the delay t_{d1} of the filter to the frequency f_1 is longer. On the contrary, the high-frequency component f_2 enters the filter after T time, and the delay t_{d2} of the filter for frequency

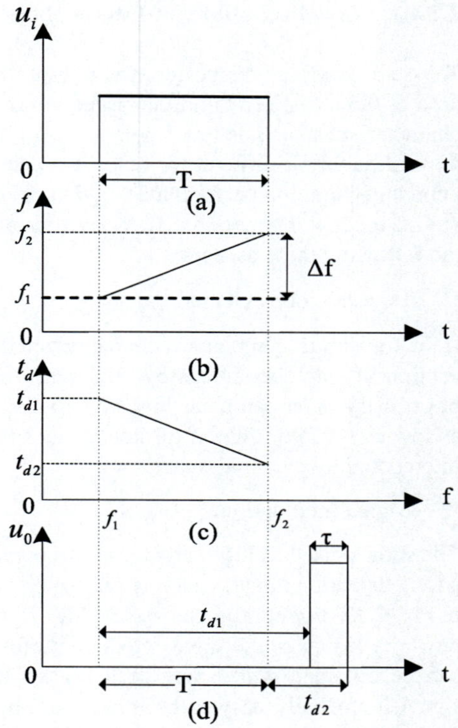

Fig. 2.16 Schematic diagram of linear frequency-modulated pulse compression

f_2 is relatively short. In this way, different frequency components in the signal arrive at the output almost simultaneously after passing through this filter to obtain a pulse signal with an increased amplitude and narrowed width. Its ideal envelope is shown in Fig. 2.16d.

2.1.4 Typical Technology of Radar Anti-jamming

Radar anti-jamming technology is also a branch of radar technology. We cannot separate radar technology and radar anti-jamming technology completely [8]. The techniques of moving target indication, pulse Doppler, constant false alarm detection, frequency agility, and pulse compression introduced above also have good anti-jamming performance. There are many anti-jamming technologies of radar; this section will mainly introduce radar anti-jamming technologies such as low probability of interception (LPI), wide band Dicke fix (WBDF) circuit, sidelobe cancelation, sidelobe blanking, automatic gain control (AGC), anti-RGPO (range gate pull-off).

2.1.4.1 Low Probability of Interception

To use a common expression, the low probability of interception (LPI) radar is a kind of radar that can "find the target without exposing itself". Usually, it has some characteristics such as low sidelobe, large bandwidth, variable frequency, and low power, making it difficult for radar reconnaissance receivers to detect. Thus, it has good anti-jamming performance, and has been widely used in radar field [9, 10]. To realize the low interception characteristics of radar signals, usually we can consider the following three aspects:

(1) Low sidelobe antenna

If the transmitting antenna is characterized by low sidelobe, the probability of interception of sidelobe signals by reconnaissance can be reduced. The sidelobe level of ordinary radar can be as low as −20 dB, while the sidelobe of LPI radar can be as low as −45Db; thus, it further reduces the possibility of sidelobe signals being intercepted by reconnaissance.

(2) Signal modulation

The short pulse has high range resolution, but it must have high peak power to have a long detection range, which is easy to be intercepted by reconnaissance, as shown in Fig. 2.17. For continuous wave (CW) signals, the ratio of average power to peak power is 1, that is, the same detection performance as that of short pulse sequences can be obtained with lower peak power. But continuous wave single frequency signal is not only easy to be intercepted by narrowband receiver but also cannot distinguish multiple targets in range dimension. LPI radar generally uses periodic-modulated continuous wave signals to acquire larger bandwidth and higher range resolution, which is very suitable for pulse compression, including linear (nonlinear) frequency modulation, phase code modulation, frequency shift keying (FSK) modulation, random signal modulation, etc. Through these modulation techniques, the lower radiation energy is dispersed over a wider frequency band, and the probability of signal being intercepted by reconnaissance is greatly reduced.

(3) Power management

Another way to achieve the low probability of interception characteristic is power management. The best LPI strategy is not to radiate signal, and the suboptimal one is to manage the radiated power. Power management is to control the antenna transmission power. On the one hand, it is controlled in space, that is, constraint the transmitted energy within the necessary space by range in order to reduce the probability of interception by reconnaissance; on the other hand, it is controlled in time, which equals to limit the transmitting time of the signal (short dwell time) to reduce the possibility of interception.

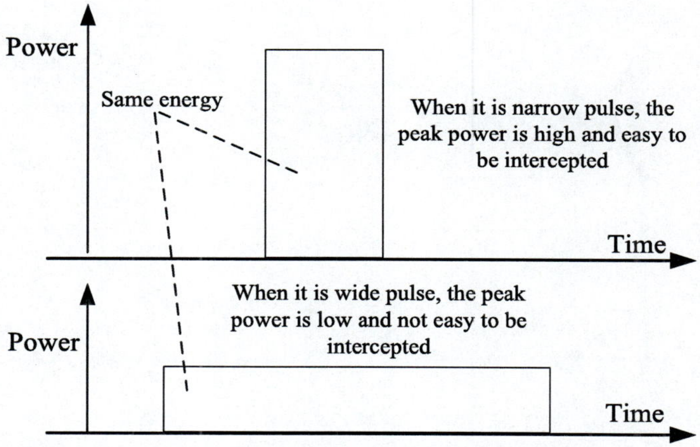

Fig. 2.17 Schematic diagram of wide pulse and narrow pulse

2.1.4.2 WBDF Circuit

Wide band Dicke fix (WBDF) circuit is a traditional anti-jamming measure, which consists of a wideband amplifier, a limiter, and a narrowband amplifier, as shown in Fig. 2.18. It has a good effect in suppressing noise jamming.

The main parameters affecting the noise suppression effect of WBDF circuit are the bandwidth of wideband amplifier, the level of limiter, and the bandwidth of narrowband amplifier. After passing through the wideband amplifier, the noise signal will become the intermediate frequency discrete random pulse signal. In order to ensure that the noise signal through the wideband amplifier is a discrete random pulse sequence, the bandwidth of the wideband amplifier should be greater than the reciprocal of the average interval of the discrete random pulse.

The level of the limiter should be equal to the level of the input signal, because the level of the input signal is changing, so the level of the limiter should also change, generally using automatic gain control (AGC) circuit, automatic control of the gain of the broadband amplifier, so that the output signal level is basically stable. In order to ensure that the output of the limiter is still a discrete random pulse train, the bandwidth of the limiter should be larger than that of the wideband amplifier.

The bandwidth of the narrowband amplifier should ensure no or little loss of the target signal, that is, the bandwidth of the narrowband amplifier should be equal to that of the signal. The noise signal is further suppressed by narrowband amplifier, and the signal to noise ratio of output is further improved.

Fig. 2.18 The composition diagram WBDF circuit

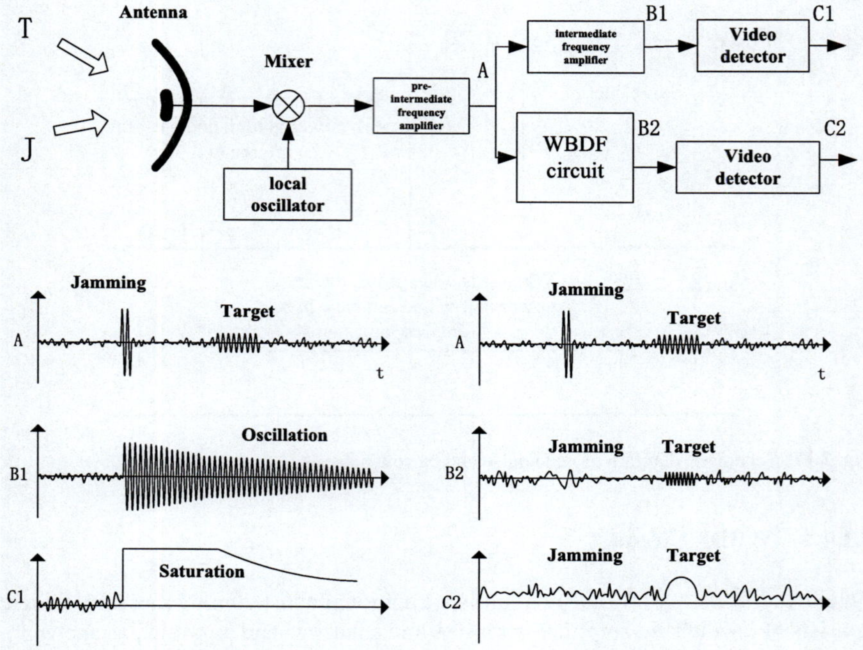

Fig. 2.19 Schematic diagram of jamming suppression for WBDF circuit

Figure 2.19 shows the process of a WBDF circuit processing a mixed signal, which consists of a noise signal and a target signal. In order to show the comparisons, two branches are given in the figure, of which A-B1-C1 branch is the signal that is not processed by the WBDF circuit, and A-B2-C2 branch is the signal that is processed by the WBDF circuit. As can be seen from the figure, the noise of the A-B1-C1 signal causes oscillation in the first stage of the amplifier, which makes the radar receiver saturated and unable to effectively detect the target signal. But the A-B2-C2 signal, by means of the wideband amplifier, limiter, and narrowband amplifier processing, suppressed the noise jamming signal well.

2.1.4.3 Sidelobe Cancelation

Sidelobe cancelation uses the signal received by several auxiliary antennas to replace the signal received by the sidelobe of the main antenna. By processing the signal samples received by the auxiliary antenna and the main antenna properly, the interference signal copy with the characteristics of the interference signal received by the sidelobe of the main antenna is obtained, and then the interference signal copy is canceled with the signal received by the main antenna, so as to suppress the jamming signal received from the sidelobe and retain the target echo signal. Sidelobe cancelation can suppress the interference with high duty ratio and similar noise received by the sidelobe.

There are generally two processing methods for sidelobe cancelation systems: open-loop processing and closed-loop processing. A significant advantage of the open-loop processing method is that the convergence time is very short, which can meet the requirements of real-time processing; the biggest disadvantage of the closed-loop processing method is that the convergence time is too long.

Figure 2.20 is a schematic diagram of sidelobe cancelation with N channels, where Y represents the signal received by the main antenna; X_1, X_2, ..., X_N stand for the signal received by the auxiliary antenna; and W_1, W_2, ..., W_N are the weighting coefficients. The function of the weighting coefficient is to adjust the amplitude and phase of the auxiliary antenna to the jamming signal received by the main antenna, so that the jamming signal received by the auxiliary antenna can be completely canceled with the jamming signal received by the main antenna.

The process is as follows: the jamming signals which received by the sidelobe of the main antenna and the auxiliary antenna are sent to the sidelobe cancelation processor at the same time. According to the corresponding algorithm to calculate the optimal weight W, after cancelation, in the output signal, the jamming signal has been suppressed.

In fact, sidelobe cancelation is equivalent to the concave point formed in the direction of interference by the radar antenna pattern, as shown in the schematic diagram (Fig. 2.21).

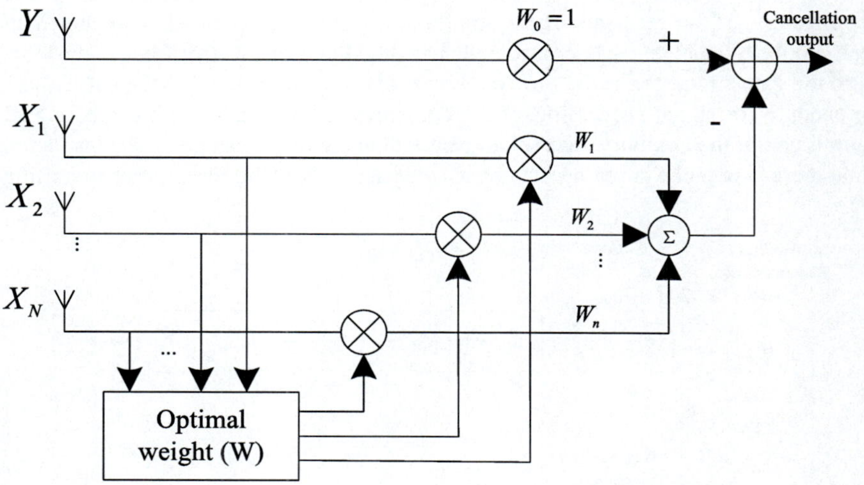

Fig. 2.20 Schematic diagram of sidelobe cancelation

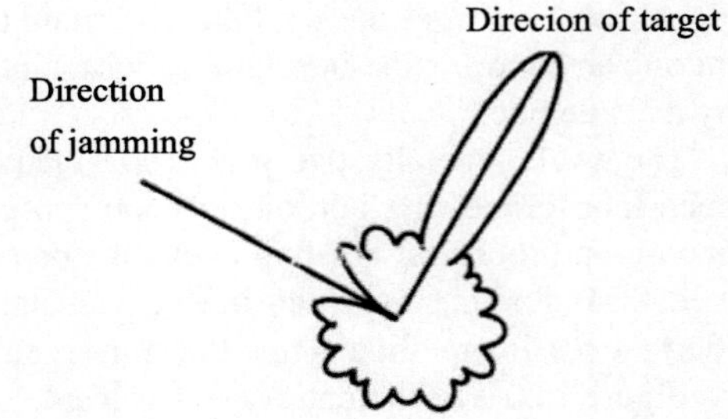

Fig. 2.21 Schematic diagram of the sidelobe cancelation antenna pattern

2.1.4.4 Sidelobe Blanking

By adding an auxiliary antenna and receiving channel in the radar system, the output signals of the main channel and the auxiliary channel are fusion processed, so that the interference signals received by the sidelobe of the radar antenna cannot seriously affect the work of the radar and can also effectively deal with the interference of the neighboring radar from the same frequency band.

The composition of sidelobe blanking is shown in Fig. 2.22, which consists of main and auxiliary channels, as well as transceiver antennas, receivers, and comparators. The main channel antenna usually boasts a high-gain main lobe and many side lobes with decreasing gain. The auxiliary antenna generally uses a weakly directional omnidirectional antenna whose gain is greater than sidelobe, but less than the gain of main lobe.

The working principle of sidelobe blanking is as follows: judging by the comparator, if the received signal amplitude of the main channel is greater than the auxiliary channel, it is considered that the signal enters from the main lobe, and the gate circuit sends the received signal of the main channel to the subsequent procedure for signal processing. If the comparator judges that the received signal amplitude of the auxiliary channel is greater than the main channel, it is considered that there is sidelobe jamming. It controls the gate to close the circuit, thus inhibiting

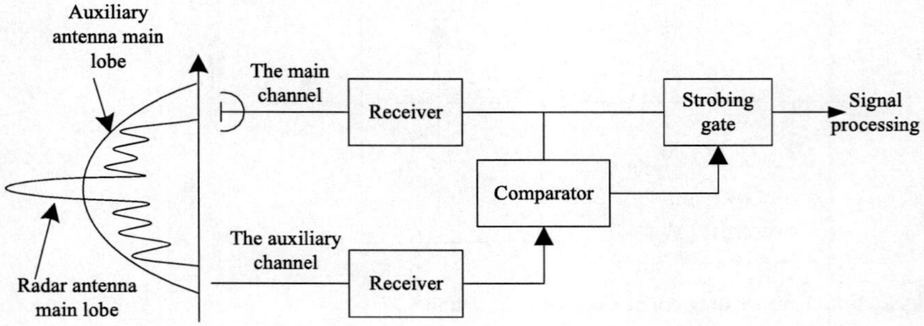

Fig. 2.22 Schematic diagram of sidelobe blanking

the normal detection of the target echo signal of the main channel by the jamming signal entering from the sidelobe.

2.1.4.5 Automatic Gain Control

Automatic gain control (AGC) is an automatic control technology in which the gain of the amplifier circuit is automatically adjusted with the input signal strength. It is mainly used to suppress the clutter of the amplitude change dramatically or the jamming of amplitude modulation.

The function of the AGC circuit is an automatic control circuit that automatically keeps the output signal amplitude within a small range when the input signal amplitude varied greatly. The composition block diagram is shown in Fig. 2.23, which is mainly composed of level detector, low pass filter, DC amplifier, voltage comparator, control voltage generator, and controllable gain amplifier.

AGC circuit is a feedback control system, its working process is as follows:

(1) When the input signal u_i is small, the amplitude of the output signal u_o is also small, and the voltage u_+ input to the voltage comparator through the feedback circuit (level detector, low pass filter, DC amplifier) is also small. In practice, u_+ must be greater than or equal to u_r. When $u_+ < u_r$, u_+ cannot change the output voltage of the comparator, so it is impossible to generate control voltage u_c to control the gain of the controllable gain amplifier. Currently, the AGC loop effectively does not work. When $u_+ < u_r$, , $u_e = u_c = 0$, in which case, u_r is called the threshold voltage of the comparator.
(2) When the amplitude of the input signal u_i increases to augment the amplitude of the output one, the corresponding DC amplifier output voltage u_+ also increases. When $u_+ \geq u_r$, the output error voltage u_e of the comparator will change, the control voltage will also change accordingly and control the gain of the controllable gain amplifier. At this time, the loop starts working, the gain of the controllable gain amplifier decreases with the increase of the output signal, so that the output signal decreases; conversely, when the input voltage u_i decreases, the output voltage u_o also decreases, and the control signal u_c generated by the loop will increase the gain of the controllable gain amplifier A_u. Through the

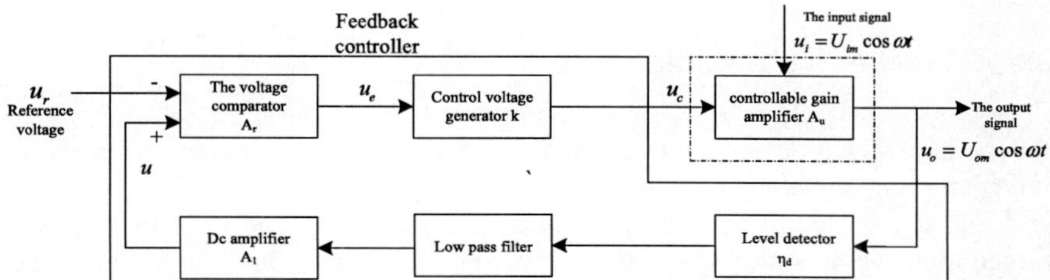

Fig. 2.23 Block diagram of the AGC circuit

control function of the loop, no matter the input voltage u_i increases or decreases, the output signal level u_o changes only in a small range, so as to keep the output signal basically stable when the input signal changes and achieve the purpose of AGC.

2.1.4.6 Anti-RGPO

Anti-range gate pull-off (RGPO) is a technology specially used to counter the RGPO jamming. There are two ways to counter RGPO jamming: the first is the front tracking of pulse echo, the second is to give different weights to the front and back gates.

The pulse front tracking for anti-RGPO schematic is shown in Fig. 2.24, which is mainly composed of receiver, differentiator and range tracking loop. The output of the receiver is sent to the differential circuit, which will eliminate the back edge of all signals that exceed a certain preset value, as shown at T_0 moment in Fig. 2.24. At this moment, only the target signal is present; when the pull-off signal appears but is not completely separated from the target signal, as shown at T_1 in Fig. 2.24, the differentiator eliminates the pull-off signal located at the back edge, leaving only the target signal located at the front edge. When the target signal is completely separated from the pull-off signal, as shown at T_2 in Fig. 2.24, although the differentiator removes the rear edge of the target signal and the tow signal respectively, leaving two signals, the pull-off signal does not play a role because the distance gate is still covered on the first echo.

The principle of the second type for anti-RGPO jamming is shown in Fig. 2.25. When there is no RGPO jamming, at T_0 moment, when $K = 1$, the weighted coefficients of the front and back wave gates are equal, and the range gate is aligned with the signal. When $K > 1$, The weighting coefficient of the front wave gate is greater than that of the back wave gate. The range gate will move forward a range ΔR, which is a predetermined amount; when the pull-off signal appears and is not completely separated from the target signal, as shown in the figure at T_1, because there is the forward movement of the wave gate, so the range tracking system is not affected by the pull-off signal; the same is true when the target signal is completely separated from the pull-off signal; as shown at the moment T_2 in the figure, the pull-off signal does not work.

2.2 Model of Radar Jamming

The model of radar detection is introduced in Sect. 2.1. The introductions of radar active jamming model are as follows.

It is assumed that the transmitted signal of radar is $s(t)$, and the target echo is formed after the action of space and target, and the transfer function of space and target is $h(t)$. It is assumed that the interference signal received by radar is $j(t)$, and without considering the influence of noise, then the radar received signal $s_r(t)$

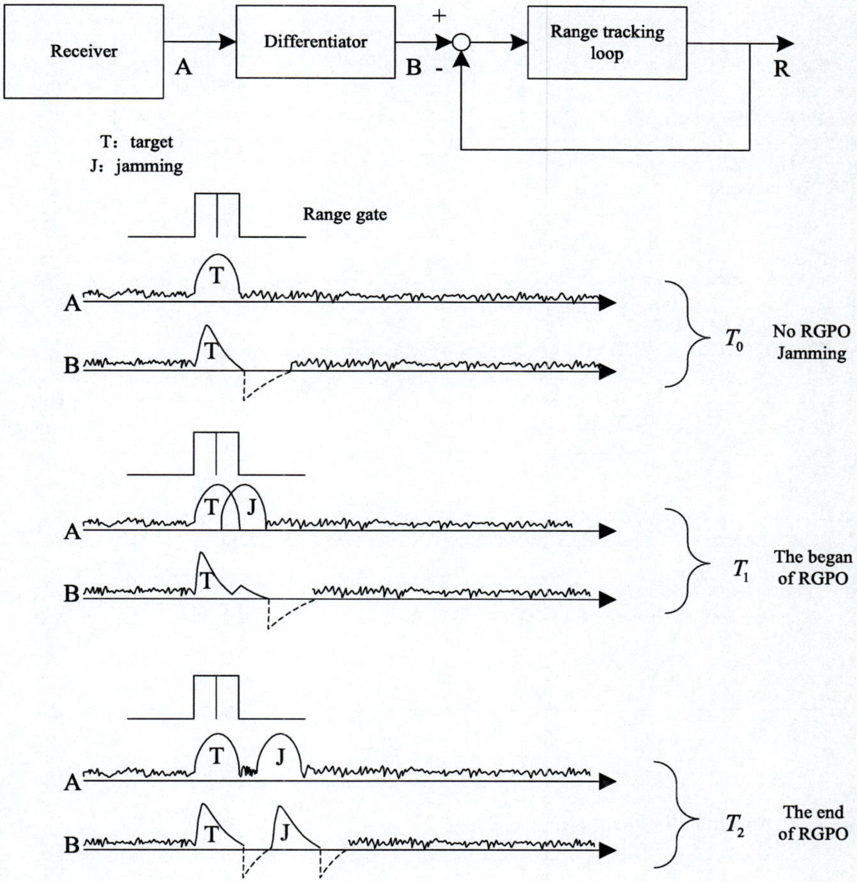

Fig. 2.24 Anti-RGPO with differentiator

represents

$$s_r(t) = s(t) \otimes h(t) + j(t) \tag{2.20}$$

In the formula, $\otimes$ represents convolution. The attribute estimation of space and target $\hat{h}_j(t)$ under jamming conditions can be obtained by formula (2.20)

$$\hat{h}_j(t) = s_r(t) \otimes s^*(-t) = h(t) \otimes p(t) + j(t) \otimes s^*(-t) \tag{2.21}$$

The first term on the right of the equation $h(t) \otimes p(t)$ in formula (2.21) is the real attributes of the target and space, which is the convolution of the real attributes of the target and the point spread function $p(t)$, including the spatial position (range and

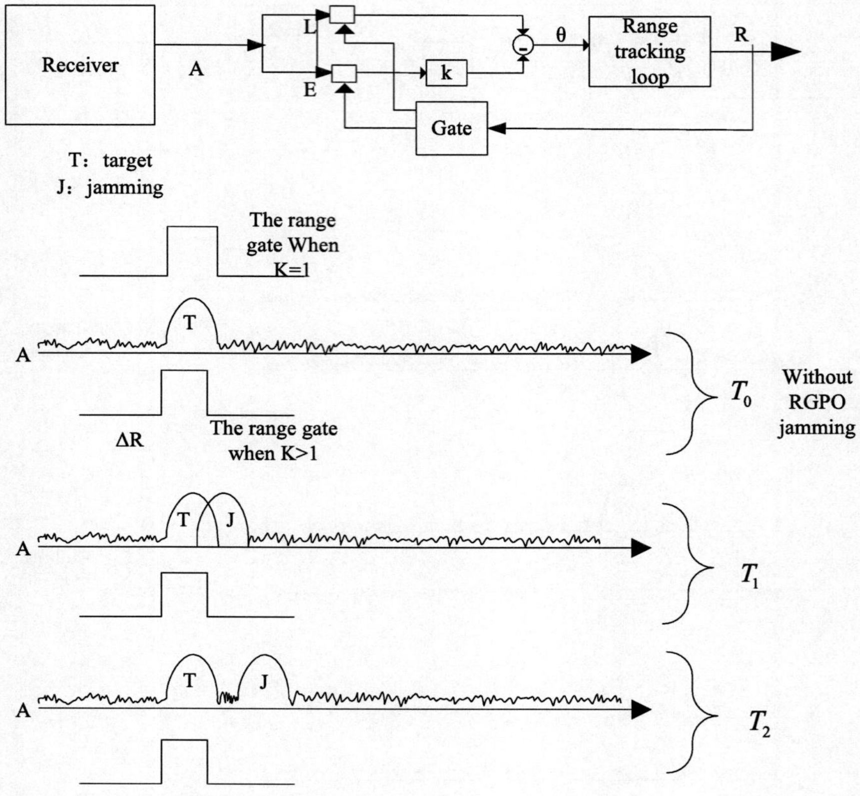

Fig. 2.25 Anti-RGPO with different weights

azimuth), scattering intensity, velocity, polarization characteristics, and so on. The second term $j(t) \otimes s^*(-t)$ is the jamming term. The jamming signal $j(t)$ is divided into noise jamming $j_n(t)$ and deception jamming $j_d(t)$. When the jamming signal is noise jamming $j_n(t)$, the output $j_n(t) \otimes s^*(-t)$ is noise. After the superposition of two terms on the right side of formula (2.21), it obscures the real target and affects the radar target detection. In addition, noise jamming will affect the radar to accurately obtain target's spatial position, scattering intensity, velocity, polarization, and other characteristics. When the jamming signal is deceptive $j_d(t)$, the output $j_d(t) \otimes s^*(-t)$ is a false target, which will confuse or deceive the real target's position, scattering intensity, velocity, polarization, and other characteristics.

2.3 Equation of Radar Jamming

Radar jamming equation mainly describes the relationship between jamming performance and many factors [11, 12].

In Fig. 2.26 [13], $F_s(\Phi, \Theta)$ and $F_j(\Phi, \Theta)$ are the normalized directional graph functions of radar jammer antenna, respectively, (R_j, Φ_j, Θ_j) are the polar coordinates of the jammer with radar as the origin, (R_j, Φ_s, Θ_s) is the polar coordinates of the radar with the jammer as the origin. *Rt* is the range from the radar to the target. The parameters of the jammer are: P_j is the jamming power of the input end of the jammer antenna, G_j is the maximum gain of the jammer antenna, Δf_j is the effective spectrum width of the jamming, γ_j is the polarization difference coefficient between the jammer antenna and the interfered radar antenna, also known as the loss coefficient of the jamming signal (in general, the jamming signal is circularly polarized, the radar antenna is linearly polarized, and the polarization loss coefficient of the jamming signal is $\gamma_j = 0.5$), σ is the radar cross section (RCS) of the target. The radar parameters are: P_s is the loss of radar power on the transmission line, G_s is the maximum gain of the radar antenna; Δf_s is equivalent noise bandwidth of radar receiver, A_s is the effective receiving area of the radar antenna, and $A_s = G_s\lambda^2/(4\pi)$.

According to the geometric relationship shown in Fig. 2.26, considering the situation where the main lobe of radar is directly facing the target region, we can get the power P_{rt} of target echo signal received by radar and the power P_{rj} of jamming signal received by radar which are respectively:

$$P_{rt} = \frac{P_s G_s \sigma A_s}{(4\pi R_t^2)^2} = \frac{P_s G_s^2 \sigma \lambda^2}{(4\pi)^2 R_t^4} \tag{2.22}$$

$$P_{rj} = \frac{P_j G_j G_s \lambda^2 \gamma_j}{(4\pi)^2 R_j^2} \frac{\Delta f_s}{\Delta f_j} F_s(\Phi_j, \Theta_j) F_j(\Phi_s, \Theta_s) \tag{2.23}$$

Where $F_s(\Phi_j, \Theta_j)$ and $F_j(\Phi_s, \Theta_s)$ are related to the relative position of the jammer and the radar and also related to the antenna pattern of the radar and the jammer.

Thus, the interference to signal ratio (ISR) between the jamming signal received by radar and the target echo signal received by radar is

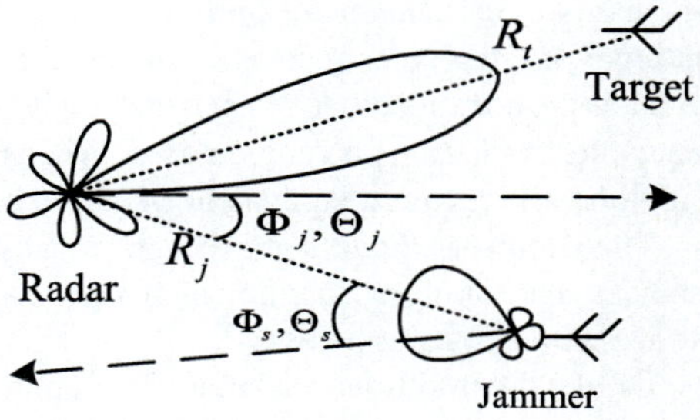

Fig. 2.26 Model diagram of radar jamming space

$$JSR = \frac{P_{rj}}{P_{rs}} = \frac{P_j G_j R_t^4 4\pi}{P_t G_s R_j^2 \sigma} \gamma_j \alpha F_s(\Phi_j, \Theta_j) F_j(\Phi_s, \Theta_s) \tag{2.24}$$

In the formula, $\alpha = \frac{\Delta f_s}{\Delta f_j}$ is the jamming signal bandwidth loss coefficient. In order to achieve effective jamming, the jamming signal ratio must be greater than the suppression coefficient K of the radar receiver, that is:

$$\frac{P_j G_j R_t^4 4\pi}{P_t G_s R_j^2 \sigma} \alpha \gamma_j F_s(\Phi_j, \Theta_j) F_j(\Phi_s, \Theta_s) \geq K \tag{2.25}$$

(1) The effective jamming range of the jammer is

$$R_t^4 \geq K \frac{P_t G_s R_j^2 \sigma}{P_j G_j 4\pi} \frac{1}{\alpha \gamma_j F_s(\Phi_j, \Theta_j) F_j(\Phi_s, \Theta_s)} \tag{2.26}$$

From this, we can see that the jamming range is not only related to the transmitting power of the radar and the jammer but also related to the range between the jammer and the radar, the radar cross section of the target, the gain of the radar and the antenna, the pattern of the radar and the antenna, and the jamming signal bandwidth loss coefficient. In order to achieve a larger range of jamming effects, we can start from several aspects like reducing the radar cross section of the target, increasing the equivalent radiated power of the jammer, making the jamming signal frequency band the same as the target signal, making the direction of the jammer's antenna as close as possible to the radar, making the required suppression coefficient as small as possible through a better jamming mode.

(2) The equivalent radiated power of the jammer

$$P_j G_j \geq K \frac{P_t G_s R_j^2 \sigma}{4\pi R_t^4} \frac{1}{\alpha \gamma_j F_s(\Phi_j, \Theta_j) F_j(\Phi_s, \Theta_s)} = P_{j0} \tag{2.27}$$

To achieve the effect of shielding jamming, the equivalent radiation power of the jammer is required to be greater than P_{j0}. When the distance R_t between the target to be covered and the radar decreases, the required equivalent radiation power increases and is inversely proportional to R_t^4. When the distance R_j between the jammer and the radar increases, the required equivalent radiation power increases, and it is proportional to R_j^2. When the target radar cross section increases, the required equivalent radiation power increases. When the jamming signal enters from the radar sidelobe, the required equivalent radiation power increases.

The jamming equation above is the radar jamming equation under normal circumstances. The jamming equation of synthetic aperture radar (SAR) is a little bit special, so we will analyze it below.

Unlike the traditional radar, SAR is an imaging radar. The main purpose of SAR jamming is to reduce the quality of SAR images, thereby to jam with the detection and identification of targets from the images. The imaging algorithm of SAR and how it works are quite different from traditional radars, and the jamming to SAR has its particularity. Assuming that the bandwidth of the jamming signal in the azimuth direction is much larger than the pulse repetition frequency of the radar, and the azimuth power spectrum of the jamming signal is rectangular. The jam-to-signal ratio of the synthetic aperture radar received signal is:

$$JSR = \frac{P_{rj}}{P_{rs}} = \frac{P_j G_j R_t^4 4\pi}{P_t G_s R_j^2 \sigma} \gamma_j \alpha F_s(\Phi_j, \Theta_j) F_j(\Phi_s, \Theta_s) \tag{2.28}$$

The jam-to-signal ratio after imaging processing is:

$$JSR_i = \frac{P_j G_j R_t^4 4\pi}{P_t G_s R_j^2 \sigma} \frac{1}{G_r G_a} \gamma_j \alpha_a \alpha F_s(\Phi_j, \Theta_j) F_j(\Phi_s, \Theta_s) \tag{2.29}$$

Among them, G_r and G_a are, respectively, the range and azimuth gain during the imaging process, and α_a is the power loss coefficient of the jamming signal in the azimuth direction. When the azimuth bandwidth of the jamming signal is greater than pulse repetition frequency (PRF), therefore, $\alpha_a = \frac{B_d}{PRF}$ because the azimuth is directly sampled and processed; when the azimuth bandwidth of the jamming signal is smaller than PRF and greater than B_d, $\alpha_a = \frac{B_d}{PRF}$; when the azimuth bandwidth of the jamming signal is smaller than B_d, $\alpha_a = 1$.

We can see that the jam-to-signal ratio in the SAR image has some lost compared with it at the radar antenna, the reason is that: jam-to-signal ratio loss caused by jamming signals in range and azimuth exceeding the system bandwidth, while the loss of jam-to-signal ratio caused by imaging processing gain. In order to concealing the target, jam-to-signal ratio is required to be greater than the intensity ratio K_{tb} of the target and the background. So, we can get:

$$\frac{P_j G_j R_t^4 4\pi}{P_t G_s R_j^2 \sigma} \frac{1}{G_r G_a} \gamma_j \alpha_a \alpha F_s(\Phi_j, \Theta_j) F_j(\Phi_s, \Theta_s) > K_{tb} \tag{2.30}$$

In order to achieve a better jamming effect, the jamming side can take measures like: point the jammer antenna's maximum gain direction at the enemy radar; the bandwidth of the jamming signal is as close as possible to the bandwidth of the radar signal; reduce the radar cross section of the target; reduce the distance between the jammer and the radar, etc.

2.4 Principle of Radar Active Suppression Jamming

Suppressing jamming in general is to cover or submerge useful signals with noise or noise-like jamming signals to prevent radar from detecting target information [14].

With the continuous development of radar jamming technology, dense false target suppression jamming has appeared.

Noise suppression jamming is to allow the noise signals which is strong enough to enter the radar receiver and to suppress and submerge the useful signal; reduce the signal-to-noise ratio as much as possible; make it difficult for radar to detect; recognize and track the target [15]. Dense false targets suppression jamming is to generate many false targets with the same or similar parameters as the real target, which saturate the target data processing of the radar or confuse the operator, making the operator unable to find the real target from many false target traces [16].

2.4.1 *Noise Blanket Jamming*

According to the jamming signal mode, noise blanket jamming can be divided into radio frequency noise, noise amplitude modulation, noise frequency modulation, and noise phase modulation, which are not detailed here, please refer to relevant references.

According to the relationship between the center frequency f_j and spectrum width Δf_j of the jamming signal relative to the center frequency f_s and bandwidth Δf_r of the radar receiver, noise blanket jamming can be divided into spot jamming, blocking jamming and frequency-swept jamming.

2.4.1.1 Spot Jamming

Generally, the spot jamming satisfies:

$$f_j \approx f_s,\ \Delta f_j = (2 \sim 5)\Delta f_r \tag{2.31}$$

Spot jamming must measure the radar signal frequency f_s first and then adjust the jammer frequency f_j to the radar carrier frequency to ensure that the narrower Δf_j can cover Δf_r. As shown in Fig. 2.27. The advantage of spot jamming is that the jamming power in the radar Δf_r is strong, which is the first choice of covering interference. However, the requirements for frequency guidance are relatively high and sometimes difficult to achieve.

2.4.1.2 Blocking Jamming

Blocking jamming generally satisfies:

$$\Delta f_j > 5\Delta f_r, f_j \approx f_s \in \left[f_j - \Delta f_j/2, f_j + \Delta f_j/2\right] \tag{2.32}$$

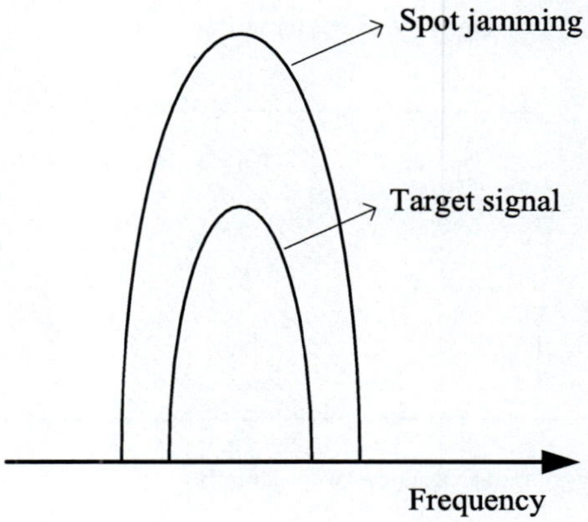

Fig. 2.27 Schematic diagram of spot jamming

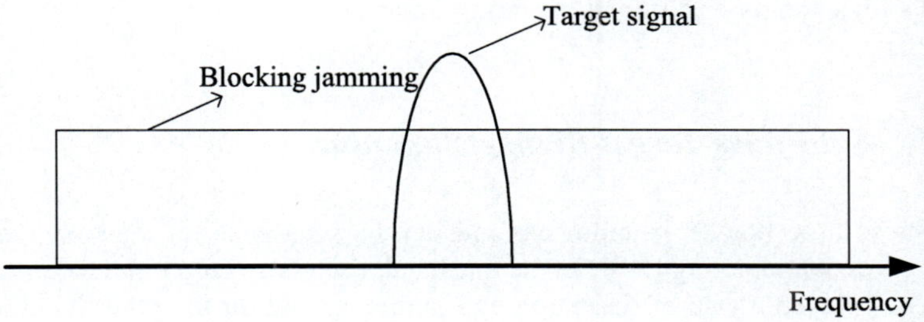

Fig. 2.28 Schematic diagram of blocking jamming

Blocking jamming has a wider Δf_j. On the one hand, the accuracy requirements for frequency guidance are reduced, making the frequency guidance equipment simple; on the other hand, it is also convenient to jam with frequency diversity radar, frequency agility radar, and multiple radars with different operating frequencies at the same time. However, the jamming power density of blocking jamming is low, as shown in Fig. 2.28.

2.4.1.3 Frequency-Swept Jamming

Frequency-swept jamming generally satisfies:

$$\Delta f_j = (2 \sim 5)\Delta f_r f_j = f_s \pm kt, t \in [0, T], k \text{ is constant} \tag{2.33}$$

That is, the center frequency of the jamming is a continuous function with T as the period. As shown in Fig. 2.29 frequency-swept jamming can cause periodic

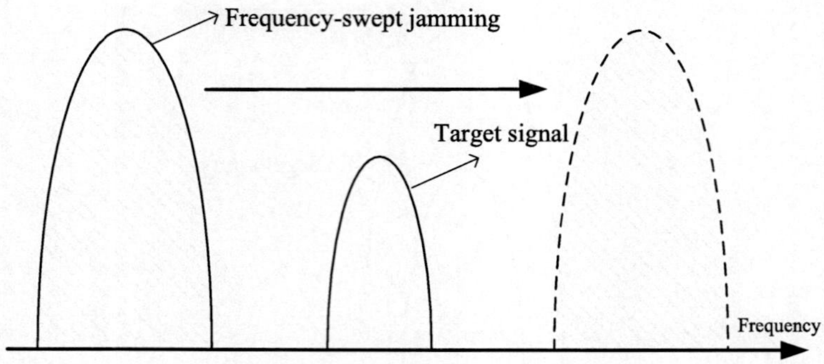

Fig. 2.29 Schematic diagram of frequency-swept jamming

and intermittent strong interference to the radar. When the frequency-swept range is wide, it can also jam with frequency diversity radars, frequency agile radars, and radars with a variety of different operating frequencies.

2.4.2 Multi-false Target Blanket Jamming

Multi-false target blanket jamming has both the characteristics of deception jamming and suppression jamming [17]. On the one hand, each false target in the dense false target has played a role of deception and jamming, and on the other hand, many dense false targets entered the radar receiver, making the radar processing system in a saturated state, which plays a role in suppressing jamming.

DRFM dense repetitive forwarding is a common way to generate dense false targets. The jammer intensively repeats the signals stored in the DRFM within a certain period and according to certain rules, which can form a dense false target. If the radar signal sample intercepted by the jammer is $s(t)$, DRFM modulates the sample signal to a certain extent and repeats it intensively according to needs. The resulting dense false target jamming can be expressed as:

$$j(t) = \sum_{n=1}^{N} A_n s(t - \tau_n) \tag{2.34}$$

Among them, N is the number of dense false targets; A_n is the amplitude of each false target; τ_n is the time interval between false targets. Normally A_n and τ_n do not change, then the radar side receives dense false targets of equal amplitude and equal interval. τ_n can also change according to a certain law or randomly, and on the radar side, the distance between false targets exhibits a certain law or random change.

Formula (2.34) only writes amplitude modulation and time domain modulation, and Doppler frequency shift modulation can also be performed as required.

2.5 The Principle of Radar Active Deception Jamming

The principle of deceptive jamming is to use false target information to act on the radar's target detection and tracking system, so that the radar cannot detect correctly the real target or measure the parameters of the real target, so as to achieve the purpose of radar confusing and disturbing the detection and tracking of the real target [18, 19].

Assuming that V is the detection space of the radar to the target, and target T is an element in the detection space V, or $T = \{R, \alpha, \beta, f_d, S\} \in V$, where x is the range, azimuth, elevation angle, Doppler frequency, and echo power of the target, respectively. According to the difference of the parameter information of the false target t and the real target T in V, the deception interference can be divided into the following categories. According to the difference of the parameter information of the false target T_f and the real target T in V, the deception jamming can be divided into the following categories.

2.5.1 *Range Deception Jamming*

The range deception jamming can be defined as:

$$R_f \neq R, \alpha_f \approx \alpha, \beta_f \approx \beta, f_{df} \approx f_d, S_f > S \tag{2.35}$$

Among them, $R_f, \alpha_f, \beta_f, f_{df}, S_f$ is the range, azimuth, elevation angle, Doppler frequency, and echo power of the false target T_f. Range deception jamming means that the false target is in range different from the real target, and the energy is generally stronger than the real target, and the rest of the parameters are generally the same as the real target.

The range gate pull-off jamming means that by changing the time delay of the range deception jamming continuously, the radar range tracking is gradually dragged away from the target, and the radar range tracking system is induced to track the false target. The schematic diagram of the jamming principle of range gate pull-off (RGPO) is shown in Fig. 2.30.

Range gate pull-off jamming is generally divided into four phases: capture, pull-off, staying, and stopping. As shown in Fig. 2.31, in the capture phase, after the jammer receives the radar emission pulse, it forwards an amplified copy signal immediately and keeps it for a period, so that the jamming signal and the target echo signal act on the same range gate; in the pull-off phase, the jammer gradually increases or decreases the forwarding delay time, so that the radar range tracking gate gradually deviates from the target. Figure 2.31 shows the situation where the jamming delay

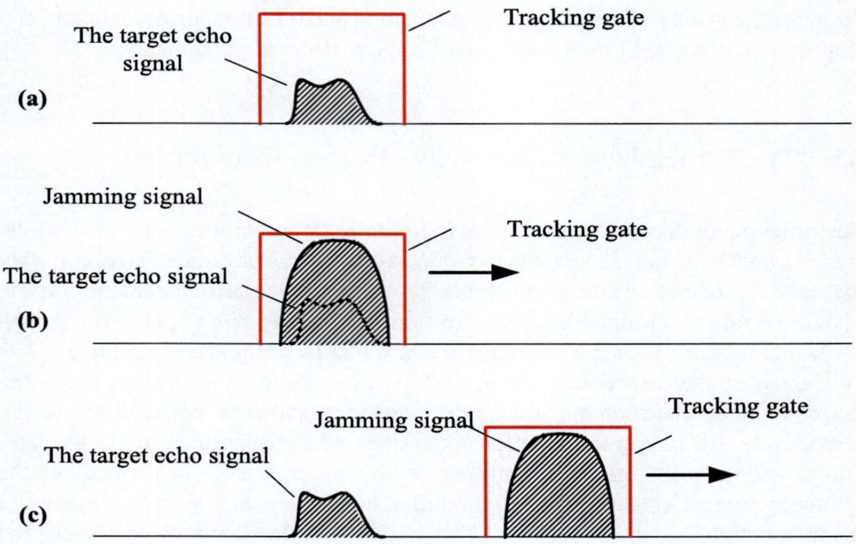

Fig. 2.30 Schematic diagram of range gate pull-off jamming

time gradually increases; in the staying phase, the delay time of the retransmission pulse no longer increases, so that the radar can track the false target stably; in the stopping phase, the jamming is turned off, the radar enters in the state of searching state, and a pull-off cycle is completed. To ensure the continued effectiveness of the pull-off jamming, the jammer usually repeats the above-mentioned pull-off process. In practical applications, some range gate pull-off jamming process may not have a staying or stopping phase.

2.5.2 *Velocity Deception Jamming*

Velocity deception jamming can be defined as:

$$R_f \approx R, \alpha_f \approx \alpha, \beta_f \approx \beta, f_{df} \neq f_d, S_f > S \tag{2.36}$$

Velocity deception jamming means that the false target is different from the real target in Doppler frequency shift, the energy is stronger than the real target, and the rest of the parameters are generally the same as the real target, as shown in Fig. 2.32.

Similar to the range gate pull-off jamming, the Doppler frequency shift of the jamming is deceived by continuously changing velocity, and the radar velocity tracking gate is gradually dragged away from the target to induce the radar speed tracking system to track the false target. This is the velocity pull-off jamming. The velocity deception jamming is mainly aimed at the tracking phase of the terminal

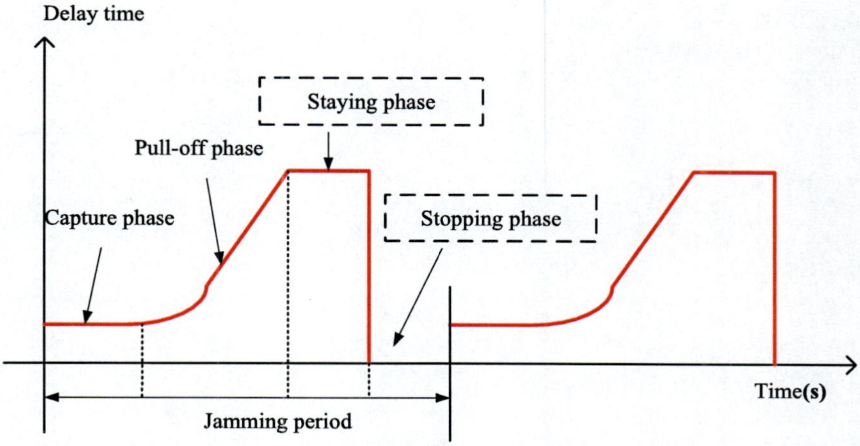

Fig. 2.31 Timing diagram of range gate pull-off jamming

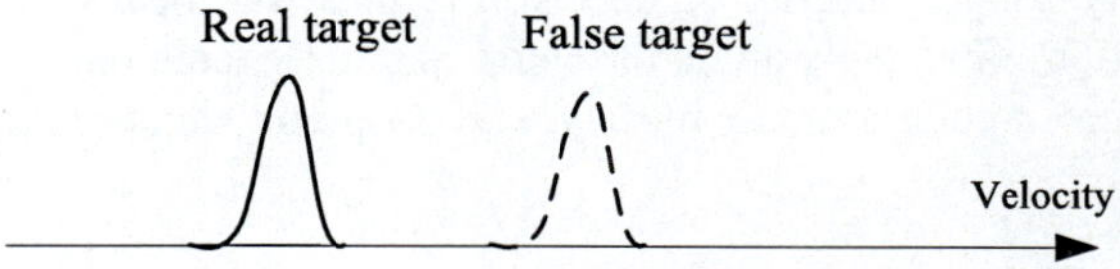

Fig. 2.32 Schematic diagram of velocity deception jamming

guidance radar. The principle and process of the velocity pull-off jamming are similar to the range pull-off jamming, so it would not be repeated here.

2.5.3 Angle Deception Jamming

Angle deception jamming can be defined as:

$$R_f \approx R, \alpha_f \neq \alpha or \beta_f \neq \beta, f_{df} \approx f_d, S_f > S \tag{2.38}$$

Angle deception jamming means that the false target is different from the real target in azimuth or pitch and has stronger energy than the real target, while other parameters are generally the same as the real target as shown in Fig. 2.33.

For the traditional cone scan tracking radar, the use of phase-inversion jamming inversely proportional to the amplitude of the received signal can deceive the radar's angle measurement. With the development of radar technology, modern radars have basically adopted a monopulse angle measurement system, and it is

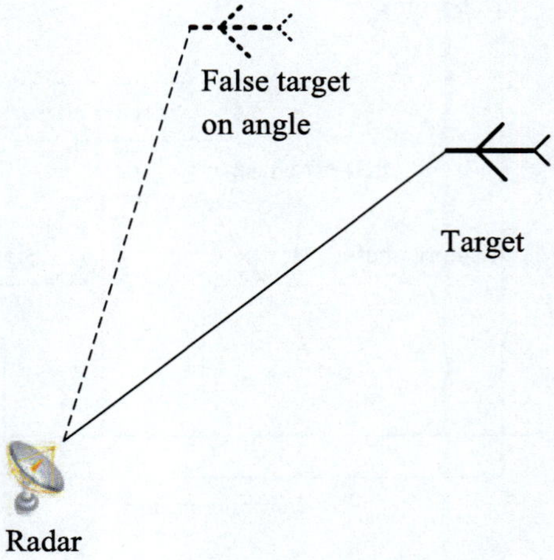

Fig. 2.33 Schematic diagram of angle deception jamming

no longer possible to deceive the radar by means of jamming amplitude modulation. The newly angle deception jamming technologies include cross-eye jamming, cross-polarization jamming, and off-platform jamming.

2.5.4 Multi-parameter Deception Jamming

Multi-parameter deception jamming means that the false target has two or more than two parameters in V that are different from the real target. With the continuous development of radar technology, many radars have the ability to resist single-parameter deception jamming. The traditional single-parameter jamming effect is greatly reduced. Thus, Multi-parameter deception jamming will be an important development direction in the future.

Radar active jamming is to disrupt or block the target detection and tracking of enemy radar by generating radio signals from electronic equipment. It has the advantages of flexible and controllable jamming power, jamming mode, and jamming effect and is an important radar countermeasure. The principle of radar active jamming technology is the theoretical basis for the design of radar active jamming system.

This chapter introduces the basic principles related to radar and radar jamming. Section 2.1 introduces the basic principles of radar detection, including the physical model of radar detection, radar equations, typical radar technologies, and typical radar anti-jamming technologies; Sect. 2.2 introduces radar jamming models; Sect. 2.3 introduces the radar jamming equation; Sect. 2.4 introduces the principle of radar blanket jamming, which not only analyzes noise blanket jamming, but also introduces the newer dense false targets suppression jamming; Sect. 2.5 introduces the principle of radar deception jamming.

References

1. Kolnik MI (2008) Radar handbook (3rd edn) [M]. McGraw-Hill Professional
2. Lufei D, Jianchun C (2009) Radar theory [M], 4th edn. Publishing House of Electronics Industry, Beijing
3. Zhaowen Z, Shunping X, Xuesong W (1999) Radar polarization information acquisition and and its application [M]. National Defense Industry Press, Beijing
4. Xianda Z (2003) Modern signal processing [M]. Tsinghua University Press, Beijing
5. Zheng B, Mengdao X, Tong W (2005) Radar imaging [M]. Publishing House of Electronics Industry, Beijing
6. Yongtan L (1999) Radar imaging technology [M]. Harbin Institute of Technology Press, Harbin
7. Chengbo Z (1989) Principle, system analysis and application of synthetic aperture radar [M]. Science Press, Beijing
8. Debao C (2002) Modern radar anti-countermeasures technology [M]. Aviation Industry Press, Beijing
9. Pace PE (2004) Detecting and classifying low probability of intercept radar [M], 2nd edn. Artech House Inc., Norwood
10. Taylor JD (2001) Ultra-wideband radar technology [M]. CRCPress, London
11. Guoqing Z (2003) Principle of radar countermeasure [M]. Xidian University Press, Xi'an
12. Yongshun Z, Ningning T, Guoqing Z (2005) Principle of radar electronic warfare [M]. National Defense Industry Press, Beijing
13. Rongbing G (2006) Research on synthetic aperture radar countermeasure and target detection technology [D]. Doctoral Dissertation of University of Electronic Science and Technology of China
14. De Martino A (2012) Introduction to modern EW system [M]. Artech House Inc., Norwood
15. Adamy DL (2001) EW101: a first course in electronic warfare [M]. Artech House Inc., Norwood
16. Adamy DL (2004) EW102: a second course in electronic warfare [M]. Artech House Inc., Norwood
17. Schleher DC (1999) Electronic warfare in the information age [M]. Artech House Inc., Norwood
18. Neri F (2001) Introduction to electronic defense system [M]. Artech House Inc., Norwood
19. Li XQ et al (2008) Integrated electronic Warfare—Killer Mace in information war [M], 2nd edn. National Defense Industry Press, Beijing

Chapter 3
System Design of Radar Active Jamming

The system design of radar active jamming is the process of designing a system which can meet the requirements of use on the basis of system analysis. The work content of radar active jamming system design mainly includes: analysis of military demands such as the combat capability and operational environment of the jamming system, and put forward the functional index requirements and technical index requirements of the system; select and determine the overall design plan; propose the system architecture, workflow and main methods of jamming; decompose the radar active jamming system into several subsystems, and determine the functions, technical indicators, and interface relationships of each subsystem.

This chapter mainly introduces the overall design method of radar active jamming system, which serves as the general outline and basis for the subsequent design of other subsystems. In this chapter, Sect. 3.1 mainly introduces the concept of radar active jamming system, including information model, basic composition, workflow and technical indicators. Sections 3.2 and 3.3, respectively, describe common design criteria and general design process for radar active jamming system. Section 3.4 introduces the design method of radar active jamming system and proposes design ideas from the aspects of requirement analysis, system model, capability analysis, choice of key technological regimes, design and decomposition of key technical indicators, engineering and product design, etc. Section 3.5 introduces several typical radar active jamming systems, and the last section introduces the development trend of radar active jamming systems in the future.

G. Tang et al., *Techniques and System Design of Radar Active Jamming*,
https://doi.org/10.1007/978-981-19-9944-4_3

3.1 Overview of Radar Active Jamming System

3.1.1 Information Model

Electronic warfare is a typical information warfare. From the perspective of information theory, it can be considered that radar jamming is a negative impact on the exchange of external information of radar and other equipment, which may include blocking, delaying the process of information transmission, and weakening, tampering, and deleting the transmitted information to destroy the integrity and accuracy of the information [1]. The information in the radar active jamming system mainly comes from protected targets, background radiation, and other electromagnetic environments on the battlefield. The information processing flow includes information acquisition, information analysis, and response to decision-making, which can be explained by OODA Loop.

The OODA Loop was originally proposed by the US Air Force Colonel John Boyd [2]. Its name comes from the four phases including Observe, Orient, Decide, and Act, as shown in Fig. 3.1. The OODA Loop theory is that armed conflict can be seen as a rivalry between the two sides to compete for who can complete the "Observe–Orient–Decide–Act" loop faster and better. The OODA Loop is the model of the human thought process, which starts with the observation of the target and the environment, and then classifies the observation objects according to the existing knowledge, and decides the corresponding action to deal with the situation. After the action is adopted and executed, the environment will be changed or responded to in some ways, and these changes will be observed again, making the loop go back and forth [3, 4].

The competition between radar jamming system and radar system is a dynamic game. The information processing OODA Loop of radar active jamming system is shown in Fig. 3.2, which mainly includes the following parts:

(1) Environmental observation. Serving as the input of the information system, it mainly collects and obtains information. The radar active jamming system uses antennas, receivers, and so on to collect information and data of battlefield

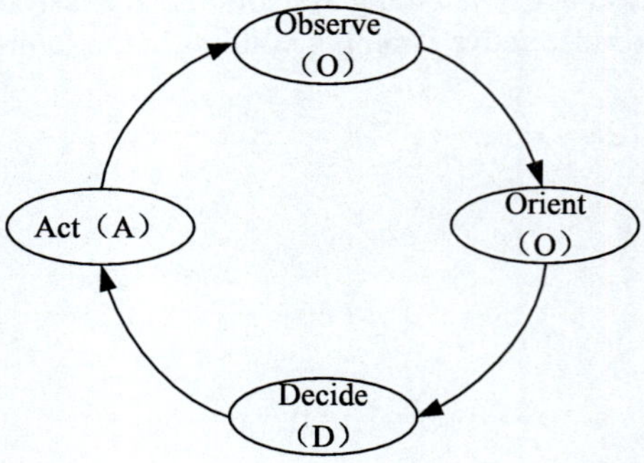

Fig. 3.1 OODA loop

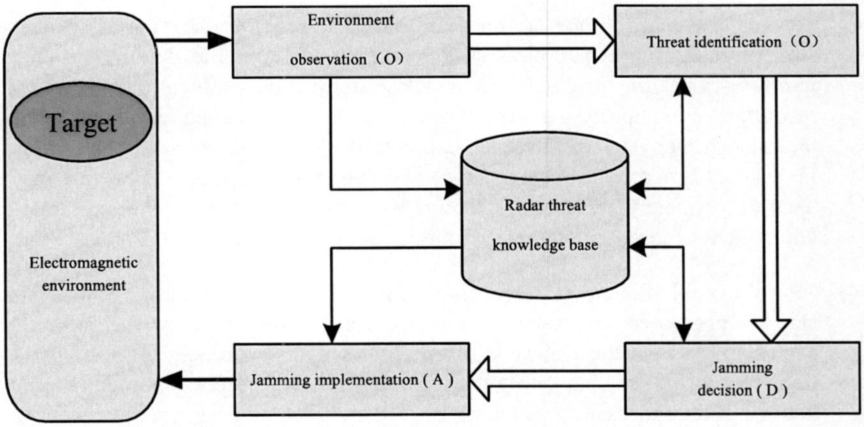

Fig. 3.2 OODA model of radar active jamming system

electromagnetic environment. The information can be directly obtained by the receiver measuring the time domain, frequency domain, spatial domain, polarization domain, energy domain, and other multi-dimensional parameters of the radiation source signal. It can also be formed by attribute reduction and dimensionality reduction to further extract the characteristic parameters. The unqualified data that affects the observation during the information acquisition process also need to be repaired or deleted to provide accurate and appropriate information for subsequent processing. The observed information and the enemy's information receiver are both potential targets for jamming system.

(2) Threat identification. It means identifying interested objects and events by comparing newly acquired information with existing information. The existing information is that has been previously observed and can be understood by the system; the existing information can be used to analyze and identify the newly acquired information. If the information cannot be identified, a new element will be created for it and associate the related information with it according to the individual's understanding, and store it in the knowledge base for the next identification process. The radar active jamming system detects, separates, selects, and recognizes the target radar and the information of battlefield electromagnetic environment through a series of signal and data processing, which helps to recognize the object and judge the current situation. Advanced systems can also use known knowledge to detect and identify unknown things through deductive reasoning and other methods when the information is incomplete.

(3) Jamming decision. It mainly according to the previous information and identification results, in order to destroy the normal transmission of enemy system information, form the method and parameters of subsequent information processing,

develop the interference action plan and task scheme. The jamming decision of radar active jamming system mainly includes the determination of jamming target, the formulation of attack strategy, and the allocation of jamming resources. When the system needs to jam multiple radar targets at the same time, jamming strategy and resource allocation become more important. It is reflected in reasonable distribution of sequence of reconnaissance interference, fast guidance of frequency and angle, adaptive tracking of jamming target, automatic adjustment of jamming waveform and jamming power, etc.

(4) Jamming implementation. According to the decision information of the previous step, it processes and releases the information obtained by the jamming system to prevent the enemy's information system from acquiring, transmitting, processing, or controlling the information. The jamming implementation actions of the radar active jamming system include jamming signal generation, jamming signal amplification and jamming signal emission, etc., so that the jamming signal and the real target echo will act on the threatening radar and the electromagnetic environment. Notice that since OODA is a circle, on the one hand, the jamming signal generated by the radar active jamming system may be observed by the receiver of the jamming system along with the radar signal, so it requires measures such as transceiver isolation and cancelation to reduce the influence of small leakage jamming signals on the jammer's receiver to continue to observe the radar and the environment as much as possible [5]. On the other hand, after the jamming is implemented, the jamming signal will have a certain impact on the threat radar and the electromagnetic environment. The receiver processor of the radar active jamming system can analyze the jamming effect by observing the changes of the target radar and the environment [6, 7].

Through the OODA Loop, all the processing procedures of the radar active jamming system constitute a complete cycle. According to the above-mentioned OODA cycle theory, shortening the cycle time as much as possible and obtaining and controlling information more efficiently and accurately is the key to victory, which corresponds to the two important requirements of "Fast Reaction" and "Accurate Processing" in electronic warfare. The two features, respectively, represent two important aspects of radar active jamming technology research [8].

3.1.2 Basic Composition

Radar generally transmits electromagnetic signal actively. By receiving and analyzing the reflected echo of the target, it can find the target and measure the parameters of the target. The basic function of the radar active jamming system is to transmit an appropriate electromagnetic signal to hinder or destroy the normal operation of the radar and achieve the purpose of protecting specific targets or specific areas. Therefore, the work of one simplest radar active jamming system should include the

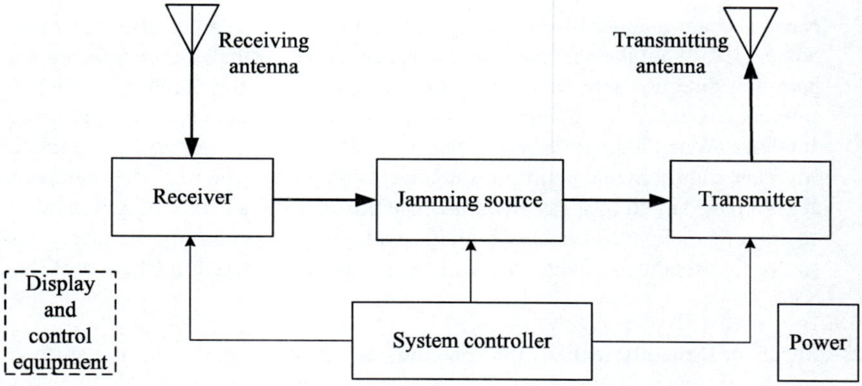

Fig. 3.3 Block diagram of radar active jamming system

following steps. At first, the signal modulator generates the radar jamming signal, then the transmitter amplifies the signal and feeds it to the antenna, and then the electromagnetic wave is radiated by the antenna. In order to realize the function of "transmitting appropriate electromagnetic wave signals", two conditions are needed. One is that the signal modulator is controllable and can generate jamming signals that meet the demands of different tasks. The other is that a receiving sensor can detect and identify threatening radar signals. To maintain the normal operation of this process, the system controller is also essential.

The radar active jamming system can be divided into the hardware system and the software system. From the functional structure, the radar active jamming system can be roughly divided into three parts: reconnaissance, jamming, and control.

The schematic diagram of the composition of a typical radar active jamming system is shown in Fig. 3.3, which mainly includes the following basic functional units: [9–15]

(1) Antenna. It acts as a transducer between the electromagnetic energy transmitted in space and the signal on the radio frequency transmission line and is an interactive channel between the radar active jamming system and the threat radar or the external environment. According to the design needs, the radar active jamming system can adopt independent receiving antenna and transmitting antenna where the receiving antenna is used to receive radar signals, and the transmitting antenna is used to transmit jamming signals. It can also use the same antenna to achieve signal receiving and jamming emission by using circulators or switches, etc. Broadband antenna is usually used in the design of radar active jamming system to meet the requirements of covering multiple frequency ranges as much as possible. According to the division of the main beamwidth, the antennas of the active jamming system can be divided into wide-beam antennas and narrow-beam antennas. Wide-beam antennas can instantaneously cover targets in a relatively larger space, but the antenna gain is low.

Narrow-beam antennas have high gains and can obtain larger effective radio power (ERP), but the space instantaneous coverage is small and more accurate jamming direction guidance is required. According to the number of antenna elements, it generally divided into single antenna, multiple antennas, or array antennas. When a single antenna cannot meet the space or frequency domain coverage requirements, multiple antennas can be used. The receiving antennas are often measured by array antennas. The transmitting antenna adopts an array antenna to achieve fast beam pointing and high-gain jamming. The principles and design methods of antennas will be introduced in detail in Chap. 4 of this book.

(2) Receiver. It mainly realizes the functions of intercepting, measuring, sorting, and identifying radar signals, and its processing results are the premise and basis of radar active jamming system for threat warning and jamming. The receiving processor mainly includes three parts: signal receiving, signal processing, and data processing. The receiver mainly completes the tasks of signal amplification, down-conversion, signal detection, and parameter measurement, while the signal and data processor complete the tasks of signal sorting, information fusion, and threat identification. The receiver of the radar active jamming system mainly includes a direction-finding receiver and a frequency-measuring receiver [10], which are used to measure the direction and frequency of the radar radiation source. By processing the intercepted radar signal, the receiver can obtain many important information such as radio frequency (RF), direction of arrival (DOA), time of arrival (TOA), pulse width (PW), pulse amplitude (PA), pulse repetition interval (PRI), signal bandwidth (BW), modulation on pulse (MOP), and pulse modulation parameters. Based on this, the signal and data processor can make radar identification and threat warning and finally guide the jammer to produce the corresponding jamming signal and enable the transmitting antenna to be aimed at the target radar. The electronic warfare receivers are generally required to have a wide frequency domain; that is, the instantaneous bandwidth is from hundreds of MHz to GHz, and the operating frequency coverage reaches a few GHz or even a dozen GHz. The working principle and design method of the receiving processor will be described in detail in Chap. 5.

(3) Jamming source. It mainly generates appropriate jamming signals based on the guidance of receiver and system controller. It is the key unit of the radar active jamming system. When jamming different radar systems, the jamming source needs to have a variety of different jamming styles. When jamming multiple radars at the same time, the jamming source must have abundant jamming resources and efficient resource scheduling capability. According to the different ways of generating jamming signals, radar jamming sources are generally divided into noise jamming, response jamming as well as repeater jamming. The noise jamming source is the most common, it produces high-power noise jamming signals, which are used to suppress target radar. The response jamming source is the jamming signal generated by frequency synthesis technology

according to the signal characteristics of threat radar measured by the receiver, such as multi-frequency synthesis technology and direct digital frequency synthesis technology (DDS). The repeater jamming source directly forwards or modulates the received radar signal to form the jamming signal. Generally, it adopts the working mode of radio frequency memory (RFM) or fiber delay device [16], which can usually accurately simulate the frequency and complex intra-pulse features. It is difficult for the target radar to identify or eliminate jamming. The working principle and design method of the jamming source will be introduced in detail in Chap. 6.

(4) Transmitter. The functional requirements of the transmitter of the radar active jamming system are almost the same as that of the radar's transmitter [17], which is mainly to amplify a low-level controllable jamming signal to the required power level without distortion as much as possible, and finally feed into the antenna for transmission. The main component of transmitter is the power amplifier, which is an important guarantee for the radar active jamming system to effectively jam the target radar. According to the type of device that the transmitter realizes the final power amplification, it can be divided into two types: electrovacuum transmitter and solid-state transmitter. The most used electrovacuum device is traveling-wave tube, which has many advantages such as high gain, large bandwidth, and output signal coherence, but also has disadvantages like that the required working voltage is higher and the volume is larger. The advantages of solid-state transmitter are high reliability, long life, large working bandwidth, low working voltage and can be turned on and off instantaneously, which is convenient for jamming control, while the disadvantage is that the output power is relatively small. The working principle and design method of the jamming transmitter will be introduced in Chap. 7.

(5) System controller. It mainly completes the sequence and parameter control of the whole active jamming processing process, such as radar signal detection, jamming signal generation, and system working mode switch. It is the brain center of the radar active jamming system. In the reconnaissance part, the system controller mainly controls the receiver to receive radar signals and carry out parameter measurement, signal sorting, and threat identification. In the jamming part, the system controller mainly determines the jamming style and parameters, so as to produce the appropriate jamming signal at the appropriate time. In terms of human–computer interaction, the system controller is responsible for monitoring the state of the system and receiving the commands, so as to realize the information interaction between the operator and the radar active jamming system. In radar active jamming systems that are partially miniaturized or require only simple jamming control, the system controller may be integrated into the jamming source. The working principle and design method of system controller will be introduced in Chap. 8.

Other components of the radar active jamming system mainly include power supply and display-control equipment, etc. The power supply provides the required electric energy for the entire system. Common operating voltages include 28, 110, 220, 270 V. The display-control equipment realizes the function of human–computer interaction, which can help electronic warfare operators obtain threat warning information and jamming system status information quickly and intuitively, as well as continuous manual command operations.

It should be noted that radar active jamming systems in different applications generally have different composition patterns. For example, in a repeater active jamming system, the jamming source may be replaced by signal storage and modulators; on a large manned combat platform, the display-control equipment generally has a strong functional configuration, while the functions on platforms such as unmanned air vehicle will be simplified, and may not be available on small platforms such as unmanned aerial vehicles.

When the system needs to have reconnaissance and jamming simultaneously with multiple different radar targets in a complex electromagnetic environment, it may need to use multiple receiver combinations of different frequency bands or different systems, or be equipped with multiple different jamming sources or multiple sets of antennas at the same time.

When the system needs to detect and jam multiple targets at the same time, the active jamming system may need to adopt multiple receivers of different frequency bands and may also be equipped with different jamming sources or multiple antennas.

3.1.3 Processing Flow

The radar active jamming system receives radar signals, after complex signal and data processing, the corresponding jamming signal is generated for the threat target. The typical radar active jamming system's processing flow is shown in Fig. 3.4, which mainly includes the following parts [9–15, 18, 19]:

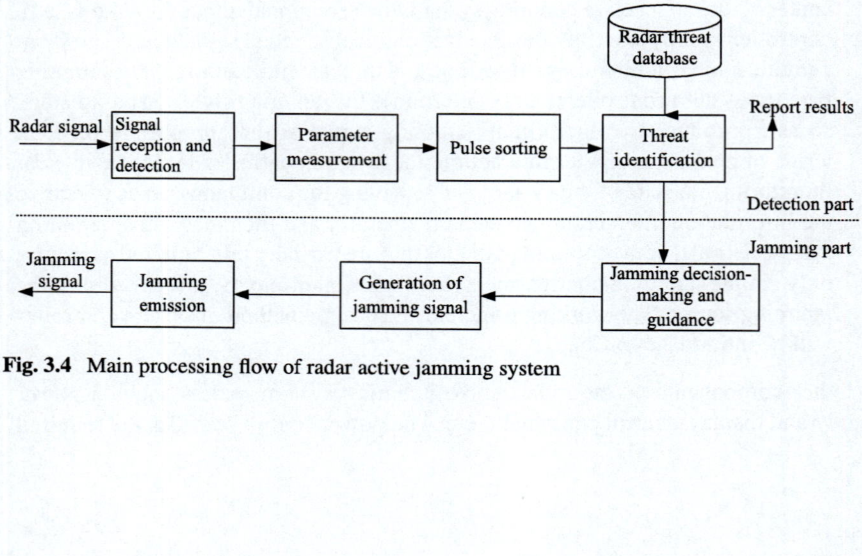

Fig. 3.4 Main processing flow of radar active jamming system

(1) Signal reception and detection
When the radar signal enters the receiver of the jamming system, it usually passes through the RF channel and down-conversion processing first. Among them, the RF channel mainly includes limiter, filter, and low-noise amplifier. The gating characteristics of the system can be adjusted by adjusting the performance of the limiter and band-pass filter, and the amplifier can improve the ability to detect small signals, while down-conversion is the conversion of radar signals from radio frequency to intermediate frequency, which is conducive to the signal's sampling, processing and storage.

Similar to the radar signal processing, the signal detection of the radar jamming system mainly uses the energy of the target signal, which can be divided into time domain detection and frequency domain detection. Among them, the time domain detection is divided into two algorithms according to the different signal objects. One algorithm is to detect the pulse video waveform, and the other is to accumulate and detect the intermediate frequency signal. However, in order to improve the detection efficiency and distinguish the signals that arrive at the same time in the instantaneous receiving bandwidth, the frequency domain signal detection method of the instantaneous coverage airspace is generally adopted, and the received signal is transformed to frequency domain according to the time slice, the signal detection is performed in the frequency domain. A detection threshold is determined according to the prior knowledge of the radiation source and the receiving sensitivity requirements of the jamming system. When the received signal exceeds this threshold, it is considered that a radiation source signal is detected. It is also very important to detect the continuous wave signal, and it often needs to be processed separately after determination; otherwise, it will occupy many resources for the jamming system.

(2) Parameter measurement
Parameter measurement is mainly based on frequency and direction measurement. For pulse signals, typical measurement parameters include radar signal carrier frequency, direction of arrival, pulse arrival time, pulse width, pulse amplitude and signal bandwidth, intra-pulse modulation type, and intra-pulse modulation parameters, etc., the specific meaning of each parameter is shown in Fig. 3.5. Continuous wave signals are represented by parameters other than pulse width and arrival time. The pulse description word (PDW) is formed by parameters above, which can characterize the radar signal. The data stream formed by continuous PDW can be called full pulse data, which can describe the electromagnetic radiation environment of the radar jamming system's reconnaissance receiver as shown in Fig. 3.6.

(3) Pulse sorting
Pulse sorting is also called signal separation. It classifies the intercepted signals according to the principle of belonging to the same radiation source based on PDW and extracts the description information of this type, which is used for

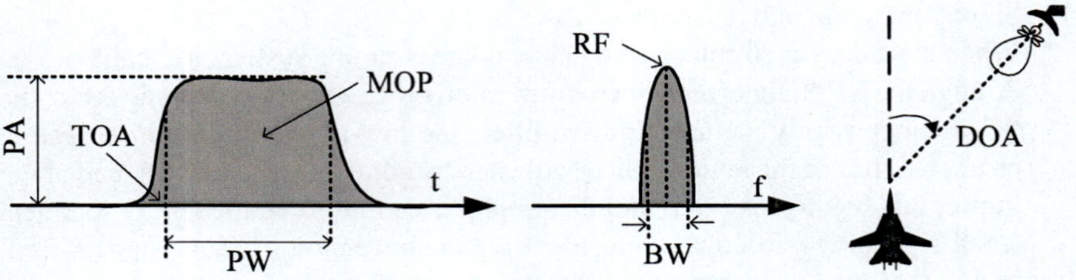

Fig. 3.5 Schematic diagram of pulse measurement parameters

Fig. 3.6 Schematic diagram of PDW [19]

TOA/us	RF/MHz	PW/ns	PA	DOA/du
8969	9470	1000	161	253
8976	9470	1000	161	253
8983	9470	1000	161	253
8990	9470	1000	161	253
8997	9470	1000	161	253
9004	9470	1000	161	253
9011	9470	1000	161	253
9018	9470	1000	161	253
9025	9470	1000	161	253
9032	9470	1000	161	253
9039	9470	1000	161	253
9046	9470	1000	161	253
9053	9470	1000	161	253
9060	9470	1000	161	253
9067	9530	1000	160	254
9077	9530	1000	160	254
9087	9530	1000	160	254
9097	9530	1000	160	254
9107	9530	1000	160	254
9117	9530	1000	160	254
9127	9530	1000	160	254
9137	9530	1000	160	254

subsequent threat identification. The basis of sorting is to analyze the relationship between pulses so that to find and determine the possibility of the existence of each radiation source signal. Pulse sorting requires the use of partially or fully of the parameters of PDW. The most common ones are RF, DOA, and PW. In addition, it is often of vital critical of the pulse repetition interval (PRI) parameter calculated from the TOA of a set of pulses with the same characteristics (Fig. 3.7).

(4) Threat identification

The received radar signal parameters or other characteristics can be matched with the radar threat database to identify the radar type, as well as determine the

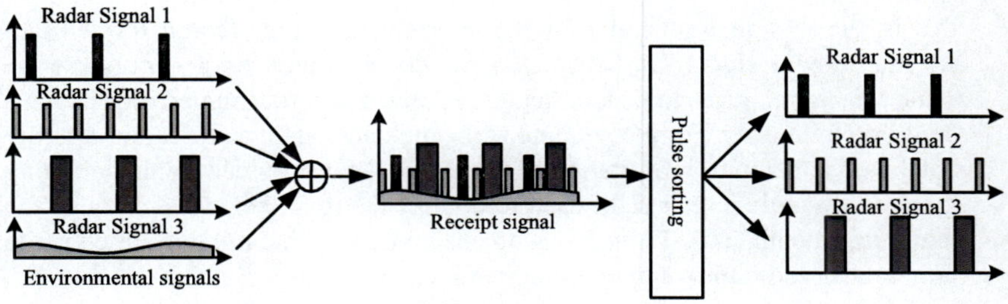

Fig. 3.7 Schematic diagram of pulse separation

target's friend–foe attribute and threat level. According to the degree of identification, it can generally be divided into radar signal identification, type identification, individual identification, and state identification [20–29]. The degree of precision in radar identification is gradually increased, and the required discrimination information is sequentially increased too. The radar threat database is a formatted description of known radar information, including pulse description words and corresponding radar model, platform category, threat level, and other information. The information in the radar threat database mainly comes from the investigative results of the electronic intelligence system and other intelligence data.

(5) Jamming decision-making and guidance
Different jamming decisions are completed mainly according to the results of threat identification, that is to determine the target and priority of jamming, allocate system jamming resources and issue jamming parameters. Among them, jamming resources include time resources, frequency resources and space resources of radar jamming system, determine jamming style and jamming time length, and guide jammers to accurately track targets and implement jamming.

(6) Generation of jamming signal
The jamming source generates the corresponding jamming signal according to the parameters such as jamming style and jamming timing. Jamming sources generally include noise signal generator, RF memory, optical fiber delay line, etc. The generation process of jamming signal is controlled by the system controller. The control of this part is often called electronic countermeasures (ECM) control, which mainly includes the working mode switching of jamming system, reconnaissance/jamming time window control, jamming signal style, and jamming modulation parameter control.

(7) Jamming emission
It generally includes up-conversion, power amplification, and antenna transmission. In order to meet the requirements of jamming power and improve the jamming effect, it is usually necessary to amplify the jamming excitation signal. Before power amplification, up-conversion processing is generally required;

that is, the IF signal generated by the interference source is transformed into RF interference signal. Up-conversion and down-conversion generally use the same LO signal, so as to ensure that the received and transmitted signals have the same frequency. For the antenna with adjustable antenna direction or array antenna, the azimuth information of the target radar provided by the jamming guidance module is generally used to control the main beam direction of the jamming antenna in real time, so as to ensure that the jamming is aimed at the target radar and improve the jamming efficiency.

3.1.4 Technical Indicators

According to their specific mission requirements and equipment form, different radar active jamming systems may have different indicators. Here are only some common functional indicators and technical indicators of radar active jamming systems [9–15, 30–32].

(1) Jamming frequency range
The jamming frequency range refers to the frequency range in which the radar active jamming system can effectively implement jamming. For the jamming system whose jamming working frequency is continuous, it can be characterized by the maximum frequency and the minimum frequency. Instead, for the jamming system whose jamming working frequency is discontinuous, segmentation is required, for example, " jamming frequency range X MHz to Y MHz" means a single frequency range, " jamming frequency: X MHz to Y MHz, W MHz to V MHz" means there are two jamming frequency ranges.

(2) Jamming airspace range
The jamming airspace range refers to the spatial angular range within which the radar active jamming system can effectively implement jamming. Generally, it includes two dimensions of azimuth and elevation, for example, "jamming airspace: azimuth X to Y degree, elevation W to V degree".

(3) Minimum jamming range
The minimum jamming range, also known as the burn-through range, refers to the minimum range at which the radar cannot find the protected target when the jammer interferes the radar. This is a very important tactical indicator of the radar active jamming system.

(4) Effectively jamming sector
The effective jamming sector refers to the range of the azimuth sector where the radar cannot find the protected target when the radar active jamming system interferes the radar. The combination of this indicator and the "minimum jamming range" can characterize the effective jamming space area of the jammer to the radar. When the jamming system implements suppression jamming, the

indicator “effective jamming suppression area” can be used, which means the area where the radar loses its normal detection ability.

(5) Types of radar that can be jammed
This indicator refers to the types of radar that the radar active jamming system can effectively jam.

(6) Adapt to the signal environmental density
This indicator refers to the maximum number of pulse signals that the radar active jamming system can correctly measure, sort, and identify in a unit time, usually “ten thousand pulses per second” as the unit.

(7) Number of simultaneous jamming targets
This indicator refers to how many target radars can be jammed by the radar active jamming system at the same time in a specific time, which represents the multi-target jamming capability of the jamming system. Generally, it can also be divided into multi-target jamming capability in the beam and multi-target jamming capability under different beam coverage.

(8) System response time
The system response time refers to the time when the radar active jamming system starts to receive the radar signal (the radar is located within the instantaneous airspace coverage of the jammer receiving antenna), after processing such as measurement, sorting, and identification, then release the rated power jamming signal to the radar.

(9) System receiving sensitivity
This indicator refers to the signal power density on the antenna port surface where the radar active jamming system can normally complete signal interception and parameter measurement, usually expressed in dBm/m^2. Usually, for the convenience of system design, the system sensitivity is equivalently converted to the signal power received by the antenna gain of 0 dBi, which is represented by dBm. Different from the receiving sensitivity of the receiver, the system receiving sensitivity is not only based on the sensitivity of the system receiver, but also the gain of the receiving antenna should be considered.

(10) System instantaneous bandwidth
The system instantaneous bandwidth refers to the signal bandwidth that the radar active jamming system can process simultaneously, which mainly includes the instantaneous bandwidth of the receiver and the bandwidth of the jamming spectrum. Generally, the receiving instantaneous bandwidth should be greater than or equal to the jamming spectrum bandwidth while the instantaneous bandwidth of the system is generally smaller than the operating frequency range of the system.

(11) Jamming pattern
The jamming pattern refers to the jamming strategy and the type of jamming modulation signal that the radar active jamming system can provide. Common jamming pattern includes noise blanket jamming, dense false target suppression jamming, multi-false-target deception jamming, range gate pull-off jamming, velocity gate pull-off jamming, angle deception jamming, and combined jamming.

(12) Jamming equivalent radiated power
The jamming equivalent radiated power refers to the product of the output jamming signal power of the transmitter of the radar active jamming system and the transmitting antenna gain, which can represent the maximum jamming signal strength that the radar active jamming system can provide.

(13) Parameter measurement accuracy
Parameter measurement accuracy refers to the error between the radar signal parameter values measured by the receiver of the jamming system and the corresponding real values, which is generally expressed by the root mean square. This indicator characterizes the accuracy of radar signal parameter measurement by the jamming system, including frequency measurement accuracy, angle measurement accuracy, pulse width measurement accuracy, repetition frequency measurement accuracy, and positioning accuracy.

(14) Frequency aiming accuracy
Frequency aiming accuracy, also known as frequency guidance accuracy, refers to the difference between the center frequency of the jamming signal emitted by the radar active jamming system and the frequency of the jammed radar signal, usually expressed in root mean square.

3.2 Design Criterias of Radar Active Jamming System

In the system design, especially when the overall scheme is demonstrated and carried out, the establishment of correct system design guidelines in advance can help designers grasp some of the key factors and achieve a multiplier effect with less effort.

Academician Sun Jiadong, a well-known aerospace scientist in China, the "father of satellites", and the chief designer of the lunar exploration project, summarized the overall system design as follows: the so-called overall system is to use the most reliable technology, the least cost, the shortest time, and the most beneficial cooperation, the most effective adaptability, and the most farsighted forward-looking to formulate the most feasible plan to ensure the best results.

The design criterias of radar active jamming system mainly include the following:

(1) Necessity criterion.
It is the most fundamental condition and basis for the design of radar active jamming system, which should base on the basic operational principles and missions of radar active jamming, highlight the differences between the designed active jamming system and previous similar products in terms of application requirements, system capabilities, technical methods, or performance indicators. The main considerations include: analyzing the current and future changes in combat objects, battlefield environment, and other changes in the formation of new requirements for the jamming system; analyzing the ability and performance of the active jamming system to be designed, and whether it complies with the electronic warfare equipment development plan and the possible role in combat.

(2) Advancement criterion.
The vitality of the product lies in its advanced nature, including the adoption of new technologies and the reasonable integration of the system, etc. The focus of radar active jamming system design is to improve the mission system's ability to radar signals reconnaissance and radar jamming in a complex electromagnetic environment, so consideration should be given to changes in the future threatening environment and operational requirements, and full consideration should be given to the capacity growth and the performance improvement of new technologies and new systems to ensure that the designed system has advanced performance as a whole and can adapt to the requirements of a relatively long period of time. However, the application of new technologies is not the more the better, the newer the better, so the adopted new technologies must be proved feasible and reasonable through pre-research.

(3) Inheritance criterion.
It means that in system design, especially in product development, it is necessary to inherit and integrate existing research results, use mature technologies as much as possible, learn from experience data and design guidelines, and proceed step by step to reduce design risks. The overall scheme design of the system should not only consider the advanced standards, aim at the world's advanced level, and innovate boldly, at the same time, it must also grasp the balance between innovation and inheritance, and pre-evaluate the possible technical risks. We should also consider adopting an open architecture as much as possible to ensure that the system is upgradeable, expandable, and partially replaceable.

(4) Feasibility criterion.
Ensure that the functional performance and technical indicators of the designed radar active jamming system can be realized, and the weight, volume and use mode of the system can meet the use conditions. During the demonstration, the favorable conditions, technical difficulties, solutions, and constraints shall be fully studied, the feasibility conclusions shall be given through comprehensive analysis, and the system design shall be carried out on the premise that it is basically feasible.

(5) Reliability criterion.
The reliability of the system is closely related to the design of the system. The failure of any subsystem may lead to the failure of the whole system. At the beginning of radar active jamming system design, reliability should be considered as a design index to ensure that the system and all subsystems can work stably, reliably, and normally. For example, conduct comprehensive analysis and system design on electromagnetic compatibility and power supply and distribution network to ensure that the active jamming system is compatible with other electronic information systems in the platform.

(6) Economy criterion.
It is another dimension relative to technology, occupying an increasingly important position in the overall system design. It mainly refers to the cost-effectiveness ratio of the designed system when designing the scheme. The fee is mainly reflected in the development costs, development personnel, and development cycle, etc. The cost of the whole life cycle includes development costs, production costs, test costs, and maintenance and repair costs in later use. The development personnel and the development schedule should be combined for optimal configuration. Among the various components of the radar active jamming system, the cost of the transmitter is often higher. We can't blindly pursue the most advanced technology, maximize system resources, and the most powerful functions, etc.

(7) Normalization criterion.
It means that scheme design shall conform to relevant national standards, military standards, and design specifications. Specifically, it includes: follow the requirements of serialization, standardization, combination, formulate the outline of the task system, development specifications and design requirements, and attach importance to ergonomic design, and improve system's usability.

(8) Compatibility criterion.
In most cases, radar active jamming systems do not work independently, but are mounted on aircraft, ship, satellite, and missile platform. The criterion of compatibility means that the volume, weight, shape, and power supply of the active jamming system should be consistent with the capability of the platform as much as possible, and the active jamming system should be integrated with other electronic equipment as much as possible.

In the specific design, some other criteria may also be included.

In summary, the design of radar active jamming system shall be comprehensively demonstrated according to the specific requirements and the main tactical and technical indicators proposed by the user, and in combination with various factors such as the cost, schedule, and risk of the system. The design principles of standardization, serialization, generalization, and modularization shall also be fully implemented in the scheme design, so that the overall scheme of the system can not only ensure the progressiveness of the technology, but also have the ability of reliability, maturity, stability, and scalability, so as to ensure that the developed radar active jamming system has good jamming efficiency and meets the needs of the user.

3.3 Design Process of Radar Active Jamming System

The general process of radar active jamming system design is shown in Fig. 3.8.

Firstly, according to the needs of users, the operational application scenarios, jamming objects, operational requirements, and tactical technical indicators of radar active jamming system are defined, the capability catalog of radar active jamming system is constructed, and the final function and performance indicators are determined.

Then, the overall design of radar active jamming system is carried out. It includes selecting the appropriate active jamming technology system, determining the software and hardware architecture and composition of the jamming system, decomposing the functions and technical indicators, determining the subsystem functions

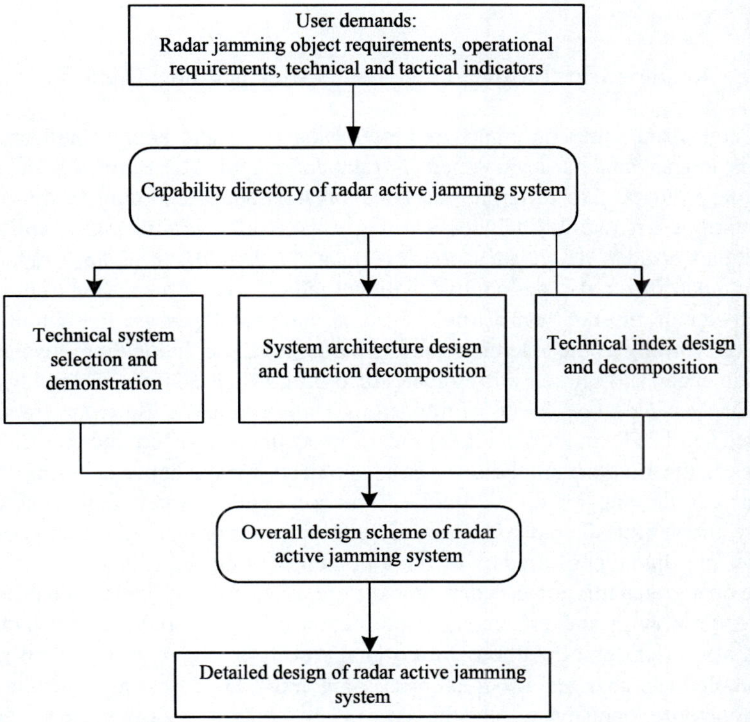

Fig. 3.8 Design flowchart of radar active jamming system

and index requirements, selecting the appropriate subsystem system, and forming the overall scheme of the radar active jamming system.

Finally, according to the overall scheme of the system, the detailed design of the radar active jamming system is carried out.

3.4 Design Method of Radar Active Jamming System

3.4.1 Requirement Analysis from Radar

The core task of radar active jamming system is to effectively jam the radar. Therefore, the main demand for jamming system design comes from the radar that needs jamming. The type, working frequency, carrying platform type, and technical index of the jammed radar will affect the design of radar active jamming system to varying degrees. Table 3.1 lists the operating frequency band, carrying platform type, technical system, and other characteristics of several typical radars (based on common terms, inaccurate classification) [33–47].

3.4.1.1 Requirements for the Functional Type of Jamming Radar

Radar countermeasures only appeared after radar came into being. The history of radar is longer, and its development is relatively rapid. There are a wide range of military radars with different classification methods. According to the operational purpose, it mainly includes warning radar, early warning radar, command and guidance radar, height measurement radar, battlefield surveillance radar, fire control radar, fuze radar, and multi-functional radar. Divided according to the form of radar signals, most of them are pulsed radars, and a small part are continuous wave radars. According to the antenna scanning mode, it can be divided into mechanical scanning radar and phased array radar. According to the adopted technology and signal processing mode, it can be divided into conventional pulse radar, frequency agile radar, pulse Doppler radar, MTI radar, and synthetic aperture radar (SAR), etc. Besides, in the modern battlefield, the combat objects of radar active jamming system also include the integrated electronic information combat system composed of air defense and antimissile radar system and radar networking system, which generally includes multiple types of radars working in multiple frequency bands.

The operational mission of radars with different functional types is often different. The design criteria and effectiveness evaluation criteria of radar active jamming should also be different. Although the signal waveform patterns of modern radars are complex and changeable, signal processing and data processing algorithms are becoming more sophisticated, and the types of anti-jamming measures are various, as far as a certain radar is concerned, they are still designed around the corresponding functional types and technical systems of the radar. Therefore, for different types of radar objects, it's only after in-depth research on working principles, technical systems, working modes, processing procedures, anti-jamming measures, etc., as well as summarize the characteristics of radar signal patterns and find weaknesses in radar processing technology that more targeted electronic warfare reconnaissance method design and jamming method design can be carried out, which is conducive to the design of a better radar active jamming system.

Table 3.1 Several typical radars and their characteristics

Radar types	Platform types	Working frequency bands	Main feature
Over-the-horizon radar	Ground	HF-band	Most of them are two coordinate radars, mainly responsible for early warning and tactical warning. Linear frequency modulation (LFM) or continuous wave (CW) signals are often used, and the operating range is relatively long. Moving target display (MTI), low sidelobe antenna, adaptive frequency conversion, long-time coherent accumulation, and other technologies are often used
Search and warning radar	Ground	V/UHF-band L-band S-band	Mechanical scanning antenna or phased array antenna can be used. There are two coordinate systems and three coordinate systems, mostly for long-range air defense and missile early warning. LFM waveform and large bandwidth signal are often used. MTI, sidelobe zeroing, interpulse frequency agility, sidelobe blanking, and other techniques are often used

(continued)

Table 3.1 (continued)

Radar types	Platform types	Working frequency bands	Main feature
Early warning aircraft radar	Airborne	UHF-band L-band S-band	Mechanical scanning antenna or phased array antenna can be used. When low pulse repetition frequency (PRF), pulse compression, phase center offset antenna (DPCA), digital moving target detection (DMTD), and other technologies are often used. When high PRF, technologies such as pitching electronic scanning, ultra-low sidelobe, adaptive sidelobe cancelation, frequency agility, and MTI are often used
Tracking guidance and fire control radar	Airborne Ground Shipborne	S-S-S-S-band C-band X-band	Mechanical scanning antenna or phased array antenna can be used. Monopulse angle measurement technology can improve the accuracy of target angle tracking. MTI, sidelobe cancelation, and other technologies will also be adopted. The working process of radar generally includes search, interception, tracking, calculation of launch conditions, initial guidance, guidance, etc.

(continued)

For warning, early warning and other search radars, the main task of the radar system is to find targets from a long distance and detect targets with high probability. Generally, after the formation of target traces, target tracking is realized through track generation and track filtering. The ground warning and searching radar generally adopt azimuth mechanical scanning, which often has rules to follow. The signal waveform is mainly characterized by intra-pulse modulation and frequency agility.

Table 3.1 (continued)

Radar types	Platform types	Working frequency bands	Main feature
Imaging surveillance radar	Airborne satelliteborne Shipborne	L-L-L-L-band S-band C-band X- band Millimeter-band	SAR or ISAR imaging adopts linear frequency modulation signal and pulse compression technology, which has the characteristics of large bandwidth. SAR generally has a variety of scanning and mapping modes. Imaging radar can generally support target classification and recognition
Active seeker radar	Missileborne	X-band Ku-band Millimeter-band	Pulse Doppler system and monopulse angle measurement technology are often used, and medium PRF or high PRF is used. It has single target active tracking and passive tracking jamming source modes. The effective working time of radar is short
Airborne multi-function radar	Airborne Helicopterborne	X-band Millimeter-band	Mechanical scanning antenna or phased array antenna can be used. It has low PRF, medium PRF, and high PRF working modes. It has multi-functional features such as air-to-air, air-to-ground, and air-to-sea and can usually achieve ultra-high-resolution imaging and ground moving target detection (GMTI)

(continued)

Table 3.1 (continued)

Radar types	Platform types	Working frequency bands	Main feature
Shipborne multi-function radar	Shipborne	S-band	Mechanical scanning antenna or phased array antenna can be used. It has many functions such as detection, search, tracking, and guidance. MTI clutter suppression, frequency agility, PD, and other technologies are often used

The processing algorithm usually uses MTI or MTD to suppress low-level clutter and performs one-dimensional constant false alarm rate (CFAR) detection in the range dimension. When the early warning aircraft monitors the ground or sea, the radar is easily affected by ground clutter and sea clutter, so the radar of the early warning aircraft generally has a strong clutter suppression ability. For these radar systems, the radar active jamming system should mainly focus on noise jamming and dense false target jamming, and destroy the processing processes of radar, such as automatic gain control (AGC), clutter suppression, target detection, ambiguity resolution, etc. If active decoy jamming is to be realized, it is necessary to consider the problem of long-time decoy association to ensure that the formed active decoy can be tracked continuously and stably during track processing.

The most important characteristic of ground guidance radar, precision tracking and measurement radar, terminal guidance radar, and other tracking systems is to keep tracking the target continuously and stably. Generally, they have tracking loops such as range tracking, speed tracking, or angle tracking. Therefore, the radar active jamming system should focus on range towing jamming, velocity towing jamming and angle deception jamming, so as to reduce the tracking accuracy of the radar to the real target, and even destroy its tracking loop. In addition, the startup time of terminal guidance radar is generally relatively short, so the requirements for signal interception and jamming response time of active jamming system are high.

For the ground air defense and antimissile radar system and other integrated radars that include both target search and tracking guidance, the radar active jamming system often needs to have a variety of jamming styles at the same time, and adopt different jamming methods in different working stages such as search, tracking, and guidance, so as to achieve effective jamming to the radar as early as possible.

The spaceborne radar is far from the ground, so it is difficult to implement active jamming to it, including jamming power and jamming pattern. When the active jamming equipment is on the ground, it is necessary to ensure sufficient power supply and increase the transmitting power of the jammer. The transmitting antenna of jammer generally adopts large-diameter parabolic antenna or array antenna. Generally, the jamming pattern is mainly coherent jamming.

Imaging surveillance radar generally has a large signal bandwidth, and the receiver instantaneous bandwidth and jamming signal bandwidth of active jamming system are also highly required. For radars with low probability of intercept waveform design or power management, the sensitivity requirements of radar active jamming system will be further improved.

The multi-functional phased array radar generally contains a variety of different working modes and states. The waveform of the radar is complex and changeable, the beam is agile, and the target search, tracking, and guidance can be realized at the same time. Therefore, the radar active jamming system also needs to have a certain ability of adaptive reconnaissance and adaptive jamming.

For other radar systems, such as continuous wave radar, bistatic radar, some special considerations corresponding to the technical system and characteristics of radar should also be considered in the design of radar active jamming system.

3.4.1.2 Requirements for Jamming Radar Operating Frequency

In terms of operating frequency, the current radar has almost covered all frequency ranges from short wave, meter wave, decimeter wave to millimeter wave and even THz and laser, which is the result of the rapid development of radar technology and electronic devices. However, in practical application, according to the operational performance requirements of radar and the difficulty of engineering implementation, most military radars in active service work in the microwave frequency range of 0.3 ~ 18 GHz and are developing toward millimeter wave and terahertz. When the frequency is higher than 18 GHz, water molecules and oxygen molecules in the air seriously absorb electromagnetic waves, and the propagation loss increases. With the further improvement of the frequency, the processing difficulty of the antenna and RF channel increases, the internal and external noise of the receiver increases, and the transmission power is difficult to improve. These frequency bands are mainly used in high-resolution radar, outer space radar, and close-range radar. Figure 3.9 shows the main operating frequency distribution range of radar and other electronic equipment.

If the radar active jamming system only needs to jam one radar or radar in a certain frequency band, the operating frequency of the active jamming system can only cover this frequency band. For example, millimeter wave active jamming system designed for Ka band terminal guidance radar, X-band radar active jamming system designed for X-band airborne fire control radar, etc.

Generally, however, a jammer often needs to deal with multiple radars at the same time, so in system design, most radar active jamming systems have a large frequency range. Typical frequency bands include 0.1 ~ 0.3, 0.3 ~ 2, 2 ~ 6, 8 ~ 12, 6 ~ 18, 18 ~ 40, 2 ~ 18, 0.1 ~ 18, and 0.1 ~ 40 GHz. Therefore, the design of antenna and RF channel should meet the corresponding frequency coverage requirements.

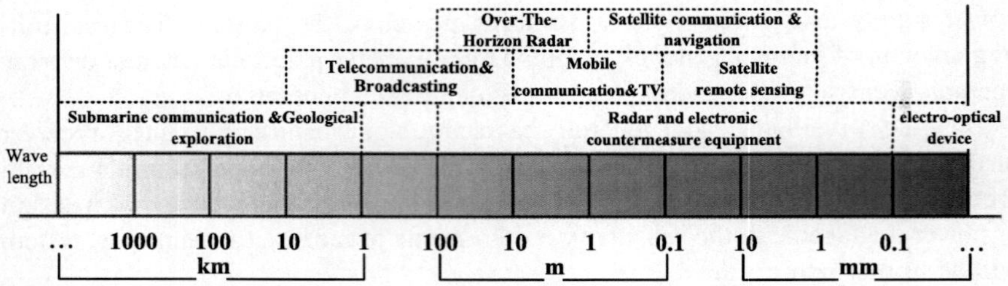

Fig. 3.9 Schematic diagram of radar main operating frequency distribution

Specifically, the typical operating frequency of early warning radar includes multiple frequency bands such as UHF, L, and S. According to different needs, SAR may also have different working frequency bands such as L, S, C, X, millimeter wave. Therefore, in the design of active radar jamming system, it is necessary to clarify whether it is a specific radar or all this type of radar that needs to be interfered, so as to put forward the appropriate operating frequency index in the design of active radar jamming system.

Due to the development of the fourth-generation stealth fighter, countries have developed meter wave radar one after another, but the antenna size of the V /UHF-band is large, radar active jamming systems are limited by the platform installation space, and the jamming power is difficult to improve. Therefore, in the design, it is generally necessary to focus on technologies such as miniaturized low-frequency antennas and high-efficiency meter wave power amplifiers.

Millimeter wave radar has the characteristics of narrow beam and short operating range, which puts forward higher requirements for signal search, target interception, jamming guidance, and other functional indicators of radar active jamming system. In addition, it is difficult to manufacture frequency conversion modules and power amplifiers.

3.4.1.3 Requirements for Jamming Radar Technology Systems

The range of radar system mentioned is generalized, including two aspects: the processing technology system of radar to achieve target search and tracking and radar anti-jamming measures. In radar signal processing, pulse compression, moving target indication, and moving target detection, pulse Doppler (PD), synthetic aperture imaging, monopulse angle measurement, and digital beamforming (DBF), etc., are commonly used. Common anti-jamming measures of modern radar include ultra-low sidelobe, sidelobe hidden (SLB), and sidelobe cancelation (SLC), frequency agility, frequency diversity, and PRF jitter. Different technical systems and anti-jamming measures all have the important impact on the design of jamming strategies and jamming modes in active jamming systems.

The radar technology system mentioned here is generalized, including two aspects: one is the radar processing technology for target search and tracking, and the other is the radar anti-jamming technology. Commonly used in radar signal processing include pulse compression, moving target indication, and moving target detection, pulse Doppler (PD), synthetic aperture imaging, monopulse angle measurement, digital beam forming (DBF), etc. Common radar anti-jamming technologies include ultra-low sidelobe, sidelobe blanking (SLB), sidelobe cancelation (SLC), frequency agility, frequency diversity, PRF jitter, etc. The technical system and anti-jamming measures of radar directly affect the design of jamming strategy and jamming style of radar active jamming system.

Pulse compression technique and pulse Doppler technique take use of accumulation gain of radar signal pulse compression and interpulse coherent accumulation gain, respectively, but common noise suppression jamming signal is difficult to obtain coherent processing gain consistent with radar echo signal. Therefore, it is necessary to consider increasing the power of noise jamming signal or using coherent jamming mode to combat effectively.

Pulse compression technology and pulse Doppler technology obtain high gain of signal through pulse compression and coherent accumulation between pulses. Noise jamming signal is difficult to obtain the same high gain as target echo signal. Therefore, to effectively jam pulse compression radar and pulse Doppler radar, it is necessary to increase the power of noise jamming signal or adopt coherent noise jamming.

The frequency agility technology makes the central working frequency of the radar change rapidly within a certain range. It requires the jamming system to master the changes of the radar working signal and quickly update the jamming signal.

The signal bandwidth of synthetic aperture imaging technology may reach up to GHz. To intercept all radar signals, the instantaneous bandwidth of the active jamming system receiver is required to be no less than the radar operating frequency range.

The signal pulse width of MTI radar is relatively wide, and the number of coherent pulses is relatively small. It is usually used together with frequency agility and pulse width agility technology. This requires that the signal sorting and identification of radar active jamming system must adapt to the characteristics of MTI, and the signal sorting and jamming response speed should be faster.

Multi-functional and phased array antenna technology makes the radar function mode diverse, waveform complex agile, and beam burst irradiation, which brings great challenges to the signal sorting, radiation source recognition and jamming fast response of the radar active jamming system, so more advanced algorithms are needed.

The trend of radar network application is obvious, which can cover a wider airspace, shorten the warning time, and enhance the anti-jamming ability of radar network by means of multi-station fusion processing such as point track association and track association. The single jammer is difficult to meet the demand, so it is necessary to consider distributed cooperative jamming.

Low sidelobe antenna technology has a higher requirement for the receiver reconnaissance sensitivity of radar active jamming system in sidelobe reconnaissance. The monopulse angle measurement technology can ensure that the radar can directly use the jamming signal to measure and track the angle under the condition of strong pressure jamming, so the jamming method of angle deception should be considered in the design. SLB and SLC are two typical anti-jamming measures for radar sidelobes, which require the high quality of system design for side lobe remote support jamming.

With the continuous development of radar technology, space-time adaptive processing (STAP) early warning radar [48, 49] netted radar [50, 51], low probability of intercept (LPI) radar [52], multiple input multiple output (MIMO) radar [53], cognitive radar [54, 55], the new systems radar have adopted the more advanced technical system, whether the signal mode, the signal and the data processing or the anti-jamming measurement are more complex, which will directly affect the design of radar active jamming system.

3.4.2 *Requirement Analysis from Operation*

The operational requirements mainly include the loading platform of jammer, the operational application mode of jammer and the electromagnetic environment of battlefield, which have some influence on the antenna design, receiver design, jamming power, and so on.

3.4.2.1 Jamming System Loading Platform

Similar to radar, the carrying platform of radar active jamming system can also be divided into airborne, shipborne, satelliteborne, missileborne, vehicleborne, ground, and portable. According to the relationship with the protected platform, it can also be divided into internal type, external type, and separate type. The different carrying platforms and specific installation positions of the jamming system will determine the antenna size, structural size, mechanical strength, jamming power, power supply, weight, stability, and other indicators of the system and will further affect the working frequency band, structural composition, calculation and processing capacity, jamming style and other design of the system.

Satellites, missiles, unmanned aerial vehicles (UAVs), balloons, and other platforms have high requirements on the size, weight, power supply, and other aspects of radar active jamming system. In the design of jamming system, it is necessary to focus on the miniaturization of equipment. Generally, the jamming style is mainly deceptive jamming with small energy demand. Spaceborne equipment is difficult to maintain and costly. Generally, mature software and hardware structures and reconnaissance and jamming technologies are used in the design of jamming system. Shipborne equipment is usually faced with serious seawater moisture, so it is necessary to protect the hardware of the jamming system against moisture, salt fog, and mold. Generally, there is no problem in the power supply and spatial layout of shipborne and ground jamming systems. The jamming power can be designed to be larger, the signal and data processing algorithms can be relatively complex, and the working frequency can cover a wider range. In addition, the radar cross section (RCS) of different protected platforms is different. For example, the RCS of a typical aircraft platform is about a few tenths of a square meter to tens of square meters, while the RCS of a ship is generally several thousand square meters or even tens of thousands of square meters. Therefore, the equivalent radiated power of the active jamming system protecting a ship is generally required to be greater.

Taking the airborne platform as an example to discuss the advantages and disadvantages of several typical active jamming systems using different physical layout design in practical applications, as shown in Table 3.2 [12–15]. Airborne carrier platforms generally include aircrafts, helicopters, and UAVs, among which electronic warfare aircraft can provide the radar active jamming system with the most adequate layout space, system resources, and power supply, while the resources of UAV platform are the scarcest. For fighter aircraft, especially the new fighter, the radar active jamming system is mainly internal; that is, the main equipment is installed inside the aircraft, and the antenna is attached to the surface of the aircraft as much as possible or adopts stealth design. The advantage of the internal jamming system is that it can be customized according to the requirements of the aircraft and can also share antennas or processing equipment with radar, navigation and other equipment. After installation, it has little effect on the aerodynamic and stealth of the aircraft. But the disadvantages are that the transmission cable is generally long, the maintenance of equipment is complex, and it is difficult to replace the jamming equipment flexibly according to the need of single operation.

The typical representative of the external mounted radar active jamming system is the airborne pod equipment, which is to put the jamming equipment into a streamlined container and hang it on the belly or wing of the aircraft with a hanger. The greatest advantage of the externally mounted pod is that it can work independently, so that a single equipment can be used on a variety of aircraft, and it only needs to be hoisted when the aircraft performs a specific task. The maintenance, repair, and upgrade are relatively simple, reducing the total cost. In addition, it also has a short internal transmission line, so that the delay of interference processing is small. Its disadvantage is that it affects the maneuverability and stealth performance of the carrier and occupies the missile or fuel tank attachment point of the aircraft, and in the harsh environmental conditions, the influence of temperature and vibration put forward higher requirements for the system design.

The separated radar jamming system includes active and passive jamming equipment, mainly from the angle domain to jam the target radar, and mostly adopts the deceptive jamming style. The separated radar active jamming system, also known as radar active decoy, mainly includes towed, accompanying and throwing types, and can also be divided into recoverable and non-recoverable types. The advantage of the separate jamming system is that it can implement angular jamming and adjacent distance unit jamming to protect the target from direct attack. The disadvantages are

Table 3.2 Different deployment modes of airborne jamming equipment

Placement	Advantages	Disadvantages	Important considerations	Typical applications
Internal type	Can be customized according to aircraft requirements. Have larger layout space. Can share components with other electronic equipment. Generally, do not affect the performance of the aircraft	Long transmission cables; Complex equipment maintenance and repair; Need to be installed on each aircraft; It is difficult to load and unload flexibly	Antenna structure and layout; The weight and volume of the system; Cost	Electronic warfare equipment of fighter, special electronic warfare aircraft, etc.
External type	Short internal transmission cables; Work independently; Can be used on a variety of aircraft; Loading equipment only when required; Maintenance, repair, and upgrade, are relatively simple; Does not occupy the cost of the carrier	Limited space and weight; Affect the flight and stealth performance of the carrier; Occupying the aircraft's missile or auxiliary fuel tank attachment point System performance may become worse in harsh environments	Shape and structure; Temperature and vibration performance in harsh environments; Weight and volume of the system	Airborne EW pod, etc.
Separation type	Low cost, some recyclable; Generally small, not afraid of being destroyed; Can implement angle jamming; Can resist radar tracking jamming source, protection platform	Severely limited in volume and weight; Limited jamming function; Limited jamming power; It is also necessary to solve the problem of equipment delivery	Method of towing or projecting equipment; Power supply remote control, recovery, and other issues; shape, structure, weight, and volume	Radar active decoy

that the volume and weight are severely limited, and the jamming function is limited. Generally, the small power deceptive jamming is the main interference. In addition, other problems such as towing or projection methods, remote control, power supply, materials, recycling, and so on need to be solved.

3.4.2.2 Jamming Operation Applications

As already described, the main applications of radar active jamming can be divided into support jamming and self-defense jamming, while support jamming includes long-range support jamming, team support jamming and close in support jamming. Under different jamming applications, there are some differences in the design of main combat objects, jamming transmitting power, jamming strategy, and jamming mode of radar active jamming system.

The main combat objects of support jamming include airborne early warning radar, ground or sea search warning radar, target indication radar, guidance radar, and airborne fire control radar. More use of support jamming is suppression jamming mode, system design requires larger jamming power. Among them, the remote support jamming is generally outside the defense area, which is very far from the radar, and is also located in the antenna sidelobe area of the radar; therefore, it is necessary to greatly improve the reconnaissance sensitivity of the receiver of the radar active jamming system to ensure the interception of the weak radiation source signal. It requires a large radiation jamming power to effectively suppress the radar and cover the target. Flexible jamming mode design is needed to resist the anti-jamming measures such as radar sidelobe blanking or sidelobe cancelation. The jamming position layout escort-support jamming is more favorable, which may realize the main lobe jamming, reduce the jamming power demand, and obtain better jamming effect. The jammer of the proximity support jamming is closer to the radar, which can effectively reduce the jamming power requirements of jamming system, if by reducing costs, increasing the number of jammers, can also make the means of warfare more flexible. In addition, two important problems need to be paid attention to in the design of support jamming. One is that the jamming system must deal with multiple threat radars at the same time in the process of shielding penetration, which leads to "dilution" of the available jamming resources allocated to each threat target. The other is that the antenna coverage, main lobe width, and beam pointing of the support jamming system need to be carefully designed to ensure that the jamming can stably and continuously cover the area to be protected when the platform is cruising.

The main combat objects of self-defense jamming mainly include airborne fire control radar, guidance radar, and terminal guidance radar. Self-defense jamming generally has both repressive jamming and deceptive jamming. Its design needs to be fully combined with the characteristics of the platform. For example, the ship platform is bulky and slow to move, and the implementation of self-defense jamming is mostly combined with the active jamming inside the ship platform and the active and passive jamming outside the ship. For fighters, it is generally necessary to carry out a full range of threat warning and missile defense. Therefore, it is generally necessary

Table 3.3 Different applications in radar active jamming systems

Applications	Main technical features	Key issues to consider
Self-defense jamming	Deception jamming is mainly used, and generally it is required to have the ability of deception jamming and suppression jamming. Jamming enters from the main lobe of radar antenna	Mainly against fire control, guidance radar Jamming power In general, angle jamming and passive jamming should be considered
Escort-support jamming	Using noise, dense false targets to suppress jamming and other styles; Jamming enters from the radar antenna main lobe	Mainly against fire control guidance radar Large jamming power Multi-target jamming capability
Long-range support jamming	Noise, dense false targets, and other suppression jamming. Jamming enters from radar antenna sidelobe	Design very large jamming power Design very high system receiving sensitivity Multi-target jamming capability

to use omnidirectional antennas or set multiple antennas in multiple directions of the aircraft. In addition, because many modern radars use monopulse angle measurement technology and have the ability to track the jamming source (also known as home on jamming, HOJ). Therefore, it is necessary to consider how to avoid radar tracking the jamming source in the design of self-defense jamming system.

The main features of different jamming applications of radar active jamming systems are shown in Table 3.3.

3.4.2.3 Complex Electromagnetic Environment

Radar active jamming system must adapt to the battlefield electromagnetic environment. From the working process of active jamming system, the influence of electromagnetic environment on the design of jamming system is mainly reflected in the reconnaissance part; that is, the signal reception, sorting, and recognition of radar active jamming system are affected.

In modern information warfare, the number of radars is increasing, the spectrum is expanding rapidly, and the electromagnetic environment of the battlefield is becoming more and more complex. Taking the design of shipborne radar active jamming system as an example, Fig. 3.10 shows the main threat radar and other electromagnetic environment faced by ships in all directions. It can be seen that the electromagnetic environment [56–59] of the sea battlefield is very complex, including electromagnetic waves from space, air, sea and underwater, and electromagnetic waves from radar, communication, navigation, identification friend of foe (IFF), and

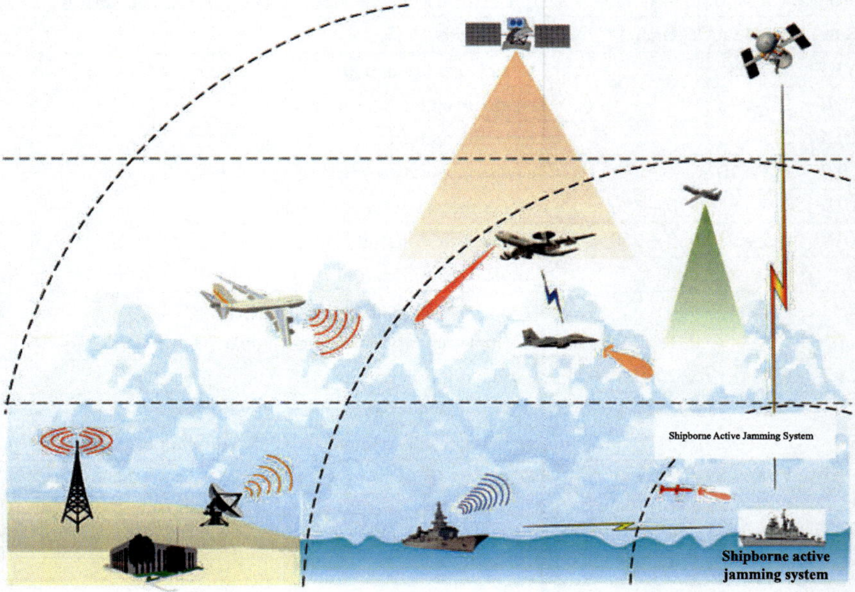

Fig. 3.10 Diagram of complex electromagnetic environment faced by shipborne active jamming system

other electronic equipment. The electromagnetic environment of an offshore battlefield may also contain electromagnetic waves radiated from terrestrial electronic equipment. These electromagnetic waves have a wide distribution range in space, time, frequency, and energy. They overlap and affect each other, forming a complex electromagnetic environment of the battlefield.

In the civil field, a large number of non-radar electromagnetic signals will seriously affect the radar signal reception of electronic warfare equipment, especially in the low-frequency band, the industrial frequency equipment with a large number, high duty ratio of continuous wave, spectrum congestion, strong power, sometimes will seriously affect the performance of radar active jamming system (Table 3.4).

In order to intercept all threat signals more effectively, radar active jamming system has the characteristics of high sensitivity and wide frequency domain. Therefore, the complex electromagnetic environment will lead to great pulse density entering the reconnaissance receiver, such as more than 2 million pulses/second, and there may be continuous wave signals, etc. This requires that the receiver of the jamming system can select continuous wave signals and cope with the sharp increase in pulse data. In addition, in order to adapt to the rapid changes of target and environment, the receiver and processor need to quickly complete the processing of parameter measurement, sorting, identification, threat judgment, jamming guidance, resource scheduling, and jamming generation. Therefore, the jamming system needs

Table 3.4 List of low-frequency band industrial radio electromagnetic spectrum occupation

Band	Frequency (MHz)	Applications
VHF	88–108	FM and data broadcasting
VHF	167–223	Television and data broadcasting
UHF	223–443	Television and data broadcasting
UHF	443–870	Television and data broadcasting
UHF	825 ~ 835/870 ~ 880	CDMA mobile communications
UHF	885 ~ 915/930 ~ 960	GSM mobile communications
L	1710 ~ 1755	Mobile communications
L	1805 ~ 1850	Mobile communications
L	1755 ~ 1785/1850 ~ 1880	3G supplementary operating frequency band
L	1880 ~ 1920/2010 ~ 2025	3G time division duplex (TDD) mode
L	1920 ~ 1980/2110 ~ 2170	3G frequency division duplex (FDD) mode
S	1980 ~ 2010/2170 ~ 2200	Frequency band of satellite mobile communication system
S	2300 ~ 2400	TDD
S	2400 ~ 2483.5	Wireless LAN, wireless access system, Bluetooth device, point-to-point or point-to-multi-point communication system

fast computing speed, large storage capacity, and strong processing capacity, so that it can make rapid response to the complex electromagnetic environment and ensure good jamming effect.

Radar active jamming systems under different loading platforms often have different electromagnetic environment adaptability requirements. For example, the ground jamming system mainly considers the electromagnetic signals that adapt to the adjacent areas and air radiation, but there are many types of civil signals in the low-frequency band and the power is strong, as well as the multipath of radar signal and the multipath of jamming signal are usually more serious. The shipboard jamming system also needs to consider the communication, navigation, and other signals emitted by many fishing vessels on the sea. The jamming system in the air is more complex, and various electromagnetic signals from the ground, sea, and air need to be considered.

In the complex electromagnetic environment, the radar active jamming system can eliminate the impact of the complex electromagnetic environment on the receiver through the following ways, such as working in different frequency bands to suppress the interference outside the frequency band, reduce the sensitivity of the receiver, filter small signals, and suppress strong interference by receiving pattern notch.

3.4.2.4 Multi-Target Jamming Capability

A radar active jamming system often needs to jam multiple radars at the same time, especially in some typical operational scenarios and complex electromagnetic environment, such as long-distance penetration, facing enemy netted radar, formation operation or saturation attack, the jamming system generally needs to have the ability to reconnaissance and jamming more than two radar targets at the same time. For supporting jamming, radar active jamming system is required to counter more radar targets at the same time.

The ability to jam multiple radars at the same time mainly affects the resource demand, resource scheduling strategy and jamming mode of radar active jamming system. The resources mentioned here are generalized, including reconnaissance and jamming resources of jamming system in frequency, space, time, energy, and other domains, such as system instantaneous bandwidth, instantaneous space-time domain coverage, and jamming transmission power.

It is easy to understand that the resources of single radar active jamming system are always limited. When multiple targets need to be jammed simultaneously, reconnaissance and jamming resources will be "diluted", and different radars generally require different jamming modes to have good jamming effects. Therefore, under the requirements of multi-target jamming capability, it is necessary to carefully design the instantaneous working bandwidth, instantaneous airspace coverage, jamming power, and system control timing of the system.

Improving the instantaneous working bandwidth of radar active jamming system is helpful to solve the problem of multi-target jamming capability. For example, most search warning radars work in S-band. When faced with multiple S-band search warning radars, the jamming system can increase the instantaneous bandwidth to 2 GHz or more and then can receive and interfere all S-band search warning radars at the same time.

In terms of instantaneous airspace coverage, the simultaneous multi-target jamming capability of radar active jamming system is mainly limited by antenna installation conditions. For airborne self-defense jamming equipment, it is often necessary to install antennas around the aircraft to achieve 360° coverage. For jamming pod equipment, due to its own linear shape characteristics, generally, the forward and backward antennas are arranged separately. If there is a requirement of full coverage in airspace, the layout of side antennas needs to be increased. For small jamming systems such as decoys, omnidirectional antennas are generally used. In the pitch dimension, if it is necessary to jam both air and ground (or sea) radar, the pitch beam width of the jamming system antenna must be large enough or its pitch direction can be quickly adjusted to ensure that all the threat targets can be covered.

The compromise design of jamming strategy and jamming style is carried out to ensure that the jamming system can track and effectively jam multiple radar threat targets at the same time. It is important to make the jamming system having rich hardware and software resources and excellent jamming resource scheduling ability, such as phased array antenna can enhance the beam agility of the antenna, digital jamming source can realize the ability of flexible jamming signal generation, superior performance processor can realize fast and intelligent resource scheduling ability and so on.

Jamming power management technology is an important technology in multi-target jamming. After the threat of environmental signals is identified, according to the battlefield situation and the total power, frequency range and other indicators of active jamming equipment, the jamming transmitter and antenna beam are controlled in the time domain, frequency domain and airspace through operations research, and the corresponding jamming signals are transmitted to different target radars with the required jamming frequency and the best jamming pattern in the appropriate time window.

3.4.3 System Modeling and Capability Analysis

Around the combat mission, combat object, and combat environment of radar active jamming system, the overall design requirements of the system are analyzed. On this basis, around the typical combat use, the capability requirements of radar active jamming system are proposed, and the requirement-capability mapping view is constructed. This process can be completed with the current popular and standardized Department of Defense Architecture Framework (DoDAF).

3.4.3.1 Introduction of DoDAF

DoD is the abbreviation of US Department of Defense; AF is the initial abbreviation of Architecture Framework. The United States Department of Defense defines AF as the structure of the components of the system, the relationship between them and the guiding principles and guidelines governing their design and evolution from beginning to end.

DoDAF is a method to analyze complex information systems by using the idea of multi-view modeling. DoDAF provides rules, guidelines and product descriptions for developing and expressing system structure. DoDAF describes different aspects of the system structure framework from different angles in the form of graphs and charts, so as to form a description of the overall system structure. DoDAF is developed on the basis of C4ISR architecture framework and has experienced multiple versions such as 1.0, 1.5, 2.0, 2.01, and 2.02.

DoDAF 1.0 came out in 2003. It was first used to guide the architecture description of defense command and control system (C^4ISR) and commercial operation process. After continuous improvement, it was applied to all business areas of US Department of Defense. DoDAF 1.0 defines four viewpoints, including all viewpoint (AV), operational viewpoint (OV), systems viewpoint (SV) and standards viewpoint (TV), a total of 26 models. DoDAF 1.0 describes the system from three aspects: operational requirements, system implementation, and technical support. DoDAF 1.5 came out in 2007, is a transitional version, and is developed in response to US Department of Defense to the network center war transition, emphasizing the architecture data.

DoDAF 2.0 came out in 2009, forming and revising some viewpoints based on version 1.5. DoDAF 2.0 includes 8 viewpoints, namely all viewpoint (AV), capability viewpoint (CV), data information viewpoint (DIV), operational viewpoint (OV), project viewpoint (PV), services viewpoint (SvcV), standards viewpoint (StdV), and systems viewpoint. Each viewpoint contains different models, a total of 52 models, as shown in the Table 3.5.

The structural framework description method in DoDAF is an effective method to obtain different operational capabilities from the top-level operational requirements, realize the internal and external interconnection, interoperability, and interoperability of information systems, and improve the operational efficiency and acquisition efficiency of weapons and equipment. DoDAF is the most widely used and mature system structure framework. It gives the framework standard to describe the system structure and can also be used as a reference method for the overall design of radar active jamming system.

3.4.3.2 System Modeling

Of the eight viewpoints mentioned above, the four most important viewpoints are all viewpoint, operational viewpoint, capability viewpoint, and systems viewpoint.

All viewpoint provides overall information about the system structure description, such as the scope and background of the system structure description. The scope includes the professional field and time frame of the description of the system structure. The background is composed of various interrelated conditions, including: ordinance, tactics, technology and procedures, descriptions of related objectives and ideas, concept of operations, scenarios, and environmental conditions. Corresponding to the design of the radar active jamming system, all viewpoint includes the position of the active jamming system product in the entire combat system, the position of the product in the mission system, the main assembly object and the method the product is used, the purpose of the enterprise, the budget, the constraints of development environment, related specifications, terminology, abbreviations, and other work content.

Operational viewpoint is used to describe the information requirements and the interrelationship among missions and actions of operations, operational elements, and operational organizations. Operational viewpoint often serves top-level design and is the basis of other view design of system structure, and it describes the operational concepts to be supported. The main modeling projects are the activities to complete the combat task and the mutual information exchange between people or organizations, revealing the requirements of operational capability and interoperability, which mainly includes the construction of combat tasks (OV-1), the determination of participating objects (OV-2), the determination of the relationship between participating objects (OV-4), the establishment of operational tasks specific execution process (OV-6C), and so on. Corresponding to the design of the radar active

Table 3.5 8 viewpoints and 52 models of DoDAF2.0

All viewpoint	AV-1 Overview and summary information	Operational viewpoint	OV-1 High-level operational concept graphic
	AV-2 Integrated dictionary		OV-2 Operational resource flow description
Capability viewpoint'	CV-1 Vision		OV-3 Operational resource flow matrix
	CV-2 Capability taxonomy		OV-4 Organizational relationships chart
	CV-3 Capability phasing		OV-5a Operational activity decomposition tree OV-5b Operational activity model
	CV-4 Capability dependencies		OV-6a Operational rules model OV-6b State transition description OV-6c Event-trace description
	CV-5 Capability to organizational development mapping	Systems viewpoint	SV-1 Systems interface description
	CV-6 Capability to operational activities mapping		SV-2 Systems resource flow description
	CV-7 Capability to services mapping		SV-3 Systems-systems matrix
Services viewpoint	SvcV-1 Services context description		SV-4 Systems functionality description
	SvcV-2 Services context description		SV-5a Operational activity to systems function traceability matrix SV-5b Operational activity to systems traceability matrix
	SvcV-3a Systems-services matrix SvcV-3b Services-services matrix		SV-6 Systems resource flow matrix

(continued)

Table 3.5 (continued)

	SvcV-4 Services functionality description		SV-7 Systems measures matrix
	SvcV-5 Operational activity to services traceability matrix		SV-8 Systems evolution description
	SvcV-6 Services resource flow matrix		SV-9 Systems technology & skills forecast
	SvcV-7 Services measures matrix		SV-10a Systems rules model SV-10b Systems state transition description SV-10c Systems event-trace description
	SvcV-8 Services evolution description	Project viewpoint	PV-1 Project portfolio relationships
	SvcV-9 Services technology & skills forecast		PV-2 Project timelines
	SvcV-10a Services rules model SvcV-10b Services state transition description SvcV-10c Services event-trace description		PV-3 Project to capability mapping
Data information viewpoint	DIV-1 Conceptual data model	Standards viewpoint	StdV-1 Standards profile
	DIV-2 Logical data model		StdV-2 Standards forecast
	DIV-3 Physical data model		

jamming system, operational viewpoint can include high-level operational concepts, operational scenarios, missions, operational unit models, operational object views, operational process diagrams, equipment interfaces and data relationships between operational units, operational activity models, and other work content.

Capability viewpoint focuses on the design objectives related to the overall vision, which refers to the ability to carry out a specific course of action or achieve the desired results under specific standards and conditions. They use a variety of means and ways to accomplish a set of tasks. In practice, it is necessary to comprehensively consider the user's demand for product capabilities and the enterprise's own demand for product capabilities, and summarize the top-level capabilities of the product, which includes capability concept models (CV-1), capability classification models (CV-2), capability segmentation models (CV-3), capability dependency models (CV-4), capability and combat activity mapping models (CV-6), etc.

It is a very important step in system design that transit the research on the operational requirements of all viewpoint and operational viewpoint to the research on capability viewpoint, realizing the quantitative mapping from "target" to "capability". Corresponding to the design of radar active jamming system, the capability viewpoint is to summarize the capability requirements, constructing the operation capability system and operational capability requirements to meet the operation requirements. For example, the top-level capabilities of radar active jamming systems usually include interception, identification, positioning, tracking, jamming, evaluation, etc., and then are refined layer by layer into radar alarm, ESM, single positioning, cooperative positioning, self-defense jamming, long-distance support jamming, accompanying support jamming, etc.

Systems viewpoint is the system and the interrelation that provides or supports operational and service functions. The system model associates system resources and operations with capability requirements. System resources support operational activities and facilitate information exchange.

3.4.3.3 Construction of System Capability Catalog

Refer to the above-mentioned DoDAF and analyze the operational requirements of the radar active jamming system to form the capability requirements of the jamming system and take it as the input for the next stage of system design.

Generally, the capability catalog of radar active jamming system may include:

(1) Radar threat warning capability: it can give warning and threat level prompt to radar signals. Radar signals include search warning radar signals, early warning aircraft radar signals, target indication radar signals, tracking guidance radar signals, fire control radar signals, terminal guidance radar signals, etc.

(2) Electromagnetic environment perception ability: it can intercept, measure, or locate radar signals in the operational area. It includes measuring and recording the waveform signal characteristic parameters of the radar emitter in the battlefield environment, such as frequency, pulse width, pulse repetition frequency, pulse modulation type, amplitude, and antenna scanning characteristics, and obtaining the angle and even geographical location of the air, ground, and sea threat emitter.

(3) Radar target recognition capability: it can recognize the type and status of radar according to the characteristics of radar signal.

(4) Self-defense jamming capability: it can implement active self-defense jamming against airborne fire control radar and missile guidance radar within the coverage of jamming airspace, reduce radar detection distance, reduce radar accuracy, and damage enemy missile launch conditions.

(5) Support jamming capability: it can jam airborne early warning radar, early warning detection radar, target indication radar, guidance radar, and airborne fire control radar. It can be divided into long-distance support jamming, team support jamming and close in support jamming.

(6) Comprehensive management and control capability: including jamming target selection capability, jamming airspace selection and control capability, jamming power control capability, and multi-target jamming capability.

(7) Cooperative combat capability: including cooperative reconnaissance, cooperative jamming, and cooperation with other combat forces.

(8) Other capabilities: including data recording, parameter loading and unloading, interaction with other avionics systems, function expansion, self-inspection, etc.

The above capabilities can be further subdivided.

3.4.4 Selection of Key Technology

According to the processing of radar active jamming system, the key technologies include signal interception technology, parameter measurement technology, signal sorting and processing technology, jamming generation technology and jamming emission technology, etc. Different technology will correspond to the selection and design of active jamming system in antenna, receiver processor, jamming source, transmitter, and other functional extensions. Of course, we can also use a variety of different systems or different types of components in a radar active jamming system according to different operating frequency bands and different application scenarios.

3.4.4.1 Technical Framework for Signal Interception

Signal interception mainly realizes the reception and detection of radar signal in electromagnetic environment, which is the premise of the radar active jamming system. The selection of the technical system will correspond to the antenna selection of the system and the design of the antenna layout, receiver and processor selection, and other functional extensions. Signal interception needs to be considered from the aspects of airspace, frequency domain, sensitivity, and multi-signal adaptability.

(1) The balance between instantaneous spatial coverage and system sensitivity
It is generally required that the instantaneous coverage of signal interception in airspace should be as wide as possible in order to intercept the threat signals in time and take jamming measures quickly. However, for a single antenna and a receiving channel, the increase of instantaneous spatial coverage means that the antenna gains decrease and the system sensitivity decreases, which restricts

each other. The formula between system sensitivity and antenna beam width is as follows:

$$P_{\min} = P_{\text{smin}} + 10 \cdot \log \frac{A}{\varphi \cdot \theta} \tag{3.1}$$

In the formula, $P_{\min}$ is the sensitivity of the receiving system, while P_{smin} is the sensitivity of the receiver, and A is constant, which is related to the efficiency of the antenna. In experience, A is 32000. φ and θ are the beam widths of antenna elevation and azimuth, expressed in degrees.

Therefore, in the design of radar active jamming system, it is necessary to balance the spatial instantaneous coverage and system sensitivity. When the sensitivity requirement is met and the spatial instantaneous coverage is not, multiple receiving systems can be used to realize high sensitivity and wide spatial coverage in space. And multiple receiving systems mean increased weight, volume, power consumption, and cost.

(2) The balance between instantaneous frequency coverage and receiver sensitivity
In order to guide jamming quickly, radar active jamming system usually needs a wide instantaneous frequency receiving range, and the best case is that the receiving system can cover the whole frequency band instantaneously. However, because the sensitivity of the receiver is affected by the instantaneous bandwidth, the wider the bandwidth of the receiver, the lower the sensitivity of the receiver that can be achieved. The relationship between receiver sensitivity and bandwidth is as follows:

$$P_{\text{smin}} = -114 + 10 \log B + \text{NF} + M \tag{3.2}$$

In the formula, P_{smin} is receiver sensitivity, NF is the noise coefficient of receiver, M is detection coefficient, and B is instantaneous bandwidth of receiver. The instantaneous bandwidth here is the bandwidth of the receiver signal detection. For example, for a channelizing receiver, the instantaneous bandwidth should take the bandwidth of a single channel, not the bandwidth synthesized by all channels; for a digital receiver, it should be the bandwidth of each frequency resolution unit detected in frequency domain.

Therefore, when designing the receiver of active jamming system, it is necessary to balance the instantaneous frequency coverage bandwidth and receiver sensitivity. If the receiver sensitivity meets the requirements, but the instantaneous frequency coverage does not, multiple receivers can be connected in parallel to achieve large instantaneous frequency coverage, but this will increase the volume, power consumption, weight, and cost.

(3) The balance between instantaneous frequency coverage and multi-signal adaptability
When the instantaneous frequency coverage is large, it means a wider instantaneous reception capacity, and more signals of the environment will enter

the receiver. The number of signals arriving at the receiver at the same time will increase, which requires stronger multi-signal processing capability. In the design of the system, it is necessary to analyze the signal environment adapted to the jamming system and the multi-signal adaptability of the receiver is considered according to the probability of multi-signals and the number of multi-signals that may occur simultaneously within the instantaneous frequency coverage.

After completing the balance of spatial domain, frequency domain, sensitivity, and multi-signal adaptability, it is necessary to select the interception technology system according to the spatial domain, frequency domain, and sensitivity.

(1) Interception system selection based on spatial domain
The interception in spatial domain is mainly used to satisfy space coverage, usually including wide-beam interception system and narrow-beam interception system. The wide-beam interception system generally adopts horn or helical antenna, which can adapt to the wide spatial range instantly, but the antenna aperture gain is low, which will have a negative effect on the sensitivity of interception, and it also faces the pressure of sorting and identifying dense pulse signals. The narrow-beam interception system usually uses reflector antenna or array antenna, which only intercepts the signals in the specified narrow airspace. Because of the high aperture gain of the antenna, it is beneficial to improve the interception sensitivity and dilute the dense signal equivalent to reduce the subsequent processing pressure. But under the condition of multi-target or initial interception, the narrow beam must have the guidance of wide airspace interception; otherwise, the effective interception probability of the target will be too low to give a normal alarm. The airspace interception can also be composed of multiple groups of antennas arranged in different directions, which will increase the complexity of the system implementation.

(2) Interception system selection based on frequency domain
The interception system in frequency domain is mainly divided according to the receiving bandwidth. It is similar with the airspace interception. It is mainly divided into instantaneous broadband interception and instantaneous narrowband interception. Instantaneous broadband interception requires a large receiving bandwidth. One receiver can cover the whole working frequency band of radar active jamming system instantaneously, or several receivers or channels can be connected in parallel to achieve instantaneous full frequency band coverage. The narrowband instantaneous interception system usually adopts superheterodyne receiving system, which is generally composed of high-speed local oscillator and frequency conversion channel covered by instantaneous narrowband. It usually needs the frequency information intercepted by broadband for guidance. Its advantage is that the performance indicators such as sensitivity and dynamic range can be improved. These are described in detail in Sect. 3.5.2 typical receiver types.

(3) Interception system selection based on system sensitivity
The system sensitivity is determined by the antenna gain and the receiver sensitivity. The receiver with high-gain antenna and small instantaneous bandwidth can achieve high sensitivity, but the spatial coverage and frequency interception ability are low, which just can be used for systems that have good frequency and space external guidance. The wide-beam antenna and the receiver with small instantaneous bandwidth can achieve medium system sensitivity and wide spatial coverage, but the interception ability is low in the frequency domain, which can be used for systems with frequency external guidance. The narrow-beam antenna and the receiver with large instantaneous bandwidth can also achieve medium sensitivity and large frequency coverage, but the spatial interception ability is low, which can be used for systems with spatial external guidance. The combination of wide-beam antenna and wideband receiver can realize wide opening of space and frequency, but the sensitivity of the system is low, which can be used when the sensitivity requirement is not high but the interception of space and frequency has a higher requirement.

3.4.4.2 Parameter Measurement

Reconnaissance mainly includes pulse parameter measurement, radiation source orientation, and location, which can also be called parameter measurement. The selection of the technical system mainly corresponds to the internal design of the receiver, in which the direction measurement is also related to the selection of antenna types.

(1) Design principles
The selection of reconnaissance and reception technology system mainly focuses on the accurate measurement of radar emitter parameters. The traditional radar pulse description parameters include frequency, pulse width, pulse repetition frequency (PRF), pulse amplitude, and angle of arrival (AOA), while the emitter parameters mentioned here also include in pulse modulation information, target distance, etc., which are required for threat target identification and positioning. The measurement accuracy of these parameters not only affects the subsequent signal sorting and threat identification processing efficiency of the active jamming system but also affects the guidance ability of the jamming source and ultimately determines the effect of active jamming. AOA is the only stable parameter among the radiation source signal parameters obtained by electronic warfare sensors in a relatively short reconnaissance time, and AOA is one of the important input parameters in target cross location algorithm.

However, the precise measurement of parameters expected by a radar active jamming system will conflict with the requirement of fast response, and often it is necessary to make a preferential choice or compromise according to the different ways of jamming applications. For alarm or self-defense radar active jamming systems, they usually require rapid identification of radar signals with high threat in order to take appropriate countermeasures in time and therefore do not require high accuracy of radar signal characteristic parameters such as frequency and angle, nor do they care about details such as intra-pulse modulation. For the long-distance support jamming system, it is not only necessary to quickly identify and jam the target radar, but also to obtain the intelligence of unknown radar signals, which should be added to the existing radar threat database after analysis and identification. Therefore, the jamming system requires not only fast but also certain accuracy for radar pulse signal parameter measurement.

(2) Receiver systems
There are mainly two types of receivers, analog receiver and digital receiver.

The advantages of analog receiver are fast response, wide bandwidth and large dynamic range. The disadvantages of analog receiver mainly include poor stability, weak ability to resist external interference and environmental changes, low sensitivity and parameter measurement accuracy; in addition, it may damage the frequency and phase modulation information of the signal. Digital receiver is an important development direction of radar active jamming system receiver. Compared with analog receiver, it has great advantages: first, it can retain more pulse information, which is convenient for long-term preservation and multiple processing of data; Second, more flexible signal processing methods can be adopted, which generally have the advantages of stability and reliability, high parameter measurement accuracy, high sensitivity, and strong simultaneous receiving ability of multiple signals. The disadvantage of digital receiver is that the receiver bandwidth and dynamic range are limited by the sampling bandwidth of ADC devices, and the response time is slower than that of analog receiver.

(3) Direction-finding systems
Different direction-finding (DF) system will determine different selection of antenna type. Common DF systems include amplitude comparison DF, time difference of arrival (TDOA) DF, interferometer DF, and digital beam DF.

The precision of amplitude comparison DF is the worst, but the implementation is the simplest. It is based on the amplitude of the incoming wave signal for angle measurement. It is insensitive to the frequency of the radar emitter. It is suitable for the situation that the platform resources are limited, and there is no need for positioning or the DF accuracy is not high.

The accuracy of TDOA DF is second. This method is insensitive to the frequency of radar emitter, but has high requirements for baseline length and time accuracy. It is generally suitable for large reconnaissance aircraft, ships, or ground systems.

The interferometer DF has high accuracy and is essentially a phase comparison DF method. The requirement for baseline length is moderate, but the requirement for phase consistency of several channels is high, and it is sensitive to the frequency of radar radiation source. Generally, accurate frequency measurement and frequency guidance are required first. It is suitable for small and medium-sized airborne, satellite, or shipborne platforms.

Digital beam DF is based on array antenna, which is consistent with the digital beam forming principle of radar. The measurement accuracy is related to the size of the array aperture. Generally, it can be up to one-tenth of the array beam width. Generally, the DF accuracy is the highest. Its disadvantages are that there are many antenna apertures and channels, and the system resource consumption is large. It is generally suitable for large airborne, shipborne, or ground platforms.

The principles of different DF methods will be analyzed in detail in Sect. 5.4.4.

3.4.4.3 Signal Sorting

Signal sorting is a very important step in the processing of radar active jamming system. The result of signal sorting will directly affect the jamming system's perception of battlefield electromagnetic environment and the jamming effect on target radar.

(1) Design principles
For the sorting algorithm, the evaluation criterion is to minimize the occurrence of "adding batch" and "missed batch". The "adding batch" means that the pulse sorting result incorrectly adds some new radar batches compared to the actual input, often because the sorting algorithm treats the complex signal of one radar as multiple radars; the "missed batch" is the opposite of the "adding batch", which means that some radars in the environment are missed in the sorting results.

In practice, radar active jamming systems may add preprocessing before sorting to filter out interfering pulses caused by receiver parameter measurement errors or complex dense environments. The post-processing is added after the sorting to correct the "incremental" and "missing" cases as much as possible.

In practical application, the radar active jamming system may add preprocessing before sorting to filter the jamming pulse caused by receiver parameter measurement error or complex and dense environment. After signal sorting, the "adding batch" and "missing batch" will be corrected as much as possible through subsequent processing.

(2) Different sorting algorithms
There are many signal sorting algorithms, and there are two common types [19, 60, 61].

The first is classical feature method. The process is to continuously extract the characteristic parameter hypothesis of the signal and perform pulse separation according to the characteristic hypothesis, including multi-parameter correlation comparison method [62, 63], time series analysis method [65], multi-parameter correlation histogram method [66–68], etc.

The second method is pattern recognition, which is characterized by the use of cluster analysis methods to achieve subset separation based on the similarities

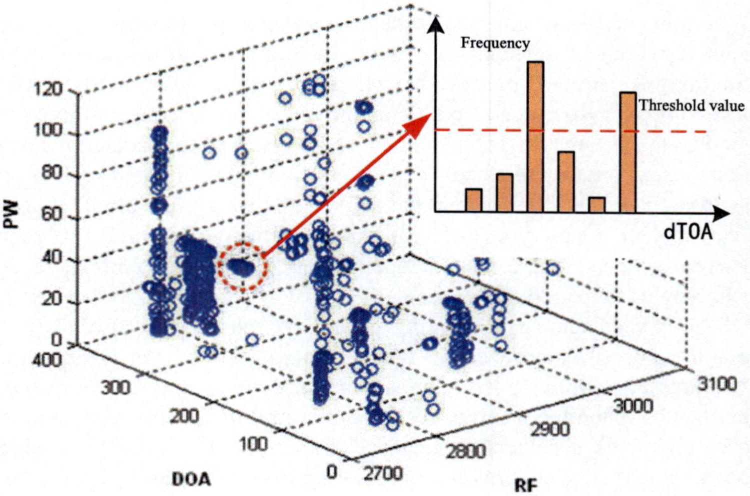

Fig. 3.11 Schematic diagram of the sorting algorithm based on clustering and histogram statistics

an64d associations existing within the PDW dataset, including K-Means cluster sorting algorithm [69, 70], weighted dynamic cluster sorting algorithm [71], and neural network sorting method [72]. The following Fig. 3.11 depicts the basic principle of a sorting algorithm combining parametric associative clustering and histogram statistical methods.

According to the implementation of sorting algorithm, it can also be divided into hardware method, software method, and comprehensive method. At present, the software method is mostly used. However, due to the increasingly complex environment faced by the radar active jamming system and the increasing requirements for sorting speed, the processing speed, storage space, and other corresponding capabilities of the processor also need to be continuously improved. In addition, the sorting method realized by hardware has gradually appeared.

3.4.4.4 Jamming Generation

Jamming generation technology refers to the different ways of radar active jamming signal generation, mainly involving the jamming source design in radar active jamming system and the jamming control design of system control unit.

(1) Design principles

The system selection of jamming generation technology is first related to the operational function and jamming pattern requirements of the radar active jamming system to be designed. For radar active jamming systems with single operational function requirements, relatively simple structure and few types of jamming styles, such as active decoys, an appropriate jamming generation technology can be selected. If a universal multi-functional jamming source with multiple jamming generation technology systems is adopted, it will bring more complex system design problems and greater power consumption, volume, weight, etc.; as a result, it is not conducive to system implementation and operational use. For large platforms and special combat equipment such as special electronic warfare aircraft, the radar active jamming system often requires that it can achieve a variety of different jamming effects on a variety of radars of different systems, so it needs to adopt a variety of different jamming generation technologies to meet the requirements.

Secondly, the choice of jamming generation technology is also related to the parameter measurement accuracy that the jamming system receiver can provide. The parameters of noise jamming signal mainly include center frequency and bandwidth. When the receiver does not measure the frequency of radar signal accurately, it is necessary to increase the bandwidth of jamming signal to ensure the effective suppression of jamming to target radar. When the receiver has high measurement accuracy for the parameters of radar signal such as frequency, pulse width and in-pulse modulation, the active jamming system can generate jamming signals independently according to these parameters. When the measurement accuracy of radar signal parameters by the receiver is not high enough, or the in-pulse modulation information cannot be obtained, the jamming signal can be generated by direct forwarding or store forwarding.

Finally, the quality of the jamming signal is also very important for the selection of the jamming signal generation technology system. For example, the spectral structure of the noise jamming signal generated in the analog way will usually be closer to the noise of the radar receiver and have better jamming effect than the noise signal generated in the digital way, but the stability and flexibility are not enough. If the coherence between the jamming signal and the radar signal is required to be high, for example, for PD radar jamming or to achieve vivid false target jamming, it is often necessary to use forwarded jamming rather than responsive synthetic jamming.

Finally, the quality of jamming signal is also an important factor affecting the choice of jamming signal generation technology. For example, the spectrum structure of the noise jamming signal generated by analog is usually closer to the noise of the radar receiver, which has better jamming effect than the noise signal generated by digital mode, but it is not stable and flexible. If the coherence between jamming signal and radar signal is required to be high, such as jamming to PD radar or lifelike false target jamming, it is often necessary to adopt the forward jamming with high coherence rather than the responsive synthetic jamming.

(2) Different jamming generation systems

One is the jamming signal generation system based on receiver measurement and parameter presetting. It can be based on the radar signal frequency and pulse-related information obtained from the measurement of the receiver of the

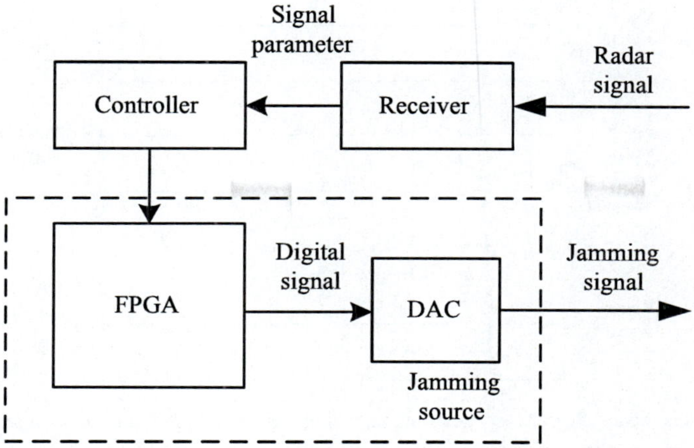

Fig. 3.12 Schematic diagram of jamming signal generation based on parameter measurement

active jamming system, and the corresponding jamming mode can be quickly generated by the system controller issuing instructions. With this autonomous jamming signal generation method, the frequency accuracy of its jamming signal is guaranteed by the measurement accuracy of the system receiver. In order to achieve fast style switching capability, it can also be achieved by loading different styles in advance and then making a quick loading based on the measurement parameters and jamming mode codes. The most common jamming signal generator in this system is the direct digital frequency synthesizer (DDS) jamming source, which can usually generate noise, single-point frequency, multi-point frequency, and other jamming mode. The following Fig. 3.12 shows a block diagram of jamming signal generation based on parameter measurement and implemented using FPGAs.

The other one is the jamming signal generation system based on the forwarding of radar signal samples. It mainly obtains samples by directly receiving radar signals and then performs different signal modulation, forwarding, etc., to form different jamming mode according to the jamming pattern code. The biggest advantage of this system is that it does not require precise measurement of radar signal parameters, it can adapt to pulse agile signals, and it can also achieve high fidelity phase replication between jamming signals and radar signals. The common jamming generating devices in this system include optical fiber delay transmitter, analog RF memories, digital RF memories (DRFM), etc., as shown in the Fig. 3.13.

Since active jamming systems are designed to jam a variety of radar objects and complex signals, it is often difficult to achieve effective jamming of all signals with a single jamming signal generation system alone. Therefore, a jamming signal generation system with the ability to combine multiple jamming

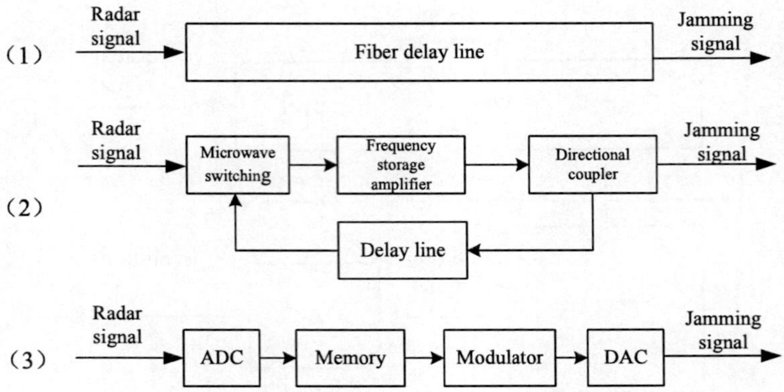

Fig. 3.13 Schematic diagram of jamming signal generation based on radar signal sample forwarding

sources is preferred. If the miniaturization and low cost of components can be achieved, it will be more widely used in practical projects.

3.4.4.5 Jamming Transmission

Jamming transmitting mainly includes two functions of jamming signal amplification and transmitting, and its technical system selection will determine the selection of transmitter and the design of transmitting antenna, etc. At the transmitting end, the transmitter and antenna are often inseparable and need to be considered together in the technical system selection and design.

(1) Selection principles

The selection of jamming transmission technology system will focus on the jamming effect of the finally transmitted jamming signal on the radar, mainly considering the following aspects.

Firstly, the quality and power of the jamming signal finally radiated into the free space should be considered. According to different tasks, jamming sources will produce different jamming signal styles. When transmitting the jamming signal, it is desirable that the signal transmission part should be a linear amplification system, which does not change other parameters of the jamming signal except the amplified power. However, in most cases, the power of the transmitter is the most important indicator, when the signal is amplified, it is difficult to ensure that the transmitter is completely linear, which will affect the spectrum structure of the jamming signal. Therefore, it is necessary to make a compromise between signal quality and amplification power in the design of transmitter.

Then, similar to the design of signal interception, it is necessary to consider the coverage and agility of jamming in time domain, spatial domain, frequency domain, polarization domain, and other dimensions in combination with specific radar objects and operational modes. Radar active jamming system is facing non-cooperative targets, so it is necessary to ensure the effective tracking of threat objects in the whole domain during the jamming process. In the time domain, the duty cycle of the transmitter and whether the transmitting antenna can be shared with the receiving antenna are mainly considered. In the spatial domain, the wide area coverage and high gain of the transmitting antenna are mainly considered. When the self-defense jamming and other jamming transmission power requirements are not high, multiple wide-beam antennas can be combined. When the support jamming and other jamming transmission power requirements are high, the narrow beam combined with digital beam forming technology or antenna servo system shall be considered as much as possible. In frequency domain, the transmitter and transmitting antenna are required to have relatively stable response characteristics in a relatively wide frequency range. In the polarization domain, the design of the transmitting antenna is mainly considered, which is required to match the receiving polarization mode of the radar. In order to meet different polarization requirements, a group of orthogonal dual polarization antennas can also be used for combination, but at least two channels of transmitters are required.

Finally, as one of the most expensive and important parts of radar active jamming system, the transmitter needs to consider the reliability, service life, maintainability, life cycle cost, volume, and weight. In addition, electromagnetic compatibility and radar stealth should be considered in the design of transmitting antenna. Solid-state transmitter is an important development direction in the future. Solid-state transmitter has no high-voltage power supply, high-power transmission line, and other easily worn parts, so its service life is longer than that of electric vacuum devices. In addition, solid-state transmitter also has the advantages of high reliability and low maintenance cost, but its deficiency is that it makes the design of jamming system more complex. Electromagnetic compatibility (EMC) mainly considers the isolation between transmitter and receiver of radar active jamming system, which often needs to be considered comprehensively, and can be solved by means of control measures, isolation design and suppression algorithm in time domain, space domain, polarization domain, and energy domain. Other EMC designs also include reducing the impact of radar active jamming system on radar, communication, navigation, and other electronic information systems on the platform. On small platforms such as missileborne platforms, conformal antennas for transmitting and receiving are usually used. On stealth combat platforms, the low RCS design of transmitting antennas and radar radomes should also be considered.

(2) Different jamming transmitter systems
According to the different ways of amplification of jamming signals by transmitters, they can be divided into two technical systems: saturation amplification and linear amplification. Saturation amplification is mainly the amplification method of vacuum electronic device, which is named because it mainly works

in the saturation amplification state and usually amplifies all the jamming signals to a uniform power value, so it may affect the spectral structure of the jamming signals, but the power utilization rate is high, which is very suitable for noise jamming signals, etc. Linear amplification generally corresponds to solid-state devices, which have a relatively good linear amplification capability, and the influence of the jamming signal is smaller. The comparison of the output signals under the two amplification methods is shown in the following Fig. 3.14.

According to the spatial coverage characteristics of the transmitting antenna, it can be divided into three technical systems: single beam, multi-beam, and phased array.

The single wave beam system adopts the working mode of single antenna and wide beam. The transmitter generally adopts electric vacuum devices. The single wave beam system is mainly used in the self-defense active jamming system. This method has the advantages of simple system and low cost. The disadvantage is that the jamming power is low, the spatial coverage requires multiple antenna combinations or antenna servo mechanisms to rotate, and the multi-target jamming can only be completed in the same beam.

Multi-beam system refers to the use of multiple electrovacuum devices for spatial power synthesis. Its core is the multi-beam antenna array, especially

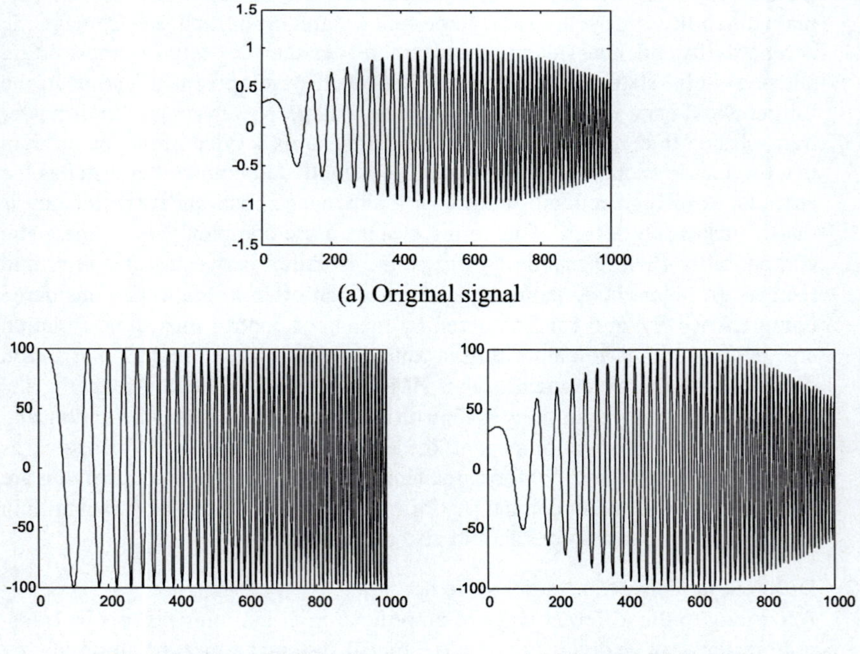

(a) Original signal

(b) Saturation amplification

(c) linear amplification

Fig. 3.14 Comparison of jamming signals under different amplification methods the Rotman lens multi-beam array is the most widely used. Through the phase adjustment of Rotman lens, spatial coverage is realized by multiple fixed beams. Its advantages are that it can form multiple beams at the same time, could interfere with multiple targets in multiple directions at the same time, and is easy to form strong effective radiation power. Its disadvantages are that the amount of hardware equipment is large, the technology is complex, and the volume, weight, power consumption, and heat dissipation requirements are large. It is mainly used in special electronic warfare aircraft or ground interference equipment.

The phased array system usually uses the solid-state transceiver module, i.e., T/R module, as the core component. It controls the direction of the transmitted beam in space through the phase shift of different units and realizes the large equivalent radiation power through the spatial power synthesis of multi-channel transmitting units. Therefore, it has the advantages of wide spatial coverage, high equivalent radiation power, and instantaneous beam direction switching. At the same time, replacing high-power electric vacuum devices with solid-state power devices can eliminate the problems of high voltage and short service life and bring about a significant improvement in reliability and service life.

3.4.5 Design and Decomposition of Key Technical Indicators

An important aspect of system design is system measurability and indicator achievability. Measurability means that it is possible to ensure that the designed radar active jamming system can meet the usage requirements through some functional indicators or technical indicators constraints. Indicator achievable means that all the tactical technical indicators required by the system should be able to be achieved through design and actual measurement. Therefore, the design needs to specialize in several of the important indicators of the radar active jamming system and decompose them into various subsystems to ensure the scientificity and realizability of the whole system indicators.

A very important aspect of system design is system measurability and indicator achievability. Among them, system measurability means that the designed radar active jamming system can meet the use requirements through some functional or technical index constraints. Indicator achievability means that all tactical and technical indicators required by the system can be achieved through design and actual measurement. Therefore, it is necessary to specially design several important indicators of radar active jamming system and decompose them into various subsystems to ensure the scientificity and realizability of the whole system.

The design principles and methods of several main technical indicators are analyzed below.

3.4.5.1 Jamming Frequency Range

(1) Determination of the jamming frequency range
The frequency range of the radar active jamming system is closely related to the frequency range of the target radars to be jammed. The jamming frequency range needs to cover the working frequency range of the target radars, which is expressed as follows:

$$F_j \supseteq (F_1 \cup F_2 \cup \cdots \cup F_i), i = 1 \cdots n \tag{3.3}$$

In the formula, F_j is the jamming frequency range of the jammer, F_i is the operating frequency range of the ith radar, n is the number of radars, and the symbols $\cup$ is the merged set and $\supseteq$ is inclusion.

In the actual determination of the indicators of jamming frequency range, if the concatenation of radar frequency range is a concentrated small range, then the jamming frequency range only needs the minimum frequency less than the minimum frequency of the radar and the maximum frequency of jamming greater than the maximum frequency of the radar. For example, the radar frequency range to be jammed is 1.1–1.4 and 1.5–1.6 GHz, then the jamming frequency range can be determined as 1.1–1.6 GHz, and considering appropriate expansion, the jamming frequency range can be determined as 1–1.7 GHz.

If the radar operating frequency range is relatively scattered, the jamming frequency range can also be a combination of more than two frequency bands. For example, if the frequency range to be jammed is 1.1–1.6, 8.5–9, 9.5–10 GHz, the jamming frequency range can be set as 1.1–1.6, 8.5–10 GHz, and the jamming frequency range can be determined as 1–1.7, 8.4–10.1 GHz by considering appropriate expansion.

(2) Decomposition of jamming frequency range
The frequency range of radar active jamming system restricts the design of antenna, receiver, and transmitter. During technical index decomposition, the operating frequency range of antennas, receivers, and transmitters shall be greater than or equal to the frequency range of the system.

If the jamming frequency range is small, the system can directly meet the jamming frequency range without dividing the frequency band. According to the traditional band division, the typical frequency range is as follows: 1 ~ 2, 2 ~ 4, 4 ~ 8, 8 ~ 18 GHz, etc. If the frequency range of the jamming system is very large, such as 0.1 ~ 18 GHz, the current signal reception, RF channel transmission, digital sampling, data storage, and signal processing will become very difficult. In the design, the full frequency band can be divided into several sections (such as 0.1 ~ 2, 2 ~ 6, 6 ~ 18 GHz). Different antenna and RF channel combinations can be used, respectively, and up-conversion and down-conversion technologies can be used to cover the whole frequency range. Some sections can be divided into several channels, for example, 6 ~ 18 GHz is divided into 12 channels such as 6 ~ 7, ..., 17 ~ 18 GHz according to 1 GHz instantaneous receiving bandwidth.

3.4.5.2 Jamming Airspace Range

(1) Determination of jamming airspace range
The jamming airspace range is jointly determined by the operational requirements and realizability of the radar active jamming system. For example, in the case of airborne self-defense, it is desired to have 360° azimuth coverage and 180° pitch coverage. However, due to the installation restrictions, the azimuth coverage and pitch coverage may be reduced to only meet the direction with the greatest threat.

(2) Decomposition of jamming airspace range
The index of jamming airspace range is ensured by the number of receiving and transmitting antennas, main lobe beam width, and beam pointing design. Jamming airspace range is usually decomposed into jamming azimuth angle range and jamming pitch angle range. It should be noted that the jamming airspace coverage here usually refers to the spatial angle range that can be reached by the radar active jamming system through antenna scanning or multi-antenna combination, while the instantaneous coverage airspace is the sum of the spatial angles that can be reached at a certain instant when the antenna is not scanned. For example, when the radar active jamming system wants to achieve 360° azimuth coverage, the omnidirectional antenna, mechanical rotating antenna, phased array scanning antenna and multiple antennas or antenna arrays can be used in the design. Here, the omnidirectional antenna can achieve instantaneous azimuth 360° coverage, and the instantaneous coverage of other antennas is less than 360°. However, the pitch coverage requirements are relatively low and vary greatly depending on different applications. Generally, the pitch coverage of airborne radar active jamming system can directly use the antenna with corresponding main lobe beam width to cover the pitch angle range instantaneously.

When a single fixed antenna is used to instantaneously cover the space range required by the jamming system, the jamming space range constrains the beamwidth of the antenna, requiring the antenna beamwidth to be greater than or equal to the coverage airspace range, the following relationship should be satisfied

$$\alpha \geq \theta, \beta \geq \phi \tag{3.4}$$

In the formula, α *is the* azimuth beamwidth of the jamming antenna, β is the pitch beamwidth of the jamming antenna, θ is the required azimuth coverage angle range, and φ is the required pitch coverage angle range.

When a single rotating antenna is used to cover the jamming airspace range, the rotation angle range of the antenna plus the antenna beam width shall be greater than or equal to the jamming airspace range; that is, the following relationship is satisfied:

$$\alpha + \mathrm{A} \geq \theta, \beta + \mathrm{B} \geq \varphi \tag{3.5}$$

In the formula, A is the antenna azimuth rotation angle, B is the antenna pitch rotation angle.

When multiple antennas are used to realize airspace coverage, the union of all antenna angle ranges should be greater than or equal to the airspace coverage.

$$\cup\alpha_i \geq \theta, \cup\beta_i \geq \varphi \tag{3.6}$$

In the formula, α_i is the azimuth beamwidth of the ith antenna, β_i is the pitch beamwidth of the ith antenna.

3.4.5.3 System Instantaneous Bandwidth

The instantaneous bandwidth of the system is generally designed to be larger than the operating frequency range of a single radar to adapt to the frequency agility. For example, modern radar can be frequency agile in the range of hundreds of MHz, and the signal bandwidth of a single pulse of high-resolution synthetic aperture radar can even reach more than 1 GHz. The instantaneous bandwidth of the system is generally less than the coverage frequency range of the system.

System instantaneous bandwidth can be divided into receiving system instantaneous bandwidth and jamming system instantaneous bandwidth. The instantaneous bandwidth of the receiving system mainly affects the design of the receiver, which is a direct requirement for the receiver; the instantaneous bandwidth of the jamming system mainly affects the design of the jamming source, which is a direct requirement for the jamming signal generated by the jamming source.

3.4.5.4 System Sensitivity

(1) Determination of sensitivity

System sensitivity refers to the minimum signal power that can be received and processed when the system works normally. This index mainly ensures that the radar active jamming system can detect the threat radar signal. High sensitivity often means a long reconnaissance range. Therefore, for a specific radar target and scene, the reconnaissance range index can also be used to characterize the reconnaissance capability of the receiver or jamming system.

The calculation of system sensitivity usually requires the following radar power density equation:

$$P = \frac{P_t G_t}{4\pi R^2} L \tag{3.7}$$

In the formula, P_t represents the peak transmit power of the target radar, G_t represents the transmit antenna gain of the target radar, R represents the distance from the radar active jamming system to the target radar, and L represents the space transmission loss.

The receiving sensitivity of radar active jamming system needs to meet the reconnaissance requirements of all different types of radars with different ranges in the operational scene; that is, the receiving sensitivity should be less than the power density of the transmitted signal of the target radar transmitted to the jammer.

But from the point of view of the overall system design, not the higher the sensitivity of the system, the better.

The first adverse effect of high sensitivity is that the isolation index of transceiver is too high, which will make it difficult for the radar active jamming system to transmit and receive at the same time or to switch quickly. Other electronic equipment will affect the signal interception of the radar active jamming system. Secondly, high sensitivity will have a great impact on the dynamic range and high-precision direction finding of the whole system, and in some applications that do not need all interception, more radar sidelobe signals or other background radiation signals will be received, which increases the difficulty of signal sorting and emitter identification. Therefore, in practical application, it is necessary to determine the sensitivity index in combination with specific reconnaissance range requirements and application methods. For example, in the application of long-distance support jamming, the radar active jamming system needs to intercept the radar antenna sidelobe signal, which requires high system sensitivity. In self-defense jamming, the radar active jamming system only needs to intercept the radar main lobe signal, and the system sensitivity is relatively low.

(2) Design of system sensitivity index
The index of system sensitivity can be decomposed into the multiplication of the receiving antenna gain and receiver sensitivity, and in the actual situation, the insertion loss also should be considered. Set the system sensitivity of the radar active jamming system as P_{sys}, the receiving antenna gain is G_{rev}, the receiver sensitivity is P_{min}, the insertion loss is L_{ins}, then the relationship between them is

$$P_{sys} = G_{rev} \cdot P_{min} \cdot L_{ins} \tag{3.8}$$

It is very important to reasonably allocate the antenna gain and receiver sensitivity based on system sensitivity requirements. In the allocation of indicators, the various types of receiver sensitivity level and advantages and disadvantages must be considered and combined with the choice of receiver type and requirements instantaneous of bandwidth needs to determine the receiver sensitivity. The antenna gain needs to be determined comprehensively according to the antenna beam coverage, installation space, and other factors.

3.4.5.5 Receiving Dynamic Range

(1) Determination of receiving dynamic range
Radar active jamming system requires simultaneous or non-simultaneous reconnaissance and reception of radar signals with different distances and powers, which requires the radar active jamming system to have the ability to adapt to different signal powers. This capability of radar active jamming system is generally expressed by receiving dynamic range, which is defined as follows:

$$\mathrm{DR} = P_{\max} - P_{\min} \tag{3.9}$$

In the formula, $P_{\max}$ represents the maximum signal power that the radar active jamming system can normally receive, $P_{\min}$ represents the minimum signal power that the system can normally receive, corresponding to the receiving sensitivity of the system.

The receiving dynamic range is determined by radar jamming scene and target radar parameters. Firstly, the maximum power and minimum power to be received are calculated, and then the receiving dynamic range is calculated according to the formula. The receiving dynamic range is mainly affected by the following factors: the radar's effective radiated power, the maximum and minimum distance between the receiver and the radar, etc.

(2) Implementation of receiving dynamic range
The receiving dynamic range of radar active jamming system is equal to the dynamic range of the receiver. The implementation of the dynamic range of the receiver will be described in detail in Chap. 5.

3.4.5.6 Equivalent Radiated Power (ERP)

(1) Determination of ERP
ERP of radar active jamming system is the decisive factor affecting the minimum jamming distance. According to the jamming equation introduced in Chap. 2, in the case of self-defense jamming, ERP is inversely proportional to the square of the minimum jamming distance. In the case of supporting jamming, when the distance and angle between the jammer and the radar remain unchanged, the ERP of the jammer is inversely proportional to the fourth power of the minimum cover distance. According to the application scenario of radar jamming system, main parameters of target radar, minimum jamming distance, RCS of protected target, suppression coefficient, and other parameters, ERP can be calculated

by using jamming equation. It should be pointed out that during the design of a specific jamming system, it is necessary to calculate the parameters of all target radars and select the maximum ERP value as the index requirement of the system. The specific calculation method of ERP is introduced in the Chap. 2 of this book.

(2) Decomposition of ERP
According to the definition, ERP is equal to the product of the jamming transmitter power and jamming antenna gain, multiplied by the insertion loss between the transmitter and the antenna.

$$\mathrm{ERP} = P_j \cdot G_j \cdot L_{\mathrm{ins}} \tag{3.10}$$

where P_j is the output power of the transmitter of the jamming system, G_j is the gain of the jamming transmitting antenna, L_{ins} is the insertion loss from the transmitter to the antenna of the jamming system.

When the ERP of the system is determined, the jamming antenna gain and transmitter output power need to be allocated reasonably. The gain of the jamming antenna is limited by the size and weight of the antenna. The greater the antenna gain, the larger the antenna aperture, and the greater the installation space and weight required. Besides, the antenna gain is also inversely proportional to the beam width. If the antenna gain is larger, it means that the jamming beam is very narrow, and more precise jamming angle guidance is required. Transmitter output power is closely related to the system power supply, the greater the transmit power means the greater the power supply demand. When the transmitter output power requirements are large, multiple amplifier units are required to be synthesized when a single amplifier piece cannot meet the requirements, which increases the manufacturing cost of the transmitter, as well as increasing the size, weight, and heat dissipation requirements. Therefore, when decomposing the equivalent radiated power into antenna gain and transmitter output power, it is required to combine the technical level of antenna and transmitter, as well as the limitations of volume, weight, power consumption, and heat dissipation, and then combine the cost factors for reasonable allocation.

After the system ERP is determined, the jamming antenna gain and transmitter output power shall be reasonably allocated. The gain of jamming antenna is limited by the size and weight of antenna. The larger the antenna gain, the larger the antenna aperture, the greater the weight and the required installation space. The antenna gain is inversely proportional to the beam width. If the antenna gain is large, the jamming beam will be narrow, and accurate angle guidance is required. The output power of the transmitter is closely related to the power supply of the system. The greater the transmission power, the greater the power supply demand. When the transmitter output power is large, a single amplifier may not meet the requirements, and multiple power amplifier units need to be combined. This method increases the manufacturing cost of the transmitter and increases the volume, weight, and heat dissipation requirements. Therefore, when the ERP is decomposed into antenna gain and transmitter output power, it needs to be reasonably allocated in combination with the specific conditions of the antenna and transmitter, as well as the conditions such as volume, weight, power consumption, heat dissipation, and cost factors.

3.4.5.7 Jamming Pattern

Jamming pattern is an important functional indicator that must be considered in the design of jamming systems. It is necessary to design the jamming pattern according to the application mode of the radar active jamming system, the radar object, and the effect to be achieved. The jamming pattern is mainly related to the design of the jamming source and the design of the jamming control system.

According to the jamming effect, common jamming styles include noise jamming, dense false target suppression jamming, range deception jamming, speed deception jamming, angle deception jamming, radar two-dimensional image jamming, etc. According to the correlation between jamming signal and radar signal, jamming pattern can be divided into coherent jamming and incoherent jamming. According to the generation mode of jamming signals, jamming pattern can be divided into noise jamming, response jamming, and forwarding jamming. According to the waveform of the jamming signal, the jamming pattern can also be divided into continuous wave jamming and pulse jamming. The jamming pattern will affect the type of radar that the radar active jamming system can jam. For example, the general active jamming system with coherent jamming style can better jam the coherent processing system radar such as PD radar, and the jamming system with angle deception jamming style can better jam the monopulse angle measurement system radar. The specific types of jamming patterns and main parameters will be introduced in detail in Chap. 7.

3.4.5.8 System Response Time

(1) Determination of system response time
In general, the shorter the response time of radar active jamming system, the better. A short response time is conducive to rapid and effective jamming of target radar. In practice, the index requirements for jamming response time are related to the jamming radar and specific scenes. For example, self-defense jamming requires fast response, while long-distance support jamming can have relatively long response time.

(2) Decomposition of system response time
The response time of jamming system can be decomposed into delay time of receiving and transmitting channel, signal and data processing time, jamming parameter adjustment time, antenna pointing adjustment, etc. The delay time of receiving and transmitting channel is the shortest, which can be basically

ignored. Signal sorting and emitter identification often occupy a lot of time, which should be considered in the index decomposition. For the automatic jamming system, the adjustment time of jamming parameters is relatively short, while for the system with manual control, the response speed of the operator should be considered, and the response time is relatively long. For the jamming system of phased array antenna, the time of phased array beam switching is very short. For the jamming system with mechanical rotation, the beam adjustment time is relatively long.

3.4.6 Engineering and Product Design

In the design of radar active jamming system, in addition to paying great attention to system composition, workflow, and functional performance indicators, it should also include three-proof design and six-performance design, as well as electromagnetic compatibility, heat dissipation, anti-vibration, software engineering, and other contents. This book only briefly introduces the basic concepts of the above aspects, for specific contents, readers can refer to relevant professional books or literatures.

3.4.6.1 Three-Proof Design

Three-proof design of radar active jamming system mainly includes damp and heat prevention, mold prevention, and salt fog prevention. Damp heat, mold, and salt fog will damage the electronic devices and system structure, causing short circuit, open circuit, structural fracture, etc., leading to the decline of indicators and loss of functions, which may even endanger the safety of products and users and must be avoided.

During the production, transportation, and use of radar active jamming system, it may be exposed to harsh environments such as high humidity, heavy salt fog, high temperature, and easy bacterial growth. Therefore, three-proof design is very important for radar active jamming system. Typical three-proof measures are as follows:

(1) Selecting excellent materials with three-proof performance.
(2) The lap joint design between metal structural members reasonably to avoid electrochemical corrosion of metals.
(3) Metal structural parts need to be structurally sealed and treated for corrosion resistance, the surface of the product, especially the circuit board sprayed with three-proof paint, etc.

3.4.6.2 Six-Performance Design

The "six" refers to six aspects of product design, including reliability, testability, maintainability, supportability, safety, and environmental adaptability.

(1) Reliability design
Reliability design is mainly to enhance the ability of radar active jamming system to work continuously and reliably. The main design tools include derating design, stability design, redundancy design, etc.

(2) Testability design
According to the division of the radar active jamming system internal module, the internal status of the main function is directly led out or through the external excitation detection, etc., to achieve comprehensive coverage of the status information to ensure that the fault can be isolated and located timely and accurately.
The main contents of testability design include: monitor the system's key functions, monitor the internal parameters of the system, such as voltage, current, temperature, humidity, and air pressure, isolate detection signals used by external test equipment, reserve debugging interfaces, etc.

(3) Maintainability design
The design of maintainability mainly includes reducing maintenance content, reducing maintenance skill requirements (e.g., structural simplification), improving interchangeability, visibility and accessibility, reducing cable and component crossings, adequate tool working space and anti-misinsertion measures and signs, etc.
The maintainability design of radar active jamming system mainly includes: the transmitter and other parts with high failure rate and easy to be damaged have good interchangeability and universality, the software module has good debugging and upgradeability, and the components that may be dangerous shall be provided with striking marks, warning lights, and audible warnings.

(4) Supportability design
Supportability design mainly includes design requirements related to support, requirements for support programs, constraints on support resources, generalization and serialization requirements for support resources, requirements related to support system survivability, and special environmental requirements.
Quantitative requirements for supportability design include indicators such as combat readiness, support scale, and cost, which can also be increased according to the project. Among them, combat readiness refers to the system's ability to start performing tasks at any time under peacetime and wartime conditions, availability, maintenance waiting time, average repair time, and maximum repair time. Support scale refers to the supportability objects that can be met by a single supportability unit. Cost refers to the overall cost required to complete the equipment supportability work.

In the supportability design, the existing support equipment and facilities shall be used as much as possible, and the general maintenance equipment shall be used as much as possible.

(5) Safety design
Safety design is to ensure the safety of system equipment, instruments, and operators.
For the radar active jamming system, the security design mainly includes: to prevent power, electromagnetic leakage, high-power microwave radiation leakage; protection and fault-tolerant design to shield the outside world from human error or intentional disorderly operation; the final power tube of the transmitter is followed by a high-power isolator, circulator, etc., for radiation protection; fault protection to prevent the system equipment overheating, the transmitter's helical overcurrent, reflected power is too large, etc. Among them, the design of fault protection is the most important.
The safety design of radar active jamming system mainly includes: prevent the leakage of high-power microwave radiation; through the protection and fault-tolerant design, avoid manual misoperation or intentional misoperation; high-power isolator, circulator, and other equipment are connected behind the final power tube of the transmitter for radiation protection; through fault protection, avoid problems such as overheating of system equipment, overcurrent of transmitter spiral, excessive reflected power, etc. Among them, the design of fault protection is the most important.

(6) Environmental adaptability design
Any product is used, transported, and stored in a certain environment. The working environment of radar active jamming equipment includes space, high altitude, sea, mountain, plateau, etc., and some environments may be very harsh. Good environmental adaptability design can ensure the normal operation performance of products. Common environmental adaptability design indicators include temperature, humidity, air pressure, acceleration, vibration, noise, etc.

The principles of environmental adaptability design include the following:

(1) The index and system design are carried out according to the specific working environment of the jamming system.
(2) Adopt advanced and mature materials, processes, and structures and have a good efficiency cost ratio.
(3) From the whole system, unit modules to components, the protection objects and protection levels shall be defined level by level.

3.4.6.3 Electromagnetic Compatibility (EMC) Design

For any electronic information system, reasonable EMC design is very critical and important, otherwise the system performance is not guaranteed, or even cannot work properly.

The EMC problem is particularly prominent because of the wide operating frequency range, many antennas and high jamming power of the radar active jamming system, and the high sensitivity and large scale of the receiving equipment. The EMC design of radar active jamming system includes the problem of the EMC itself and the problem between the jamming system and the original electronic equipment on the platform.

(1) EMC between the jamming system and the original electronic equipment of the platform
Generally, due to the limitation of the jamming system carrying platform (such as aircraft), the distance between the receiving and transmitting antennas is limited, so it is difficult to isolate the receiving and transmitting antennas.
For the above problems, the solutions follow the following main principles. The jamming system itself should first meet the relevant EMC requirements; The power supply, cable layout, antenna installation, etc., provided by the carrying platform to the jamming system should also meet the relevant EMC requirements. For the transceiver equipment that can solve the problem from the transceiver isolation, the system can work normally by increasing the transceiver isolation measures. For the equipment that cannot solve the jamming from the transceiver isolation, other technologies can be used to solve the problem. From the overall consideration of the platform, special structural design and space reservation can be made for the jamming system, etc.

(2) EMC between modules or components within the jamming system
The basic requirements of EMC design of radar active jamming system can be formulated according to the corresponding national military standard and the characteristics of specific active jamming system. Generally, the main requirements include the following aspects.

    (a) During the module design, EMC design shall be carried out for the electromagnetic jamming sensitive parts, electromagnetic jamming source parts, power supply, and control signals to suppress the radiated jamming and conducted jamming. In terms of the layout and signal flow direction of each device, the analog and digital separation method is adopted to isolate the digital circuit from the analog circuit.
    (b) High-speed transmission lines strictly require differential design and single-ended impedance design to ensure impedance matching and reduce signal reflection. Limit the distance between long-distance parallel lines to reduce signal crosstalk.
    (c) When designing the circuit board, measures such as filtering should be taken to reduce the interference caused by the power supply.

Table 3.6 Common cooling methods

Cooling method	Main parameters and design elements
Air cooling	Maximum power consumption, calorific value, inlet air temperature, outlet air temperature, air volume (kg/h), baseboard temperature, etc.
Liquid cooling	Maximum power consumption, calorific value, liquid inlet and outlet temperature, flow, pressure, etc.
Radiation cooling	Maximum power consumption, calorific value, upper and lower limits of ambient temperature, baseboard temperature, etc.

3.4.6.4 Cooling Management Design

Radar active jamming systems involve high-power amplifier devices and a lot of digital signal processing, so the thermal management of the system is also very important.

Cooling management technology generally uses a combination of computational and experimental approaches, making full use of some thermal design software to carry out modeling simulations and thermal characteristics of product prototypes. Through heat generation analysis and heat reduction design, the heat dissipation problems of high-density microwave components and large systems are solved. Generally, the following aspects can be considered.

(1) The heat generated by the heating device can be transmitted to the shell of the module through conduction, and the shell radiates the heat through thermal radiation.
(2) The module with high calorific value is filled with heat absorbing materials with high specific heat, so that the heat generated during the operation of the jammer can be absorbed as much as possible.
(3) The active jamming system adopts air-cooled or liquid-cooled heat dissipation design, which is guaranteed from the aspects of system structure and loading position.

The main parameters of cooling management design include heat transfer area, allowable maximum temperature, environmental requirements, cooling mode, etc. The common cooling methods include air cooling, liquid cooling, and radiation cooling, and the main parameters are shown in Table 3.6.

3.4.6.5 Anti-Vibration Design

Strong vibration will cause the stress of components and structural parts of electronic equipment to exceed their strength limit or yield limit, resulting in damage or permanent failure. Especially for aircraft, helicopter, missile and vehicle platforms, the anti-vibration ability of radar active jamming system is very high.

According to the electrical performance, interface, installation and other requirements of each equipment of the system, different anti-vibration designs are carried out for different parts, mainly including the following two ways:

(1) Reinforcement design
The reinforcement design is to select components with good anti-vibration performance, so that the components and the whole machine can work reliably under the specified environmental conditions without any vibration isolation buffer. The main measures include: comprehensively considering the requirements of weight and strength, optimizing the structure according to the mechanical principle on the premise of meeting the requirements of volume and weight, and improving the inherent strength of the structural unit as much as possible; the circuit board shall be of small board structure as far as possible, and the circuit board with larger size shall be strengthened in the middle; devices with large mass and cantilever shall be mechanically fixed; try to eliminate the connection gap between various parts of the equipment to reduce the amplification of vibration in the transmission process. In the design, the height dimension of each independent installation body shall be minimized to improve its stability.

(2) Vibration isolation design
Due to technical or economic constraints, equipment such as cabinets, display, and control consoles are difficult to reinforce, or the anti-vibration requirements cannot be met after reinforcement measures are taken. It is necessary to install vibration isolation buffer for this equipment to make the equipment work reliably under the specified environmental conditions. If necessary, secondary vibration isolation can be carried out for vulnerable and precision components inside the cabinet and display console.
In a word, the anti-vibration design of the equipment shall be implemented in every link of the system design, so that all equipment in the system can meet the anti-vibration requirements.

3.4.6.6 Software Engineering Management

Software system is an important part of radar active jamming system. Software engineering design and management mainly include software hierarchical management, requirement management, and quality management.

Software hierarchical management means that after the system design scheme is determined, the importance degree of software will be determined according to the degree of influence of software failure on equipment safety and function, which can usually be divided into three levels: critical, important, and general. Among them, critical software will directly affect the safety of equipment use and endanger the safety of personnel, or affect the completion of key tasks of the system, etc. Important software will not affect the safety of equipment use, but will affect the tasks completion of the system, while general software will not affect the safety of equipment use and the tasks completion of the system.

Requirement management refers to the definition method, verification method, change process, approval authority, and other requirements of software requirements. And in the process of supporting software life cycle, the software requirements and changes are effectively managed to ensure the correctness, integrity, traceability, and no ambiguity of requirements.

Quality management refers to the non-functional requirements such as security, ease of use, robustness, maintainability, expansibility, and document specification on the premise of ensuring that the software system meets the specified function and performance requirements. Among them, security refers to the supporting software should have the ability to run normally for a long time, and in case of failure, can be downgraded to use to maintain the lowest working state, to ensure the safety of the system. Ease of use refers to the human-computer interaction interface realized by software needs to have good man-machine ergonomics, with a consistent and simple display style and fast and convenient operation methods. Robustness refers to the ability of the software system to protect all kinds of irregular operations, misoperations, wrong commands, or illegal data input, so as to ensure clear and unambiguous display of interactive information. Maintainability means that the software structure design should follow the principle of modularization, and reduce the coupling degree and structural complexity between modules. Expansibility means that open architecture and unified interface standards should be adopted to meet the expansion of software system functions, but the types of programming languages should be reduced as far as possible in a software system. Document specification means that all kinds of software documents should be written well while software development and testing, and ensure the integrity, specification, correctness, and consistency of software documents.

3.5 Typical Radar Active Jamming Systems

The structure, function, and technical index of different radar active jamming systems will be different. Several typical radar active jamming systems are introduced from the aspects of airborne, shipborne, ground and small platforms [73–76].

3.5.1 Airborne Electronic Jamming Pods

The airborne jamming pod can be used as either self-defense jamming or high-power support jamming. The United States, Britain, France, Israel, Italy, and other countries have developed and produced airborne electronic jamming pods, of which the United States has the largest number and the most advanced technology.

The AN/ALQ-99 airborne electronic jamming pod has been the main combat equipment of the US Navy's EA-6B for many years, and it's modified AN/ALQ-99F (V) is currently the primary system for electronic attack missions on EA-18G

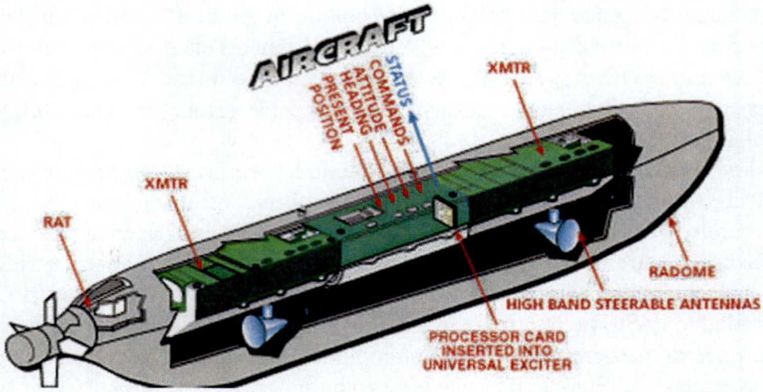

Fig. 3.15 AN/ALQ-99F (V) airborne electronic jamming pod

aircraft. AN/ALQ-99 can jam conventional radar, agile frequency radar, continuous wave radar, pulse radar, covert scan radar, and other systems. The EA-6B can be equipped with five AN/ALQ-99 jamming pods. Each pod is equipped with two ultra-high-power jamming transmitters, a tracking receiver and its associated antenna, and a ramjet turbine generator for power supply. The pod is canoe-shaped, 4.7 m long and contains two sets of antennas, low frequency and high frequency, as shown in Fig. 3.15.

The AN/ALQ-99 pod can manually select a variety of jamming methods, such as spot jamming, dual-frequency jamming, frequency-swept jamming, and noise jamming. It also can work automatically according to the environment of the theater and operational requirements. Currently, the pod can cover a frequency range from 64 MHz to 18 GHz, with a beam width of up to 30 degrees, and can jam multiple threat signals in a dense electromagnetic signal environment. It adopts power management and jamming directional targeting technology, which significantly reduces response time. According to the operational needs, EA-18G can carry 3-5 AN/ALQ-99 pods, and each pod can be flexibly configured with different transmitters to realize the free combination of various jamming bands. The equivalent radiation power of each transmitter is greater than 100 kW, and the continuous wave power generated is 1 ~ 2 kW, which can radiate to the front and rear of the aircraft, respectively [10, 77–80].

The AN/ALQ-184 electronic jamming pod is designed to provide effective jamming against ground-to-air missiles, radar-guided fire control systems, and airborne interception weapons. It developed AN/ALQ-184 J and AN/ALQ184V systems for use on F-15, F-16C/D and other aircraft. The key technologies of AN/ALQ184V jamming pod are multi-beam antenna system by Rotman lens, high-gain antenna, medium power small traveling-wave tube, crystal video receiver, and high-performance signal processor. The same jammer is installed at the front and rear ends of the jamming pod. Each jammer is equipped with 8 TWTs as power transmitters. The space power synthesis is completed by the Rotman lens antenna, which can quickly change the direction of the jamming beam, form multiple beams at the same time to jam multiple targets or concentrate energy to jam a radar. At present, the effective jamming frequency coverage of the pod is 1 ~ 18 GHz, and the jamming bandwidth can reach 1 GHz. Each jammer can provide very high equivalent radiated power and has jamming modes such as noise, response, or forwarding. Among them, the forwarding mode can also choose a variety of deceptive jamming styles [78, 81, 82].

3.5.2 Airborne Self-protection Jammers

Airborne self-protection jammer is a built-in radar active jamming system, which is the first choice of most fighters. The AN/ALQ-165 self-protection jamming system has been installed on US F-14, F-16, F/A-18E, and other fighters to jam the threat of radar terminal guided missiles and radar fire control systems. The jammer is a modular system designed with new electronic countermeasure technology, which is divided into two frequency bands to cover its frequency range. The principal block diagram of the system is shown in Fig. 3.16.

The operating frequency range of AN/ALQ-165 700 MHz-18 GHz, the high frequency can be extended to 35 GHz, the instantaneous bandwidth can be up to 1.44 GHz, and the system response time is about 0.1 to 0.25 s. Its receiver can be used in the battlefield environment with extremely dense signal distribution and can adapt to many types of pulse, continuous wave, pulse Doppler, and agility signals. Its sensitivity is about − 71 dBm, and dynamic range is about 50 dB. The peak power of the jammer is about 58 ~ 63 dBm, and the accuracy of aiming frequency is ±0.5 ~ ±20 MHz. It has two jamming modes: continuous wave noise and pulse deception. It adopts advanced power management unit, which can simultaneously jam 16 ~ 32 radar radiation sources and has the ability of reprogramming for jamming the emerging radar threat [10, 13, 83].

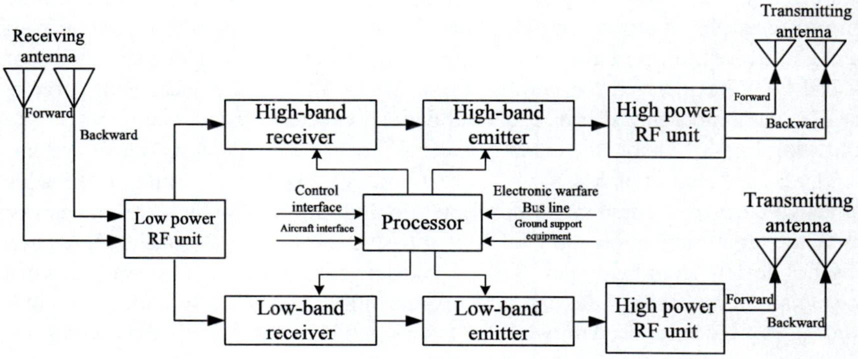

Fig. 3.16 Block diagram of AN/ALQ-165 self-protection jammer [13]

The AN/ALQ-137 is a self-protection jammer equipped with US air force FB-111 and EF-111 bombers. This is a deceptive jamming system with power management capability. The operating frequency range is composed of three sub-bands, which can cover 2 ~ 15 GHz. It has three jamming modes: continuous wave noise jamming, forwarding jamming, and response jamming. The receiver uses broadband crystal video equipment. The response time of the jammer is about 100 ns, the pulse jamming power is greater than 1000 W, and the continuous wave jamming power is about 100 W. It can produce range deception jamming, velocity deception jamming, and other types.

For advanced fighters such as the F-22 or F-35, the self-protection jamming system usually uses the Integrated Electronic Warfare System (IEWS) [84]. It can automatically select active and passive countermeasures and can adapt to a variety of radar, such as continuous wave, pulse, microwave and millimeter wave, as well as laser and infrared threats. It has a fully integrated structure and has the capability of radar warning, signal collection and analysis, passive radiation location and electronic countermeasures.

3.5.3 Naval Radar Active Jamming Systems

The ship has large radar RCS, as well as a large installation space and power supply capacity, so the shipborne radar active jamming system is a comprehensive, huge, and complex system.

The AN/SLQ-32 is a standard electronic warfare system for US Navy ships. It was developed by Raytheon Company and began to be equipped with US ships in 1977. At present, it is also installed on ships in Australia, Bahrain, Mexico, Poland, Portugal, Saudi Arabia, and other countries. It has become the electronic warfare system equipped with the largest number of large and medium-sized surface ships in the world [85, 86]. The main functions of the system include radar warning, signal interception, and electronic jamming. It can serve as a point defense task on the ship to resist the attack of anti-ship missiles. Therefore, it has the characteristics of high interception probability and short total response time. The basic model of AN/SLQ-32 has derived five models from (V) 1–(V) 5. After decades of service, the US Navy modernized it in 2003 and derived two models (V) 6 and (V) 7. The following Fig. 3.17 show the antenna and cabin console of AN/SLQ-32(V) system.

The main function of AN/SLQ-32 (V) 1 and (V) 2 is threat warning. (V) 3 adds an active electronic countermeasure system based on (V) 2, which can prevent or delay the target indication and launch of anti-ship missiles, and make the launched missiles deviate from the target. Active electronic countermeasure system can work in semiautomatic mode, and can also operate fully automatically under computer control. (V) 4 is composed of two sets of (V) 3, which can be installed on both sides of the ship respectively. The two sets are connected by optical fiber for rapid data exchange, which can realize cooperative operations. (V) 5 reduces the scale of the system, and the use of “Side Kick” active jammer can effectively combat anti-ship

Fig. 3.17 Antenna and console of c/SLQ-32 (V) [85, 87]

missiles and target indication radar. (V) 6 is developed by Lockheed Martin Company. It adopts new phased array technology, digital receiver technology and combat system interface, and adopts a modular open architecture, which greatly improves the support ability of electronic warfare and the adaptability to the platform. The system has the ability of instantaneous frequency measurement. The reconnaissance working frequency is 0.25 ~ 20 GHz, and the jamming frequency is 8 ~ 20 GHz. It has a variety of jamming modes such as noise suppression jamming, range deception jamming, angle deception jamming, etc. The maximum power is about 1 MW, and it can jam 75 ~ 80 targets at the same time. The jamming response time is about 51 ns. (V) 7 will integrate with other sensors and combat management systems, and adopt new transmitter technology and array technology to enhance electronic attack capability.

3.5.4 Ground-Based Radar Active Jamming Systems

Relatively speaking, there are fewer types of radar active jamming systems on the ground and on board. Krasuha vehicleborne electronic warfare system of Russia is a typical example.

The Krasuha electronic countermeasure and reconnaissance system uses a special vehicle chassis. The maximum road speed is 80 km/h, the maximum travel distance is 1000 km, and the crew is 3 ~ 7 people. Krasuha-2 was exhibited at the Moscow International Air Show in 2015. Its main function is to jam the S-band airborne early warning radar, with a jamming distance of up to 250 km. It can also jam various air radars such as radar-guided missiles and air-based observation radars. The system has a parabolic reflector with a diameter of 9 feet. The antenna and RF signal generator are located on a 360° rotatable platform with a maximum pitch angle of 5°. In addition, the system usually uses 2 ~ 4 sets of equipment as a group to effectively cover the search range of the early warning aircraft. Krasuha-4 is a broadband and multi-functional ground-based electronic suppression and protection system, which

Fig. 3.18 Krasukha-4 mobile electronic warfare system

can work independently. It has three large reflectors and a horn receiving antenna. It mainly adopts jamming patterns such as noise suppression. It can reduce the real-time perception and transmission ability of enemy electronic information systems through active countermeasures. It can be used to jam medium orbit and low orbit radar reconnaissance satellites, airborne surveillance radars with a distance of 150 ~ 300 km, it also can be used to jam unmanned aerial vehicles (UAV) and radar-guided weapons. Figure 3.18 is the appearance photo of Krasuha-4.

The operating frequency coverage range of Krasuha-4 system includes common L, S, C, X frequency bands. In order to deal with the complex and changeable electromagnetic threats in the modern battlefield more effectively, it adopts many advanced computer technologies, greatly improves the automation degree of the whole system, so as to have better real-time ability, self-adaptive ability, and full power management ability. The receiver of the system can accurately measure the frequency, azimuth, and other information of the threat radar in microseconds or even shorter time, and then automatically adjust the transmitter, accurately control the width and direction of the jamming beam, and conduct directional jamming or other corresponding jamming to the threat radar. The system also uses pulse repetition frequency tracking technology, which can simultaneously jam hundreds of targets. In addition, the Krasuha mobile electronic warfare system can be used in the extreme cold of the Arctic Circle, as well as in the Arabian Desert and other harsh environments.

3.5.5 Airborne and Outboard Active Radar Decoys

Active decoy is a separate radar active jamming system. It mainly includes small air-launched active decoy used by airborne platforms and outboard active decoy used by ships. MALD-J is a small air-launched decoy developed by Raytheon Company of the United States. Its shape is more like a cruise missile, as shown in Fig. 3.19. It is launched from the aircraft outside the defense area to the sky over the battlefield to jam the target radar.

Because the small air-launched active decoy can jam close to the enemy radar, the jamming range is greatly shortened and the jamming power is reduced. The core component of the MALD-J is an electronic warfare payload, and while there can be different types of payloads for different combat missions, it is currently focused on applications to deceive and disrupt enemy air defense systems. The early MALD-J system is mainly used for transmitting deception jamming signals, and its operating frequency can cover VHF, UHF, and microwave. Under the control of computer, the active decoy can simulate the radar echo of different aircraft in signal amplitude and frequency, so as to confuse the enemy radar operator. In addition, the MALD-J active decoy can also use GPS navigation system or inertial navigation system to simulate the flight trajectory of the aircraft, further weakening or destroying the enemy radar's ability to detect and identify targets. However, MALD-J system has low jamming power, poor maneuverability and can only work in a limited frequency band due to size and weight limitations [88–90].

Outboard active decoys [91–96] are usually used to jam the active terminal guidance radar of anti-ship missiles from angle or distance. At present, most of the outboard active decoys use hovering methods, including rocket hovering, parachute hovering, tethering, and UAV hovering. Figure 3.20 shows the Nulka active radar decoy system jointly developed by the United States and Australia. It uses rocket hovering and autonomous control technology and can hover briefly in the air after launch. The system is equipped with four antennas, which can generate jamming signals of 8 ~ 20 GHz and can jam a variety of anti-ship missiles. It has been equipped

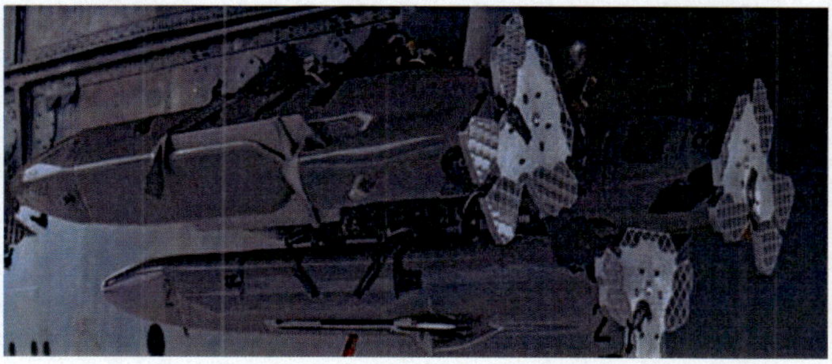

Fig. 3.19 MALD-J active decoy [88]

with hundreds of ships from the United States, Canada, Australia, and other countries. Figure 3.21 shows the C-Gem outboard active decoy developed by Israel. It is a parachute hovering device, which adopts a series of new technologies in system power supply, jamming signal transmitter, and transceiver antenna.

In addition, the US Naval Marine Research Laboratory has developed Eagle air tethering active radar decoy, which relies on one main rotor and two small auxiliary rotors to maintain the stability of the body in the air and connects with the ship through an optical cable to provide control signals and power, so that it can work for a long time.

Fig. 3.20 Narka outboard active decoy

Fig. 3.21 C-Gem outboard active decoy

3.6 Development Trend of Radar Active Jamming System

The future development of radar active jamming system mainly depends on the development of radar active jamming technology and electronic devices. “Distribution” and “self-adaption” are the important characteristics of the future electronic warfare system. Electronic devices are the basis for realizing the ability of jamming system, mainly including antennas, RF, and digital components. The development of electronic devices will be affected by many technologies such as materials, processes, and structures, such as microwave technology, integrated electronic technology, and advanced manufacturing technology. By inferring the combat form of the future war and the characteristics of electronic warfare, this paper summarizes the development trend of radar active jamming system, which may include the following aspects.

3.6.1 Distribution and Cooperation

The US Navy put forward the concept of “Network Centric Warfare” as early as 1997. After years of enrichment and development, it has become an important basis for the transformation and construction of the US military and the new joint operation theory. Through the network integration and organic synthesis of traditional combat units, network-centric warfare forms a cooperative combat system, improves the overall resultant force, and makes the combat effectiveness potential of the system far greater than that of individuals [97].

The future war will have the characteristics of cluster operation, system confrontation system, maximization of airspace coverage, and the best jamming efficiency. Therefore, the distribution and coordination of electronic warfare are the general trend [98, 99]. In the cooperative operation mode, a large amount of information can be gathered and interacted, and system resources can be allocated and used more reasonably, forming a powerful organization with system combat capability. Through time and space diversity, this organization not only has a strong ability to obtain information, which is conducive to making correct decisions quickly on the battlefield, but also can use multi-dimensional cooperation to deal with dense and complex threat environments.

This new combat style will face many new challenges. First, it is necessary to ensure the reasonable allocation and system control of distributed radar active jamming nodes, including between equipment inside and outside the platform, between manned and unmanned platforms, and between moving and non-moving resources, so as to maximize the efficiency of electronic warfare. Then, the active jamming system of each node needs more efficient reconnaissance and jamming capabilities. Finally, this system also needs to adopt the technology and algorithm of joint location, tracking and electronic attack of multiple resources.

3.6.2 *Intelligentization*

In recent years, driven by adaptive and cognitive technologies [100–102], the intelligence of radar has been continuously improved. Radar's ability of self-sensing and rapid response to the environment has become a hot spot, and there have been many advanced radar systems, such as multi-functional active phased array radar, digital programmable radar, cognitive radar, and so on. On the other hand, artificial intelligence technology has made rapid development in many fields. The rapid improvement of the computing power of modern information equipment also means that the radar active jamming system can achieve complex intelligence. Therefore, intelligent radar active jamming system with multi-function, multi-mode, and cognitive ability will be the development direction in the future.

The most important feature of this intelligent jamming system is its autonomous optimization and its ability to self-adapt to unknown radiation sources. At the reconnaissance level, the system is mainly characterized by adaptive capabilities to radiation source targets and the electromagnetic environment, such as maximizing dynamic adaptability to time-sensitive targets and special targets, and optimizing information perception of the electromagnetic environment. At the jamming level, the system is capable of autonomous decision-making, rapid response, and automatic jamming adjustment, such as identifying new and unknown threat signals through learning, automatically generating optimal jamming strategies based on battlefield posture, adaptively managing and allocating system resources, evaluating jamming effects online, and automatically adjusting jamming modes based on the evaluation results of jamming effects.

The main feature of intelligent jamming system is self-optimization and self-adaptation to unknown radiation sources. In terms of reconnaissance, it is mainly reflected in the adaptive ability to emitter targets and electromagnetic environment, such as maximizing the dynamic adaptability to time-sensitive targets and other special targets, optimizing the information perception of electromagnetic environment, etc. In terms of jamming, it is mainly reflected in the ability of independent decision-making, rapid response, and automatic adjustment of jamming, such as identifying new and unknown threat signals through learning, automatically generating the optimal jamming strategy according to the battlefield situation, adaptive management and allocation of system resources, online evaluation of jamming effects, automatic adjustment of jamming patterns according to the evaluation results of jamming effects, etc.

In addition, the improvement of the intelligent level of radar active jamming system can greatly reduce the operation and participation of personnel and promote the more application of unmanned platform of radar active jamming system [103].

3.6.3 Miniaturization

The miniaturization of the jamming system cannot only reduce the loading space and reduce the pressure on the platform, but also improve the combat effectiveness by taking advantage of the quantitative advantage. In addition, it also has the advantages of lower price, lighter carrying, more energy saving, and so on. The miniaturization of radar active jamming system is of great practical significance in airborne, spaceborne, missileborne, unmanned aerial vehicles, separated decoys, and other platforms [104].

The concept of microsystem was first put forward by DARPA. It is a micro- or small system with one or more functions, which takes the system architecture and algorithm as the core, based on microelectronics, optoelectronics, and micro-machinery, integrates microsensors, microdrivers, microactuators and signal processors through interdisciplinary and multi-disciplinary integrated design, and adopts advanced three-dimensional packaging manufacturing technology. Microsystem products are divided into two categories: function integrated microsystem and chip integrated microsystem. Functional integrated microsystem refers to a micro- or small product that takes the system architecture and algorithm as the core, based on microelectronics, optoelectronics, MEMS/NEMS, electromagnetic field, microwave/radio frequency, digital, structure, and from the perspective of system engineering, realizes a certain system function by means of interdisciplinary and multidisciplinary integration of integrated design, SOC/SIP and three-dimensional system packaging and integrated manufacturing. Chip integrated microsystem refers to a microsystem that integrates microelectronic devices, optoelectronic/photonic devices, MEMS/NEMS, and other devices through three-dimensional heterogeneous integration by combining the functional algorithm of microsystem architecture and advanced system packaging technology, forming a microsystem with certain system functions that integrates sensing, processing, communication, execution, micropower supply and other functions. The development of intelligence allows to reduce the radiation power of jammers, the rapid development of digital technology allows the miniaturization of information processing units, and the application of distribution requires the miniaturization of a large number of individuals, which objectively promotes the process of miniaturization.

The miniaturization of radar active jamming systems and microsystem mainly include: adopt small antennas and batteries to reduce the size and weight of the system, improve the efficiency of digital signal processing and RF systems, and multi-functionally integrate sensors, RF circuits, passive circuits, high-power microwave circuits, etc. The development of miniaturized systems will involve technologies such as microminiature jamming architecture, high-speed digital and analog processing, three-dimensional high-density integration, electromagnetic compatibility, and new materials. In addition, effective heat dissipation will also be a challenge due to the high component density of miniaturized system.

3.6.4 Synthesis and Reconfigurable

Future electronic warfare systems will require to deal with uncertainties as much as possible, including all traditional and non-traditional radio frequency threats. For example, radar active jamming systems will use multispectral, integrated receivers and transmitters to provide electronic warfare systems with higher index and cross-generational capabilities in terms of frequency, bandwidth, dynamics, sensitivity, transmit power, simultaneous transceiver isolation, and cooperative capabilities to distortion-free and fully intercept threat targets and accurately, efficiently make jamming over the full frequency range. The modular, open, and reconfigurable architecture will be conducive to the timely deployment and transformation of electronic warfare functions of radar active jamming system when the target and environmental state change rapidly, and the modular system will be configured according to the latest requirements in any combat action [105].

Among them, radar active jamming system based on microwave photonics is a typical representative of advanced electronic warfare system [106], which uses cross-discipline through domain fusion and solves the bottleneck problem in the microwave domain with the ideological approach and technical characteristics of the optical domain, which has the advantages of large bandwidth, high processing speed, parallelism, and small size [107].

3.6.5 High-Power and Precision Control

From the point of view of the suppression effect of the radar echo signal, higher power jamming signal is generally more conducive to successful jamming with radar, in addition, high-power jamming also provides an important guarantee for long-range combat and anti-radar low-antenna sidelobe technology. Therefore, one of the future directions of development of radar active jamming systems will continue to move in the direction of higher power jamming to ensure the successful suppression of threat targets.

In order to improve the jamming power of radar active jamming system, it is necessary to improve the equivalent radiation power of the transmitting antenna. Using phased array, GaN and other technologies [108], higher equivalent radiation power can be achieved. At present, phased array antenna and solid-state power amplifier technology are relatively mature in the field of radar. GaN is an ideal microwave power amplifier material. Compared with traditional amplifiers, GaN has higher voltage and temperature tolerance and has smaller volume, weight, and power consumption.

On the other hand, with the trend toward full digitalization of radars and the increasing refinement of signal processing techniques, the jamming signal also needs to have more accurate control in order to make effective jamming of radars [109, 110]. One of the very important aspects is the precise jamming signal power control.

It can form a more refined jamming pattern, improve the fidelity of false targets and avoid being recognized by radar, while reducing the impact on other electromagnetic actions of our side. A typical representative of this is the "Surgical Jamming" proposed by the United States in recent years.

Of course, the radar active jamming system will also have new trends in some other technologies and system designs, such as multi-mode and multi-function of the active jamming system, flexibility and accuracy of the jamming source, integration of radar communication and electronic warfare, system reliability enhancement, and so on [111–114]. We believe that the future radar active countermeasures system must be an integrated electronic information system more conducive to the use of combatants, more conducive to the improvement of jamming effectiveness, and more conducive to the application of new technologies.

References

1. Poisel RA (2013) Information warfare and electronic warfare systems [M]. Artech House Inc., Norwood
2. Boyd JA (1987) Discourse on winning and losing. Maxwell air force base [M]. Air University Library Document No. M-U43947 (Briefing slides), AL, USA
3. Brumley L, Kopp C, Korb K (1991) The orientation step of the OODA loop and information warfare [D]. Monash University, Australia, Clayton School of Information Technology
4. Yu Z, Mingzhi Z, Jingyu Y et al (2013) A modeling method of weapon equipment system based on OODA loop [J]. Acta Simulata Systematica Sinica 25(S1):6–11
5. Sang N (2012) Research on the algorithm of jammer emission leakage and multipath signal cancellation [D]. Master thesis, University of Electronic Science and Technology, Chengdu
6. Zhao Y, Xu W (2016) An online evaluation method of jamming effect based on radar state change [J]. Electron Inf Countermeasure Technol 31(3):42–46
7. An H, Li Y, Gao Y et al (2014) Preliminary study on jamming effect evaluation method of phased array radar based on combat application [J]. Electron Inf Countermeasure Technol 29(3):42–46
8. Ke Z, Xuan Z, Jiacai J (2013) Preliminary study of cognitive electronic warfare [J]. Aerosp Electron Countermeasure 29(1):53–56
9. Zhao G (1999) Principle of radar countermeasure [M]. Xi'an Electronics and Science University Press, Xi'an
10. Zhou Y, An W, Guo F et al (2009) Principles of electronic warfare system [M]. Electronic Industry Press, Beijing
11. Adamy D (2001) EW101: a first course in electronic warfare [M]. Artech House Inc., Norwood
12. Wiegand RJ (1991) Radar electronic countermeasure system design [M]. Artech House Inc., Norwood
13. Schleher DC (1999) Electronic warfare in the information age [M]. Artech House Inc., Norwood
14. Neri F (2001) Introduction to electronic defense system [M]. Artech House Inc., Norwood
15. De Martino A (2012) Introduction to modern EW system [M]. Artech House Inc., Norwood
16. Jixin C, Tao Z (2008) Retransmitted jamming of optical fiber delay line to broadband imaging radar [J]. Electron Inf Countermeasure Technol 23(6):32–35
17. Zheng X, Li W, Pan H et al (2006) Radar transmitter technology [M]. Electronic Industry Press, Beijing
18. Tsui J (2010) Digital techniques in wideband receivers [M]. Artech House Inc., Norwood

19. Li Z (2010) Full pulse processing architecture [M]. Twenty-ninth Institute of China Electronic Technology Group, Chengdu
20. Zhang G (2005) Research on radar emitter identification technology [D]. Doctoral dissertation, University of Defense Science and Technology
21. Hu G (2011) Radar signal modulation recognition technology research [D]. Doctoral dissertation, Nanjing University of Aeronautics and Astronautics
22. Yuan X, Li X, Yu H et al (2003) Radar signal recognition technology research [J]. Electron Eng 1:1–6
23. Li Y, Zhang B (2005) A radar type identification method [J]. Mod Defense Technol 33(3):65–68
24. Langley LE (1993) Specific emitter identification and classical parameter fusion technology [C]. In: WESCON'93. pp 377–381
25. Xu D (2009) Study on fingerprint mechanism and identification method of radiation source [D]. University of Defense Technology, Changsha
26. Sheng C, Qiuxi J, Jifei P (2013) Feasibility analysis of radar individual identification based on inter-pulse fingerprint characteristics [J]. Electron Inf Countermeasure Technol 27(3):6–9
27. Chaowen J, Shuilou Z (2011) Working pattern recognition of airborne radar [J]. Electron Inf Countermeasure Technol 26(1):14–16
28. Visnevski N (2005) Syntactic modeling of multifunction radars [D]. McMaster University, Hamilton
29. Liu H, Li Y, Liu Z (2010) Multifunctional radar identification method based on stochastic grammar [J]. J Aviat 31(9):1809–1817
30. Terminology for Electronic Countermeasures [S]. National Military Standard of the People's Republic of China (GJB 891A-2001)
31. Comprehensive Electronic Warfare Technical Terms [M] (2001) Twenty-ninth Institute of Electronics, Ministry of Information Industry, Chengdu
32. Tactical Technical Requirements Project Format for Radar Countermeasure Equipment [S]. National military standard of the People's Republic of China (GJB 2092–1994)
33. Skolnik MI (2008) Radar handbook, 3rd edn. [M]. McGraw-Hill Professional
34. Mingyou Z, Xuegang W (2005) Radar system [M]. Electronics Industry Press, Beijing
35. Zhou W, Jiao P et al (2008) Over-the-horizon radar technology [M]. Electronic Industry Press, Beijing
36. Guangyi Z, Yujie Z (2007) Phased array radar technology [M]. Electronic Industry Press, Beijing
37. De B, Chuanan W, Youquan L (2006) Airborne radar technology [M]. Electronics Industry Press, Beijing
38. Philippe L et al (2001) Air and spaceborne radar systems: an introduction [M]. Elsevier
39. Long M (1992) Airborne early warning system concept [M]. Artech House Inc., Norwood
40. Li S, Luo Y (2013) E-2 Hawkeye airborne early warning aircraft [M]. Aviation Industry Press, Beijing
41. Lu J, Li N, Cao C et al (2011) Introduction to airborne early warning system [M]. Defense Industry Press, Beijing
42. Dechun W, Jiahui D, Wangdong C et al (2006) Precision tracking measurement radar technology [M]. Electronics Industry Press, Beijing
43. Huang H, Qi R, Wen S (2006) Guidance radar technology [M]. Electronic Industry Press, Beijing
44. Feng G (2010) The introduction of radar seeker [M]. Electronic Industry Press, Beijing
45. Cumming IG, Wong FH (2007) Digital processing of synthetic aperture radar data algorithms and implementation [M].In: Trans HW et al. (eds). Electronics Industry Publishing House, Beijing
46. Zheng B, Mengdao X, Tong W (2005) Radar imaging technology [M]. Electronics Industry Press, Beijing
47. Xiaomo W, Yongsheng K, Zhongxian C (2008) Surveillance radar technology [M]. Electronic Industry Press, Beijing

48. Yongliang W, Yingning P (2000) Space-time adaptive signal processing [M]. Tsinghua University Press, Beijing
49. Klemm R (2004) Applications of space-time adaptive processing [M]. Institution of Electrical Engineers
50. Li D (2002) Research on networking radar technology [D]. PhD thesis, Beijing Institute of Technology, Beijing
51. Chen Y, Li X, Shen Y (2006) Analysis and evaluation of netted radar combat capability [M]. Defense Industry Press, Beijing
52. Pace PE (2004) Detecting and classifying low probability of intercept radar, 2nd edn. [M]. Artech House, Inc., Norwood
53. Li J, Stoica P (eds) (2013) MIMO radar signal processing [M]. Defense Industry Press, Beijing
54. Haykin S (2006) Cognitive radar: a way of the future [J]. IEEE Sign Process Mag 23(1):30–40
55. Xia S (2012) Cognitive Radar signal processing-detection and tracking [D]. PhD thesis of Xi'an of Electronics and Science University, Xi'an
56. Ruqun W (2006) War-field electromagnetic environment [M]. Liberation Army Press, Beijing
57. Li W (2007) Some understandings of complex electromagnetic environment in sea battlefield [J]. Inf Countermeasures 4:10–13
58. Yi R, Ruilin L (2009) Research on the development of electronic warfare technology in sea battlefield [J]. Ship Electron Warfare 32(1):9–15
59. Qisheng S, Sheng Y et al (2015) Anti-ship missile battlefield electromagnetic environment simulation and test identification technology [M]. Defense Industry Press, Beijing
60. Qiang G (2006) Theoretical research on signal sorting of unknown radar emitter in complex environment [D]. Harbin Engineering University, Harbin
61. Pu Y (2007) Research on radar emitter signal sorting model and algorithm for complex system [D]. Southwest Jiaotong University, Chengdu
62. Jun C, Ming G, Jianmin Z (2001) FPGA and CPLD in radar signal sorting preprocessor application [J]. Syst Eng Electron Technol 23(10):22–24
63. Xu X, Feng D, Zhou Y et al (2001) Research on real-time pulse deinterlacing method based on CAM [J]. J Circ Syst 6(3):94–98
64. Willey RG (1983) Electronic intelligence: the analysis of radar signals [M]. Artech House Inc., Norwood
65. Li J, Wang Y (1992) SDIF algorithm for radar pulse deinterleaving [J]. Electron Countermeasure Technol 5:35–40
66. Ma S, Wu H, Liu Z et al (2012) Radiation source time difference sorting method based on recursive extended histogram [J]. J Nat Defense Univ 34(5):83–89
67. Milojevie DJ, Popovie BM (1992) Improved algorithm for the deinterleaving of radar pulses [J]. IEE Proc F Radar Signal Proc 139(1):98–104
68. Zhengwei Z (1992) Histogram analysis of complex radar signals [J]. Electronic Countermeasures 06:6–14
69. Xin S, Huiqun H, Chengzhi Y (2010) Unknown radar signal sorting based on improved k-means algorithm [J]. Mod Electron Technol 17(3):91–93
70. Ran Z, Houpei X (2015) A new k-means clustering radar signal sorting algorithm [J]. Mod Defense Technol 43(6):136–141
71. Zhengwei Z (2005) Cluster sorting method for radar signals [J]. Electron Confrontation 105(6):6–10
72. Jiang T, He M, Zhai W (2004) Application of wavelet neural network in radar signal sorting [J]. Radar Sci Technol 2(6):314–344
73. Retrieved from http://www.raytheon.com/capabilities/ew/overview/index.html
74. Zhu H (2006) Overview of U.S. army integrated electronic warfare system equipment and technology development [J]. Radar Electron Warfare 4:1–2
75. Yan W, Jian J (2010) Analysis of US army electronic warfare system and its development trend [J]. Ship Electron Eng 30(3):16–19
76. General Staff Part IV (2008) Reference manual for information combat equipment in major countries of the world [M]. Liberation Army Press, Beijing

77. Xianyu T (2004) The status and development of airborne electronic jamming pods abroad [J]. Mod Defense Technol 32(2):66–72
78. Retrieved from http://www.fas.org/man/dod-101/sys/ac/equip/an-alq-99.htm
79. Winger (2004) “Roarer” will come---the new generation of U.S. navy electronic attacker EA-18G [J]. Mod Weapons 4:17–19
80. Lian L, Hua X, Hu C (2006) U.S. military EA-18G “roarer” electronic attacker details [J]. Defense Technol 9:11–16
81. Retrieved from http://www.fas.org/man/dod-101/sys/ac/equip/an-alq-184.htm
82. Chen J, Yi M, Wu W (2012) Research on electronic warfare system after Taiwan F-16 A/B (block 20) upgrade [J]. Ship Electron Eng 32(11):11–13
83. Jingmin L (1997) ALQ-165 (Airborne self-defense jammer) [J]. Photoelectronic Countermeasure Passive Jamming 3:33–39
84. Kun Q (2006) Electric wave titan—integrated electronic warfare system for F-35 fighter [J]. Int Outlook 22(11):62–65
85. Retrieved from http://www.fas.org/man/dod-101/sys/ac/equip/an-slq-32.htm
86. Luo Rong compile (2011) The Latest Improvement of SLQ-32 [J]. Int Electron Warfare 7:27–28
87. Retrieved from http://www.raytheon.com/capabilities/ products/slq32.htm
88. Retrieved from http://www.raytheon.com/capabilities/products/mald/
89. Ge Y, He X (2016) American small-scale air-launched decoys may change the future air combat mode [J]. Aerodyn Missiles 1:15–18
90. Naiguang C (1999) American MALD small aerial launch bait [J]. Mod Weapons 11:17–19
91. Jianjun H et al (2000) American navy weapons and equipment manual [M]. PLA Press, Beijing
92. “Shipborne weapons” Editorial Committee (2010) Shipborne Weapons [M]. Aviation Industry Press, Beijing
93. Zhiqiang L (2001) Development and tactical application of outboard jamming technology [J]. Aerodyn Missile 3:61–63
94. Cuilan Z (2004) Discussion on onboard active jamming [J]. Mine Warfare Warship Prot 3:32–35
95. Jie W, Jisong S (2011) Active jamming technology against anti-ship missiles [J]. Ship Electron Countermeasures 2:36–40
96. Shigen Z (2000) World ship electronic warfare system manual [M]. Science Press, Beijing
97. Dingding C (2005) Network center electronic warfare research [J]. Electron Sci Technol Rev 6:1–5
98. Ning L (2008) “Wolf pack” distributed system introduction [J]. Foreign Inf War 4:40–42
99. Xiujin W, Qiang W, Fan Y (2012) The concept of distributed electronic warfare system for “sea wolves” [J]. Ship Electron Warfare 35(4):8–10
100. AFRL. Cognitive jammer. RFI-PKS-0001–2010 [R/OL]. Retrieved from http://www.fbo.gov. Accessed on 20 Jan 2010
101. DARPA. Adaptive Radar Countermeasures (ARC). DARPA-BAA-12–54[R/OL]. Retrieved from http://www.fbo.gov. Accessed on 3 Jul 2012
102. Ni S, Long X, Wei X et al (2011) Analysis of the development of cognitive technology in electronic warfare equipment [J]. Electron Inf Countermeasure Technol 26(6):22–26
103. Hu J, Zhang B, Zhou S (2000) Research on the characters and survival ability of UAV in electronic warfare [J]. Aeronaut Sci Technol 2:32–34
104. Bingrong Z (2004) Micromation progress of military electronic system [J]. Electron Countermeasures 3:1–8
105. Yang C, Li Q, Zhang Y (2014) New shipborne integrated electronic warfare system [J]. Ship Electron Warfare 37(6):1–5
106. Fan B, He Z, Liu J (2014) Microwave optical links for electronic warfare systems [J]. Semicond Optoelectron 35(3):559–562
107. Ni S, Wei X, Dong C (2013) The development of high-power integrated digital-analog conversion technology in electronic warfare system [J]. Electron Inf Countermeasure Technol 28(2):30–33

108. Wu L, Sun Z (2012) Overview of the latest developments in core technology of T/R modules [J]. J. Chin Acad Electron Sci 7(1):29–33
109. Liu Y (2012) SAR active false target accurate position deception jamming technology research [D]. Master degree thesis, University of Defense Science and Technology, Changsha
110. Lei G, Li H, Li Q (2013) The coherent radar intermittent sampling smart jamming method [J]. Mod Defense Technol 43(3):111–116
111. Qing S, Jianfeng T, Ning Z (2005) Development status and trend of electronic warfare technology [J]. Aerosp Electron Warfare 21(5):28–31
112. Songhai C (2006) Shipborne electronic warfare system and development trend of American navy [J]. Nat Defense Sci Technol 3:6–10
113. Li XQ et al (2008) Integrated electronic warfare -- killer mace in information war, 2nd edn. [M]. National Defense Industry Press, Beijing
114. Rongbing G, Jindan C (2010) Discussion on the future development trend of electromagnetic space decurity and defense-technology [J]. Dynamics of Electromagnetic Space Security and Defense, Twenty-ninth Research Institute of China Electronic Science and Technology Corporation

Chapter 4
Antennas of Radar Jamming System

Antenna is a kind of energy converter, which converts the electromagnetic wave signal propagating in space and the alternating electrical signal conducted in the transmission line [1]. Transmitting antennas convert RF energy in the alternating circuit in the transmission line into electromagnetic energy radiated into space. Receiving antennas convert electromagnetic energy in space into RF energy in the alternating circuit. Normally, antennas are reciprocal.

The antenna is a very important part of the electronic warfare system. It includes various directional antennas and omnidirectional ones. If the entire radar active jamming system is compared to the human body, then the antenna is like the "ear" and "mouth" of the entire system. Comparing it to the "ear" means that the receiving antenna can collect and perceive the radar signal from the outside to convert it into an "electric" signal that can be processed and recognized and transmit it to the "brain" that processes the signal—the receiving processor. It is likened to "mouth," which means that the transmitting antenna can radiate the jamming signal transmitted from the "brain" into the space to achieve jamming to the radar [2].

This chapter focuses on the content of antenna in radar active jamming system, including antenna index system, antenna selection and design basis, typical antenna types, and design methods of radar jamming system and points out the technical development trend of antenna in active countermeasure system.

4.1 Technical Indicators of Antenna

The technical indicators of antenna in radar active jamming system are consistent with that of other electronic systems (such as communications and radar), mainly including antenna directivity and pattern, antenna bandwidth, antenna impedance, antenna efficiency, antenna gain, antenna polarization, and other indicators [3].

G. Tang et al., *Techniques and System Design of Radar Active Jamming*,
https://doi.org/10.1007/978-981-19-9944-4_4

4.1.1 Directionality and Antenna Pattern

Antenna is usually a passive device. What we often refer to as active antenna is the combination of antenna and active amplifier circuit. As far as the antenna itself is concerned, it is still passive [4]. The antenna itself does not amplify the signal, but it can either focus the signal in a particular direction (the receiving antenna) or radiate it in a particular direction (the transmitting antenna), so that the signal in a particular direction "looks" stronger than the signal received or transmitted without directionality.

The characteristic of an antenna that concentrates RF energy in a particular direction is called its directionality. Directionality reflects the degree of RF energy concentration in three-dimensional space and is also a function of the azimuth and pitch angles of the antenna relative to the non-directionality. The direction of the maximum value of this function is usually regarded as the direction of the maximum antenna gain [5].

$$D(\theta, \phi) = \frac{4\pi P_o(\theta, \phi)}{\iint P_i(\theta, \phi)\sin\theta d\theta d\phi}\begin{cases} 0 < \phi < 360^\circ \\ 0 < \theta < 180^\circ \end{cases} \tag{4.1}$$

In the formula, D represents the directionality of the antenna, ϕ refers to the azimuth of space, θ is the pitch angle of space, $P_o(\theta, \phi)$ is the radiated power in a particular direction, and the denominator $\iint P_i(\theta, \phi)\sin\theta d\theta d\phi$ refers to the power integral of the whole space.

A graphical way to show the directivity of antenna radiation in space is called a pattern. In polar coordinate system or rectangular coordinate system, the different intensities radiated by the antenna in each direction are expressed by different length vectors starting from the origin. Then, the envelope surface formed by connecting the endpoints of all vectors is the antenna pattern which is a function of the two-dimensional angle of azimuth and pitch. The pattern of different types of antennas is different. Power is usually used to characterize the pattern, so it is also called power pattern. Figure 4.1 is a typical horn antenna and its two-dimensional pattern.

In engineering, the directionality of the antenna is generally represented by two orthogonal main planes, which are called plane E and plane H. Plane E is the plane passing through the maximum radiation of the antenna and parallel to the electric field vector. Plane H is the plane passing through the maximum radiation of the antenna and perpendicular to plane E, as shown in Fig. 4.2.

According to the characteristics of antenna pattern, antennas can be roughly divided into omnidirectional antenna, semi-omnidirectional antenna, wide-beam directional antenna, narrow beam directional antenna, and so on. The selection of antennas for different types of patterns depends on the application scenarios of antennas.

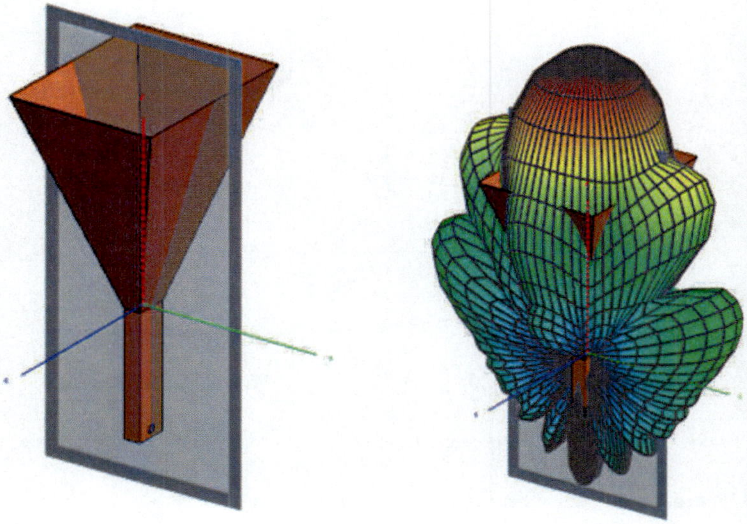

Fig. 4.1 X-band horn antenna and its two-dimensional pattern

Fig. 4.2 Diagram of antenna plane E and plane H E-plane pattern and H-plane pattern of antenna

4.1.2 Beamwidth of Antenna

The main lobe beam width of the antenna refers to the included angle between two directions of half of the maximum radiated power density on a specific plane, also known as half power beam width or 3 dB beam width, which is also a major characterization of the antenna directivity. An antenna with a stronger directivity has a higher gain and narrower main lobe beamwidth; an antenna with a weaker directivity

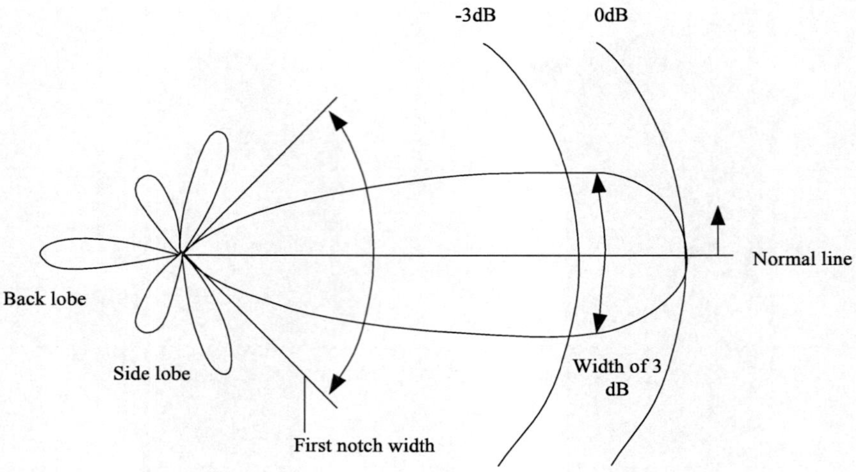

Fig. 4.3 Schematic diagram of beam-related concepts of antenna

has a lower gain and wider main lobe beam bandwidth. Besides the main one, other lobes are called sidelobe. The secondary lobe that differs 180° from the main one is called the back lobe. The energy radiated from the secondary and back lobe is in the undesired direction, so it is necessary to minimize the sidelobe and backlobe radiation in antenna design. In antenna design, sidelobe and backlobe radiation should be minimized as far as possible. The sidelobes adjacent to the main lobe are referred to as the first sidelobe and the second side one in turn. The lowest point between the main lobe and the first sidelobe is called the first zero depth, and the included angle between the first zero depth on both sides of the main lobe is known as the first zero depth angle, as shown in Fig. 4.3.

4.1.3 Gain of Antenna

Under the condition of equal input power, antenna gain refers to the ratio of the power density of the signal generated by the actual antenna and the ideal radiation unit at the same point in space. Quantitatively, it describes the degree to which an antenna concentrates the input power. Gain is closely related to antenna directionality. The narrower the main lobe and the smaller the sidelobe of the pattern, the higher the gain.

In engineering applications, the first focus is the working frequency band and gain of the antenna, and the estimation of parameters, such as beam width, can be obtained roughly by estimating the gain and efficiency of the antenna.

We usually use an ideal non-directional point radiation source as a reference for power gain. Gain is the ratio of the area of the whole sphere to the area of the sphere sector corresponding to a given antenna beam, namely:

$$G = \frac{\text{Surface area of sphere}}{\text{Spherical fan area corresponding to antenna beam}} \tag{4.2}$$

The spherical fan area can be calculated by using the two approximate graph area of the rectangular model or the ellipse (circle) model. This is shown in Figs. 4.4 and 4.5.

For the rectangular model:

$$G = \frac{\text{The ball surface area}}{\text{The rectangular area}} = \frac{4\pi r^2}{\sin\theta \sin\varphi r^2} = \frac{4\pi}{\sin\theta \sin\varphi} \tag{4.3}$$

If it uses degree as the unit of the beamwidth:

$$G = \frac{4\pi}{\sin\theta \sin\phi} \approx \frac{4\pi}{\theta\phi} = \frac{4\pi}{\theta_d \phi_d}\left(\frac{360}{2\pi}\right)^2 = \frac{41253}{\theta_d \phi_d} \tag{4.4}$$

In the formula, θ and ϕ, respectively antenna azimuth and pitch beamwidth, are specified in radians, but θ_d and ϕ_d are specified in degrees to represent them

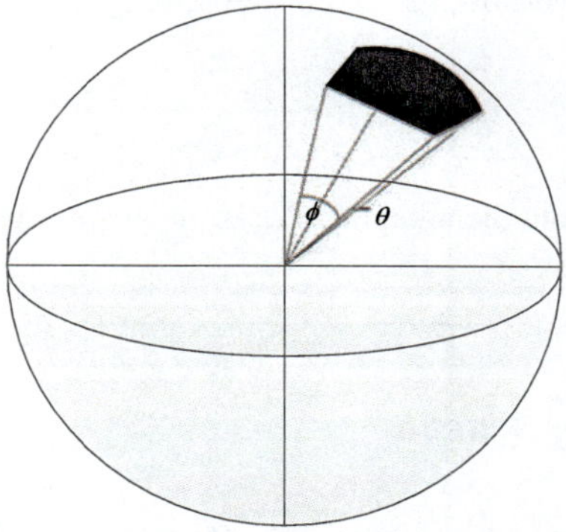

Fig. 4.4 Graphical diagram of antenna gain calculation

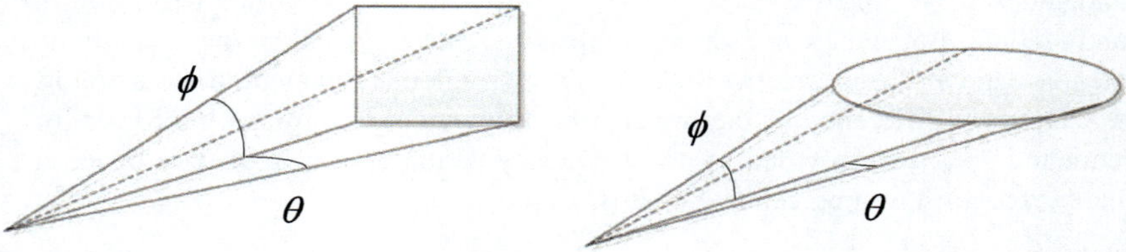

Fig. 4.5 Using rectangle and ellipse to approximate the projected area of antenna beam

respectively. Similarly, the approximate formula of antenna gain obtained by using elliptical model is $G = 52525/\theta_d\phi_d$.

What needs illustration is that the above calculation formulas do not take the antenna efficiency into account. The actual gain of the antenna is equal to the ideal gain multiply by the antenna efficiency. Taking an antenna with an azimuth beamwidth of 3°, a pitch beamwidth of 5° and an efficiency of 70% as an example, the gain calculated by the two formulas is 32.8 dB (rectangular model) and 33.9 dB (elliptical model) respectively, which are similar.

For an antenna with a fixed operating frequency, the larger the antenna aperture area, the greater the gain and the narrower the beam. For a wideband antenna, when its aperture area is certain, the higher the operating frequency, the greater the gain due to the smaller wavelength.

4.1.4 Efficiency of Antenna

Antenna efficiency is characterized by the ability of the antenna to radiate the input energy into free space, denoted by the symbol η, which is usually related to the type and design of the antenna. Since there are always losses inside the antenna, such as dielectric loss, feed irradiation mismatch, port surface wavefront phase mismatch, sidelobe and backlobe overflow, the antenna efficiency is always a value less than 1:

$$\eta = \frac{P_{out}}{P_{in}} \tag{4.4}$$

P_{in} is the power input to the antenna; P_{out} is the power radiated from the antenna port.

4.1.5 Bandwidth of Antenna

The bandwidth of the antenna usually has two meanings. One is called absolute bandwidth, which is the ratio between the upper limit f_{max} and the lower limit f_{min} of the operating frequency of the antenna. For example, the working frequency band index of an antenna is 8 GHz ~ 12 GHz. According to this definition, the bandwidth is 1.5, which is a dimensionless parameter. Another representation of antenna bandwidth is "relative bandwidth." Its calculation method is as follows: the frequency difference of the upper limit frequency f_{max} minus the lower limit frequency f_{min}, then divided by the frequency center of the frequency band, and finally expressed as a percentage, that is:

$$BW = \frac{2(f_{max} - f_{min})}{f_{max} + f_{min}} \times 100\% \tag{4.5}$$

Among EW practitioners who are not antenna professionals, there is a more popular term for bandwidth: the differential value between the upper limit f_{max} and the lower limit f_{min} of the antenna operating frequency is a dimensionless parameter of absolute bandwidth. For the above X-band antenna, we can think that its bandwidth is 4 GHz.

To avoid ambiguity, the dimensionless "absolute bandwidth" is referred to in this book as the "absolute bandwidth ratio".

4.1.6 Polarization of Antenna

Polarization of electromagnetic wave refers to the orientation of electric field vector. The direction of electric field vector, magnetic field vector, and wave propagation direction are orthogonal to each other. Vertical polarization shows that the electric field vector is perpendicular to the ground; horizontal polarization shows that the electric field vector is parallel to the ground; circular polarization refers to the projection of the endpoint of the electric field vector on the plane perpendicular to the radiation direction as a circle, which can be divided into left-handed and right-handed polarization according to its rotation direction. In radar countermeasure, different polarization forms have an all-important effect on the performance of active jamming system, so it is necessary to fully consider which polarization forms to adopt (Table 4.1).

The case where the receiving polarization and the transmitting polarization are orthogonal is called orthogonal polarization. Theoretically, under this circumstance, the signal should not be received at all (the power loss is infinite). However, in

Table 4.1 Loss of antenna combination with different polarization forms

Polarization form of radar emission	Polarization form of countermeasure antenna	Theoretical value of polarization loss (dB)
Verticality	Verticality	0
Verticality	Oblique 45°	−3
Verticality	Horizontality	$-\infty$
Verticality	Right/left-handed circular	−3
Horizontality	Horizontality	0
Horizontality	Oblique 45°	−3
Horizontality	Verticality	$-\infty$
Horizontality	Right/left-handed circular	−3
Right-handed circular	Right-handed circular	0
Right-handed circular	Left-handed circular	$-\infty$
Right/left-handed circular	Oblique 45°	−3

practice, the polarization isolation of the antenna cannot reach the theoretical infinity, and the orthogonal loss is usually about 20 dB.

It should be noted that the polarization direction of the antenna is normally defined along the propagation direction from the back of the antenna as the transmitting antenna, which is particularly important when the antenna is circularly polarized.

4.1.7 Impedance of Antenna

The impedance of the antenna is the ratio of the voltage and current at the feed point of the antenna, which is related to the type, frequency, and surrounding appendages of the antenna. The antenna impedance normally is a complex form, expressed as:

$$Z_A = R_A + jX_A \tag{4.6}$$

In the formula, Z_A is the antenna complex impedance; R_A refers to the resistance of the antenna, which corresponds to active power, dissipated in the form of thermal loss or radiation; X_A is the antenna's reactance, representing the energy in the near field, which corresponds to the reactive power.

The antenna is connected with the receiver or transmitter through the feeder or waveguide. In order to ensure the maximum transmission efficiency, it is necessary to match the antenna impedance with the feeder impedance. The matching degree of impedance directly affects the transmission efficiency. For the transmitting antenna, if the impedance mismatch occurs, the transmission energy will be lost in the feeder or the joint between the feeder and the antenna, and even the energy will be reflected to the transmitting end. In serious cases, the transmitter will be ignited or damaged.

In the RF frequency band applied in military and communication industries, the standard impedance of the feeder is 50 Ω, so the antenna impedance of the radar active jamming system should be designed at 50 Ω as much as possible, in accordance with the industry practices and the standard of general products.

4.1.8 SWR of Antenna

Standing wave ratio is short for voltage standing wave ratio, abbreviated as VSWR or SWR (voltage standing wave ratio), which is used to characterize the degree of impedance match or mismatch of the antenna input. Because the RF energy input to the antenna is not fully radiated to the free space, the reflected wave is generated, and the reflected wave is superimposed with the input signal to form a standing wave. The larger the standing wave is, the worse the match at the antenna input become, and the more energy is reflected to the antenna input. Assuming that Γ_0 represents the reflection coefficient, Z_L refers to the antenna load impedance, and Z_0 is the characteristic impedance, then

$$\Gamma_0 = \frac{\text{Reflected wave amplitude}}{\text{Incident wave amplitude}} = \frac{Z_L - Z_0}{Z_L + Z_0} \tag{4.7}$$

$$SWR = \frac{1 + \Gamma_0}{1 - \Gamma_0} \tag{4.8}$$

It can also be defined as the ratio of the maximum value to the minimum value of the voltage along the line:

$$SWR = \frac{E_{\max}}{E_{\min}} = \frac{E_i + E_r}{E_i - E_r} \tag{4.9}$$

At the feed point, $E_{\max}$ is the maximum voltage and $E_{\min}$ is the minimum voltage. E_i is the incident voltage amplitude, and E_r is the reflected voltage amplitude.

In the case of ideal matching, the impedance is perfectly matched, $Z_L = Z_0$, and the reflection coefficient $\Gamma_0 = 0$. The energy input to the antenna is completely absorbed or radiated, and there is no reflected energy, at this moment, $SWR = 1$.

In practice, the impedance of the antenna cannot be completely consistent with the characteristic impedance, so there will be partial energy reflection, resulting in SWR always greater than 1. For the transmitting system, if the standing wave is high, the reflected energy may damage the transmitter, when the standing wave ratio of the antenna is 2, about 11% of the energy is reflected to the transmitter (Table 4.2).

4.2 Requirements Analysis of Antenna

In this book, it mainly discusses the antenna design related to radar active jamming systems, including the RWR system closely related to ECM, but not involving the antenna of communication and communication countermeasures system.

Many different types of antennas are applied in radar active jamming system, and many of them are wideband antennas, which are determined by the wideband characteristics of electronic warfare system itself. In terms of the technical system of antenna, frequently used antennas in radar active countermeasures mainly include vibrator antenna, Yagi antenna, spiral antenna, horn antenna, surface antenna, waveguide antenna, planar microstrip antenna, etc. [5]. The selection and design of antenna used in radar active jamming system need to consider many factors.

Table 4.2 Conversion table of several indicators

VSWR	Reflected loss (dB)	Transmitted loss (dB)	Reflected power (%)	Transmitted power (%)
1.00	∞	0	0	100.0
1.01	46.1	0	0	100.0
1.10	26.4	0	0.2	99.8
1.20	20.8	0	0.8	99.2
1.30	17.7	0.1	1.7	98.3
1.40	15.6	0.1	2.8	97.2
1.50	14.0	0.2	4.0	96.0
1.60	12.7	0.2	5.3	94.7
1.70	11.7	0.3	6.7	93.3
1.80	10.9	0.4	8.2	91.8
1.90	10.2	0.4	9.6	90.4
2.00	9.5	0.5	11.1	88.9
2.50	7.4	0.9	18.4	81.6
3.00	6.0	1.2	25.0	75.0
4.00	4.4	1.9	36.0	64.0
5.00	3.5	2.6	44.4	55.6
10.00	1.7	4.8	66.9	33.1
15.00	1.2	6.3	76.6	23.4
20.00	0.9	7.4	81.9	18.1
30.00	0.6	9.0	87.5	12.5

4.2.1 Working Frequency Range and Bandwidth Requirements

According to the regulations of their respective industries, there are certain differences in the division and naming of main frequency bands between electronic warfare and radar, as shown in the following table (Table 4.3).

Table 4.3 Division and naming of frequency bands for EW and radar

Frequency range (GHz)	Frequency band naming system (ECM)	Frequency band naming system (IEEE)
0.5 ~ 2	Band C/D	UHF/L
2 ~ 4	Band E/F	S
4 ~ 8	Band G/H	C
8 ~ 20	Band I/J	X/Ku

According to the constraints such as the weight, size, and installation location of the equipment and considering the antenna design, frequency conversion design of the receiving channel, and amplifier design, the antenna of electronic warfare equipment usually adopts broadband antenna. The working frequency range of the antenna shall match the technical index of the working frequency range of the radar active jamming system. If the frequency range of the antenna is narrower than the requirement of the system, the signals of some frequency bands cannot be received or transmitted. If it is too wide, which has no effect on the transmission, but for the reception, the external useless signal will enter the receiver, which may affect the performance of the system. In some cases, when the frequency coverage required by the system is too wide to be meet by a single antenna, the combined one is needed to achieve goals.

For different application scenarios, the antennas in electronic warfare equipment are not only ultra-wideband antennas spanning multiple octaves, such as 0.5 GHz ~ 2 GHz, 2 GHz ~ 6 GHz, 6 GHz ~ 18 GHz, 2 GHz ~ 18 GHz, but also relatively "narrowband" antennas for specific bands, such as C/D band antennas, E/F band antennas, G/H band antennas, and I/J band antennas [6].

The central and relative bandwidths that can be realized by different types of antennas have different characteristics. For example, in the actual radar active jamming system, horn and log-periodic antenna are usually used in 0.5 GHz ~ 2 GHz frequency band, horn antenna, spiral antenna, and paraboloid antenna are commonly used in 2 GHz ~ 6 GHz and 6 GHz ~ 18 GHz frequency band, which are all broadband antennas.

4.2.2 *Gain and Beamwidth Requirements*

The requirements of EW antenna beam are closely related to the requirements of application scenarios.

The gain and beamwidth of the receiving antenna are determined by the sensitivity of the receiving system, the instantaneous coverage of the receiving space, and the accuracy of angle measurement. Given the system sensitivity requirements, the selected receiver sensitivity, and the cable insertion loss, the required gain of the receiving antenna can be determined, which is its minimum gain requirement. The determination of system sensitivity requirements refers to the analysis and calculation in Chap. 3. The sensitivity requirements of typical receivers and the calculation of receiver sensitivity are shown in the analysis and calculation in Chap. 5.

Receiving antennas usually have two application requirements; one is wide airspace interception, and the other is to meet the high-gain requirements for long-distance interception or accurate measurement. Antennas intercepted in wide airspace, such as RWR antennas, require 360° coverage. For example, using a (planar) spiral antenna, the 3 dB beamwidth is more than 90° and the gain is about 3 dB. As shown in Fig. 4.6, four antennas can be used to achieve 360°azimuth coverage. High-gain antenna usually adopts reflector antenna or array antenna, which has

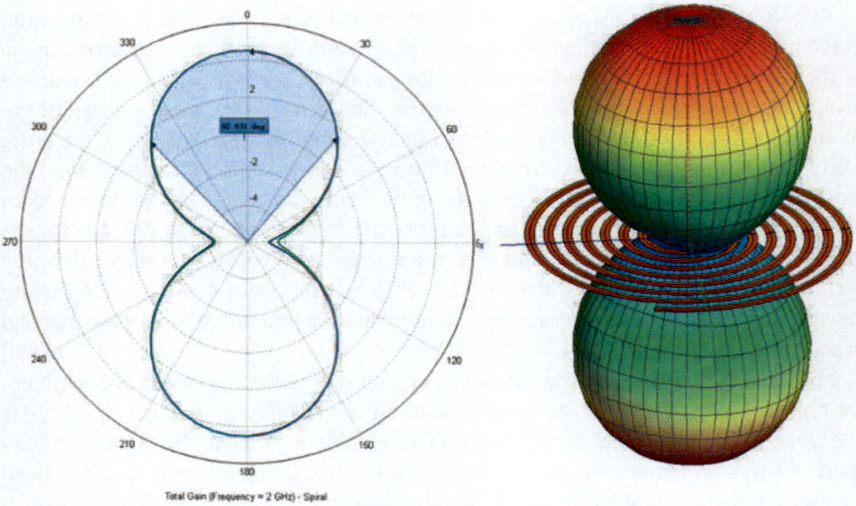

Fig. 4.6 Schematic diagram of a spiral antenna pattern and its three-dimensional pattern

small instantaneous coverage angle in space. For reflector antenna, it is necessary to receive signals from different angles by rotating the antenna. For array antenna, it is necessary to receive signals from different angles through phased array beam switching. Therefore, reflector antennas and array antennas usually work under wide open guidance.

For jamming transmitting antennas, a compromise should be made between antenna gain and pointing requirements. To meet higher ERP requirements, horn antennas, surface antennas, or array antennas with narrow beams are usually preferred. In this case, the antenna beam is narrow, and the jamming requirements from different angles can be achieved through mechanical rotation, phased array or multiple groups of antennas.

The ERP required by the transmitting subsystem can be determined according to the jamming-to-signal power ratio, operating distance, and other parameters of the jamming system. Knowing the output power of the transmitter and cable insertion loss, it is possible to determine the gain required for the transmitting antenna. The decomposition of ERP refers to the analysis and calculation in Chap. 3.

4.2.3 Sidelobe Level and Pattern Shape Requirements

The requirement of antenna shaping comes from the system's restriction on the gain of antenna sidelobe and the requirement on the shape of main lobe. As for the receiving antenna, too high sidelobe will lead to strong sidelobe signals entering the receiver and forming false targets. When the antenna without shape design cannot meet the requirements, it is often satisfied by shape design.

In some special cases, it is necessary to shape the main lobe of the wide-beam antenna. For example, for ground countermeasure equipment, in order to reduce the impact of ground clutter, it is usually required to generate a narrow beam in the azimuth plane and a pre-specified beam in the pitch plane, namely the shaped beam. In some EW equipment with direction finding requirements, in order to reduce the influence of antenna sidelobe, the antenna with wide beam is usually used to reduce the sidelobe, or the form of array non-uniform weighting is used to reduce the sidelobe.

For the shaping of array antenna, the unit antenna (equidistant or non-equidistant array element) in array antenna can be weighted by amplitude, phase, or amplitude phase.

For the shaping of reflector antenna, the methods of feed illumination adjustment and reflector edge optimization are usually adopted. The reflector is fitted in segments, and each segment points to a specific different direction. Finally, the formed beam is shaped within the specified angle range.

4.2.4 Volume Requirements

Antenna volume requirements in ECM are usually confirmed by the platform hosting the electronic warfare equipment. For ground and shipborne countermeasure equipment, the volume constraint is usually small. Wire antenna or surface antenna is commonly used. Its effective area is large enough to achieve high gain and improve jamming efficiency or reconnaissance sensitivity.

For airborne or missileborne countermeasure equipment, due to the strict size requirements, it is almost impossible to use wire antenna or surface antenna. Airborne countermeasure equipment often adopts horn antennas or phased array antennas. As for missileborne countermeasure equipment, due to the limitation of the installation size of the platform, the spin of the missileborne platform and the need to cover a wide-angle range for the countermeasure of the missileborne platform, the circularly polarized spiral antenna with wide beam is usually chosen.

4.2.5 Power Capacity Requirements

The power capacity of the antenna refers to the maximum power that the antenna can work normally. Signals exceeding the power capacity entering the antenna may cause antenna damage and even endanger the transmitter. Different types of feeding and the form of the antenna determine that the antenna has different power capacity. There are different ways to feed the antenna, such as waveguide feed, coaxial waveguide feed, coaxial feed, microstrip feed, and so on. Higher power capacity can usually be obtained by waveguide feeding, while the power capacity of microstrip line or SMA probe to microstrip line feeding is smaller.

Ground, shipboard and airborne countermeasure equipment, their transmitter power is usually in tens to several kilowatts; therefore, waveguide feed structure or N-type head feed antenna is often utilized, such as surface antenna and horn antenna. However, the self-defense jamming power of missileborne equipment is usually small, which is in the range of several watts, so the SMA-fed microstrip spiral antenna is usually used.

4.3 Typical Antennas of Jamming System

4.3.1 Spiral Antenna

The spiral antenna is an antenna with spiral shape. It is composed of helical lines of metal conductors arranged in a certain regularity [7]. The radiation of helical antennas is bi-directional, but in practical applications, the antenna is often required to have one-way radiation characteristics, so a reflection cavity is generally installed on one side of the helical antenna to reduce the radiation of the back lobe. Such spiral antennas are also called back cavity spiral antennas.

Spiral antenna can be divided into plane spiral, column spiral, and cone spiral. Among them, plane spiral antenna is one of the most widely used antennas in electronic warfare. It is characterized by wide bandwidth, wide beam, small size, light weight and can span multiple octaves. In electronic warfare system, spiral antenna is usually used as the receiving antenna of radar warning receiver or guidance receiver, but it is also used as the transmitting antenna in some jammers with low power and wide-beam transmission.

The basic form of planar spiral antenna is equiangular spiral antenna and Archimedes spiral antenna, and the latter is used most frequently. In the structure of the helical line, there are single-arm, double-arm, and four-arm, as shown in Figs. 4.7 and 4.8 [8, 9]. As the name suggests, the number of helicals extending outward from the center of rotation, the single-armed spiral antenna is 1; bifilar helix antenna is 2, and quadrifilar helix antenna is 4.

The radius of the helix increases uniformly with the change of angle, and the polar coordinate equation of the curve can be expressed as:

$$r = r_0 + a \cdot \phi \tag{4.10}$$

This is the Archimedes curve and the origin of the name of Archimedes spiral antenna. In the formula, r_0 is the initial radius, a is the helical growth rate, and ϕ is the amplitude angle expressed in radians. For a circular helicoidal surface, the perimeter is an area near a wavelength, forming the main radiation area of the planar helix. The circumference of the Nth circle is $L_n = \pi D_n = \lambda_n$. The maximum outer diameter of

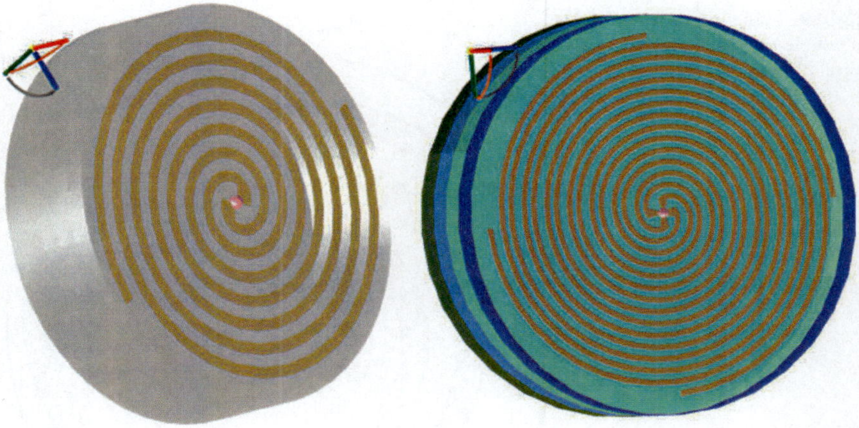

Fig. 4.7 Schematic diagram of two-arm and four-arm Archimedes spiral antenna with plane dorsal cavity

Fig. 4.8 Typical spiral antenna

the antenna (D_{max}) determines the maximum mwavelength (lowest mmfrequency), $\lambda_{max} = \pi D_{max}$. The minimum working wavelength (highest working frequency) is determined by the minimum distance D_{min} between the helical feed points (as shown in the Fig. 4.9), so the spiral antenna can achieve larger bandwidth.

In addition to the helical radiation surface, the complete antenna also includes a feed balancer and a back cavity. Among them, the input impedance of self-complements Archimedes helicoid with equal helix width and spacing is about 125 Ω. In order to match with the commonly used 50 Ω input impedance, a balanced–unbalanced impedance transformer needs to be used. At the same time, in order to suppress the influence of back valve, the form of adding back cavity reflecting surface is adopted.

Fig. 4.9 Radiating surface of double-arm archimedes spiral antenna

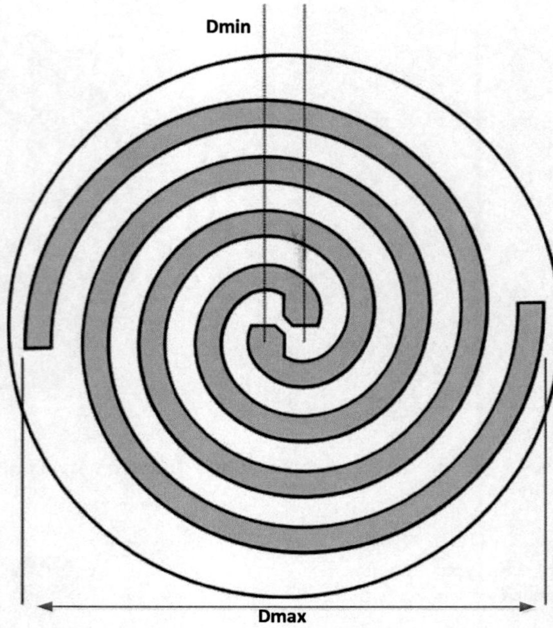

Planar spiral antenna has many advantages, such as wide frequency band, spanning multiple octaves, simple structure, low machining cost, low profile, and small size. Its typical polarization mode is circular polarization. The beamwidth is usually 60° ~ 90°, and the antenna gain is −1dBi ~ 3dBi, with low power capacity. It is suitable for all kinds of platform and boasts a wide range of applications in seeker, electronic reconnaissance, active decoy, radar warning receiver, missileborne jammer, and other fields.

4.3.2 Horn Antenna

Horn antenna is the most used antenna form in electronic countermeasures system. It is widely applied in satelliteborne, airborne, shipborne, vehicleborne and ground reconnaissance, positioning, and countermeasure systems [10]. In practical application, horn antenna can be used alone, as the feed of reflector antenna and as the unit antenna of array antenna. The horn antenna can achieve large bandwidth, high gain, and high power capacity.

The horn antenna can be divided into H-plane sector horn, E-plane sector horn, mmhorn, conical horn, and other types.

The H-plane horn antenna is formed by extending the wide side of the rectangular waveguide (H-plane, length a) at a certain opening angle, while the narrow side of

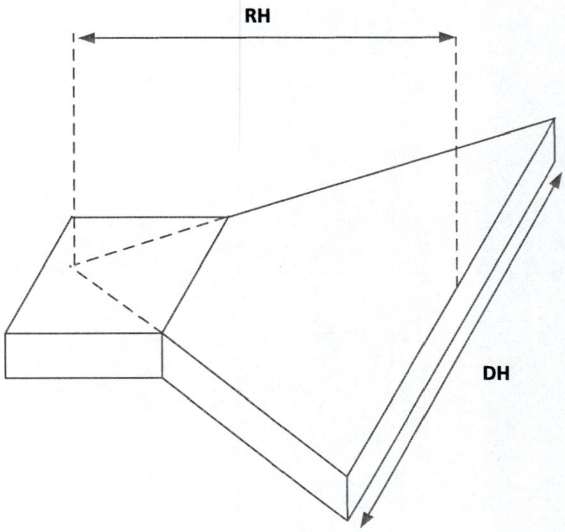

Fig. 4.10 Schematic diagram of H-plane horn antenna

the waveguide (E-plane, length b) remains unchanged, as shown in Fig. 4.10. The aperture size of the horn is $D_H \times b$, and the distance between the virtual vertex of opening angle and the center of the aperture is R_H, the difference of R_H will lead to the wave path difference of electromagnetic waves at different positions on the aperture surface, and there is an optimal size, which makes the antenna obtain the maximum gain.

Optimal size for H-plane horn antenna:

$$R_H = \frac{D_H^2}{3\lambda} \tag{4.11}$$

Under this condition, the width of the main lobe is $2\theta_{0.5H} = 80\lambda / D_H(°)$ and the efficiency is $\eta = 63\%$.

The E-plane horn antenna is formed by extending the narrow side (E-plane, length b) of the rectangular waveguide at a certain opening angle, while the wide side (H-plane, length a) of the waveguide remains unchanged, as shown in Fig. 4.11. The aperture size of the horn is $D_E \times a$, and the distance between the virtual vertex of opening angle and the center of the aperture is R_E, the difference of R_E will lead to the wave path difference of electromagnetic waves at different positions on the aperture surface, and there is an optimal size, which makes the antenna obtain the maximum gain.

Optimal size for E-plane horn antennas:

$$R_E = \frac{D_E^2}{2\lambda} \tag{4.12}$$

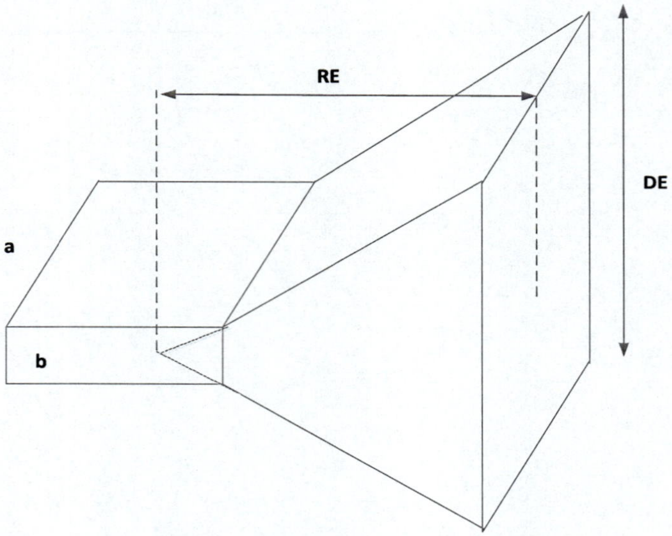

Fig. 4.11 Schematic diagram of E-plane horn antenna

Under this condition, the width of the main lobe is $2\theta_{0.5E} = 54\lambda/D_E(°)$ and the efficiency is $\eta = 63\%$.

Pyramid horn antennas are formed by expanding the narrow and wide sides of a rectangle at a certain opening angle. The aperture size of the horn is $D_E \times D_H$, and the distance between the virtual vertex of opening angle and the center of the aperture, respectively, is R_E and R_H (Fig. 4.12).

For the pyramid horn antenna, there is still an optimal size:

$$R_E = \frac{D_E^2}{2\lambda}, R_H = \frac{D_H^2}{3\lambda} \tag{4.13}$$

In such a condition, the width of the main lobe is $2\theta_{0.5E} = 54\lambda/D_E(°)$ and $2\theta_{0.5H} = 80\lambda/D_H(°)$, the efficiency is $\eta = 51\%$.

The pyramid horn is fed by double ridge waveguide, and the upper and lower symmetrical metal ridges are added to the horn body, and the transition curve of the ridge is reasonably selected. This horn antenna is called the ridged horn, as shown in Fig. 4.13. Ridged horn antenna is a typical ultra-wideband antenna, which has the advantages of small size and high gain, and it also has an extremely wide working frequency band. Most of the commonly used 2 GHz ~ 6 GHz and 6 GHz ~ 18 GHz broadband horn antennas are ridged horn antennas in electronic warfare.

The horn antenna is characterized by wide frequency band, simple structure, and easy processing. The typical polarization mode can be either circular polarization or linear polarization according to the different excitation modes. The beamwidth is usually 10° ~ 80°, and the antenna gain is 5 dBi ~ 20 dBi, with large power capacity.

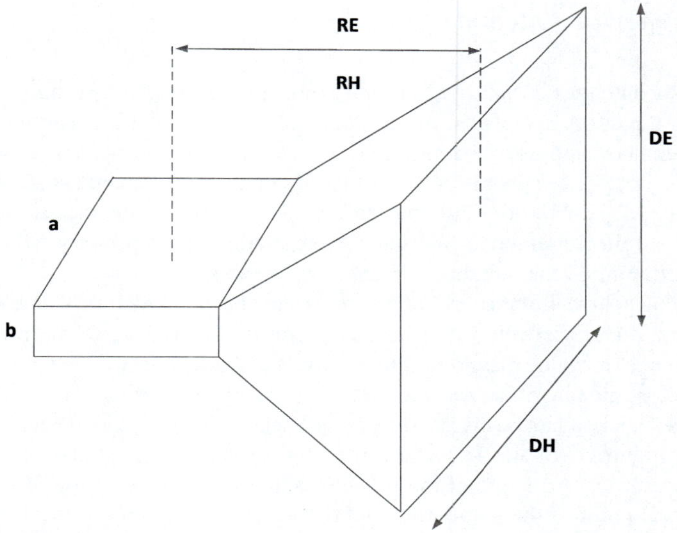

Fig. 4.12 Schematic diagram of pyramid horn antenna

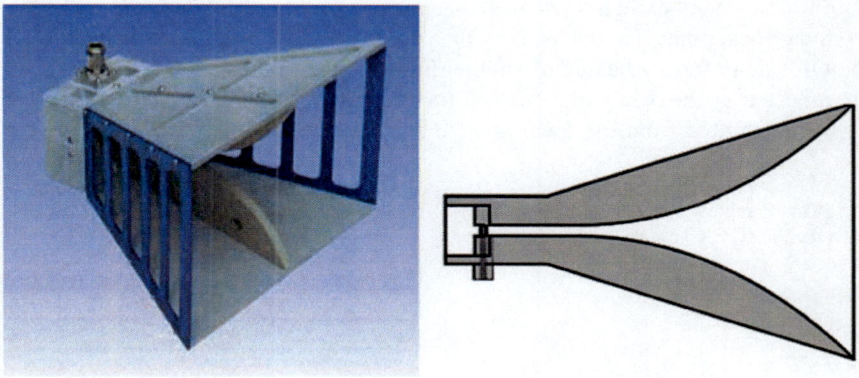

Fig. 4.13 Ridged horn and its cross-sectional schematic diagram

It is widely used in reconnaissance antennas and jamming transmitting antennas of airborne, shipborne, and ground equipment.

4.3.3 Reflector Antenna

The reflector antenna consists of a feed and a reflector. The feed is a primary irradiator with weak directionality placed on the focus of the paraboloid. It can be a single dipole antenna or dipole array, a single horn or multiple horn array, a waveguide slot array, and so on. The reflector of the antenna is composed of metal surface or grid in the shape of paraboloid. Utilizing the geometric focusing characteristics of the reflector, the reflector antenna can reflect the radiation of the primary radiator with weak directivity into the radiation with strong directivity.

According to the number of reflectors, it can be divided into single-reflector antenna and double-reflector antenna and so forth. According to the position of the feed source, it can be classified into positive feedback reflector antenna, reareed-back reflector antenna, offset reflector antenna, among other types. According to the shape of the beam, it can be divided into pencil beam antenna, shaped beam antenna, and so on. Figures 4.14 and 4.15 show different types of reflector antennas.

The geometrical principle of the reflector antenna is shown in Fig. 4.16, where F is the focal point of the paraboloid and the distance between the focal point and the origin O is the focal length of the paraboloid. 2α is the angle between the ray emitted from the focus to the edge of the paraboloid and the axis, also known as the feed irradiation angle; the diameter of the paraboloid is D.

A rotating paraboloid has two main geometric properties; one is that after the ray from the focal point F is reflected by the paraboloid the reflected ray is parallel to the OF axis to form a bundle of parallel lines. Conversely, when the parallel lines are reflected by the paraboloid, they all focus on the focal point F. The other is that each line emitted from the focal point F and reflected by the paraboloid has the

Fig. 4.14 Parabolic cylinder antenna (AN/TPS-63 radar)

Fig. 4.15 Paraboloidal and cutting paraboloidal antennas (Russian Krasuha-2/4 ground countermeasures system)

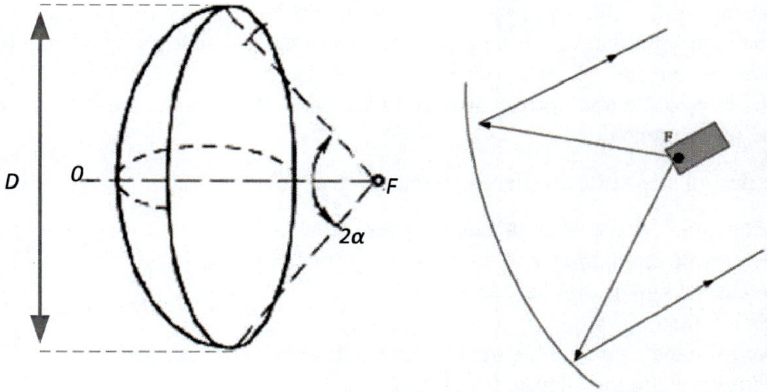

Fig. 4.16 Schematic diagram of the basic principle of a parabolic reflector antenna

same distance to the aperture surface (or any plane perpendicular to the OF axis), signifying that the spherical wave from the focal point F is reflected as a plane wave.

When the rotating paraboloid is irradiated by the feed at the focus F of the paraboloid, the spherical electromagnetic wave emitted by the feed becomes a plane wave on the aperture surface after being reflected by the paraboloid. After the superposition of the in-phase aperture field, the far-field radiation on the paraboloid will synthesize a narrow beam antenna pattern, so as to obtain the high gain of the antenna.

In the parabolic antenna, there is an important parameter called focal length aperture ratio: f/D. If the focal length aperture ratio is larger, the beam of antenna is narrower, the gain is higher, and the sidelobe is lower. As shown in Table 4.4. When the gain is given, the aperture D can be determined first and then f/D can be

Table 4.4 Relationship between focal length aperture ratio and antenna beamwidth

f/D	$2\theta_{0.5H}$	$2\theta_{0.5E}$	Side-lobe level	
			H-plane	E-plane
0.625	$61\lambda/D(°)$	$63\lambda/D(°)$	13	16
0.417	$63\lambda/D(°)$	$71\lambda/D(°)$	16	20
0.213	$70\lambda/D(°)$	$83\lambda/D(°)$	24	25
0.250	$79\lambda/D(°)$	$96\lambda/D(°)$	27	26

selected according to the requirements of the overall structure. Finally, we can take this expression $f/D = 1/[4\tan(\alpha/2)]$ to calculate the feed irradiation angle.

Requirements for feeders:

(1) The feed should have a definite phase center placed at the focal point of the paraboloid so as to obtain equal phase distribution on the aperture.
(2) The shape of the feed pattern should conform to the optimal irradiation as far as possible, and the sidelobe and the backlobe should be as small as possible, because they will reduce the antenna gain and raise the sidelobe level.
(3) The feed should have a smaller volume to reduce its shielding to the paraboloid aperture surface.
(4) The bandwidth and power capacity of the parabolic antenna are determined by the feed antenna.

The design steps of the reflector antenna are as follows:

(1) Determine the diameter of the main reflector.
(2) Determine the diameter of the secondary reflector (if any).
(3) Select the hyperboloid eccentricity.
(4) Design feed.
(5) Determine the reasonable manufacturing tolerance of reflector.
(6) Determine the installation accuracy of feed.
(7) Design the structure of reflector.

Reflector antenna is characterized by high gain and narrow beam. The typical polarization mode can be either circular polarization or linear polarization according to the different feeders. The beam width is usually a few tenths to ten degrees; the antenna gain is above 20dBi; and the power capacity is large. It is widely used in reconnaissance antennas and jamming transmitting antennas for shipboard and ground equipment.

4.3.4 Array Antenna

Array antenna is more and more used in electronic countermeasure equipment. Using array antenna can easily realize high-gain narrow beam and large transmission ERP.

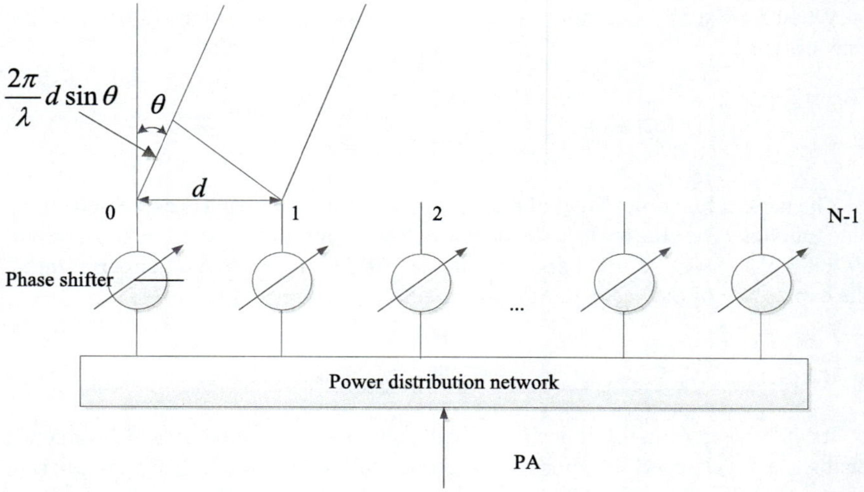

Fig. 4.17 Schematic diagram of N-element linear phased array antenna

At the same time, using array antenna can improve the flexibility of beam control and realize the functions of beam forming, low sidelobe antenna, multi-beam scanning, and so on. Among array antennas, phased array antennas and digital beam forming (DBF) antennas are most frequently used.

Assuming that the unit antenna pattern $f(\theta, \varphi)$ is wide enough to satisfy omni-directivity, for narrowband signals, the linear array antenna is shown in Fig. 4.17, and its pattern function can be expressed as

$$F(\theta) = \sum_{i=0}^{N-1} a_i e^{ji(\frac{2\pi}{\lambda} dsin\theta - \Delta\phi_{\max})} \tag{4.14}$$

In the formula, a_i is the amplitude weight coefficient, $\Delta\phi$ is the feed phase difference between units, $\Delta\phi_{\max}=\frac{2\pi}{\lambda} d\sin\theta_{\max}$ and $\theta_{\max}$ is the maximum direction of antenna beam.

Assumed that $\Delta\phi = (2\pi/\lambda)d\sin\theta$, it represents the phase difference of the signal from direction θ received by adjacent units, which can be called the spatial phase difference between the adjacent units. When $\Delta\phi - \Delta\phi_{\max} = \vartheta$, for the uniformly distributed irradiation function, $a_i = 1$, $F(\theta) = (1 - e^{jN\vartheta})/(1 - e^{j\vartheta})$ can be calculated, and the formula is simplified by Euler formula:

$$F(\theta) = \frac{\sin\frac{N}{2}\vartheta}{\sin\frac{1}{2}\vartheta} e^{j\frac{N-1}{2}\vartheta} \tag{4.15}$$

When x is small, according to $\sin x \approx x$, the amplitude pattern function of the line array is:

$$|F(\theta)| = N\frac{\sin\frac{N}{2}\vartheta}{\frac{N}{2}\vartheta} = N\frac{\sin\frac{N\pi}{\lambda}d(sin\theta - sin\theta_{\max})}{\frac{N\pi}{\lambda}d(sin\theta - sin\theta_{\max})} \tag{4.16}$$

The pattern function $|F(\theta)|$ of the linear phased array antenna is expressed by the sinc function. Then the basic performance of the linear array antenna can be acquired. When $\vartheta N/2 = 0$, $|F(\theta)|$ has maximum, $|F(\theta)| = 1$, under such circumstances, the expression of the beam pointing $\theta_{\max}$ is:

$$\theta_{\max} = \arcsin(\frac{\lambda}{2\pi d}\Delta\phi_{\max}) \tag{4.17}$$

It can be seen from the above formula that the maximum direction of the antenna beam can be changed by altering the phase shift value $\Delta\phi_{\max}$ between adjacent elements in the array. The $\Delta\phi_{\max}$ is realized through the phase shifter or optical fiber delay line set at the back end of each antenna unit.

Phase shifters are mainly classified into analog phase shifters using voltage-controlled varactor diodes or ferrites and digital phase shifters adopting PIN diodes as switching devices. In recent years, with the development of microsystem technology, digital phase shifters based on MEMS micro-electromechanical switches have also emerged.

Phased array antennas are characterized by flexible beam control and can quickly switch from one wave position to another, which is especially suitable for multi-target tracking and jamming in active countermeasures. It boasts a wide application in airborne, shipborne, and ground-based reconnaissance and transmitting antennas.

4.4 Common Antenna Design Tools

4.4.1 HFSS Software

HFSS is a full-wave three-dimensional electromagnetic field simulation software developed by ANSYS Company in the USA. High computational accuracy can be obtained through the finite element method. It is the first-class electromagnetic field simulation software in the industry and one of the benchmarks in the industry [11]. HFSS can accurately calculate a variety of antenna performance, including two-dimensional, three-dimensional far-field and near-field radiation pattern, directivity coefficient, gain, axial ratio, beam width, input impedance, standing wave, S parameter, and current distribution characteristics.

The process of antenna design and simulation by using HFSS is shown in Fig. 4.18. First, set the solution type of the antenna (driven model/driven terminal) and then

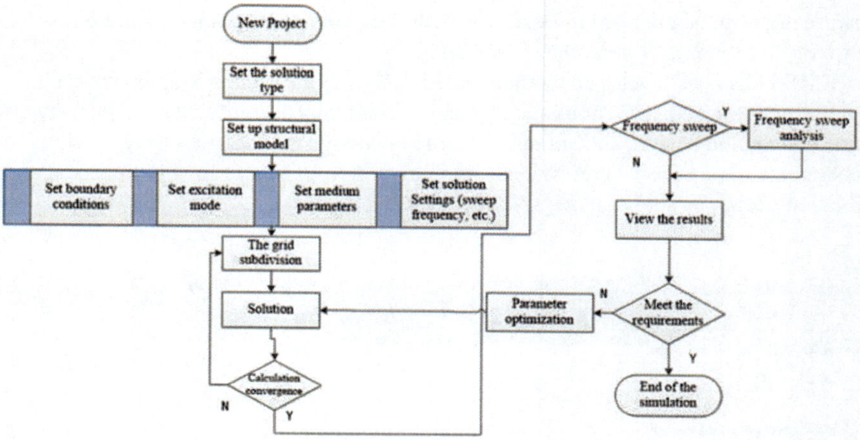

Fig. 4.18 HFSS antenna design process

create the parametric structure model of the antenna. The next step is to set the boundary conditions, the excitation mode (e.g., wave port/lumped port), the medium parameters, and the parameters of solution. After the relevant parameters are set, the structural model is meshed for solution. The solution of HFSS is usually automated. If the parameters are set correctly, a convergent solution can be obtained basically, but the solution time is different. If the obtained results do not meet the design requirements, it is necessary to carry out certain optimization, and finally get the simulation results that meet the requirements.

4.4.2 FEKO Software

FEKO is a powerful 3D full-wave electromagnetic field simulation software owned by EMSS, South Africa. Its biggest feature is the combination of moment method (MOM), multi-layer fast multipole method (MLFMM) and finite element method, which is especially suitable for solving the problem of free space coupling between structures. FEKO also includes a wealth of high-frequency calculation methods, such as physical optics (PO), large surface element physical optics (large element PO), geometric optics (GO), uniform geometry diffraction (UTD), etc., which can quickly solve super electric large-scale problems with less resources.

In electronic warfare, FEKO is commonly used for antenna analysis, array antenna design, RCS stealth analysis, and so forth.

The usage of FEKO is similar to the design process of HFSS, and its main operation interface is shown in Figs. 4.19 and 4.20. The following steps are carried out in CADFEKO module, such as model establishment, import, solution parameter setting, medium parameter setting, feed parameter setting, mesh splitting. The solved parameters can be checked in Postfeko, including near-field results, far-field results, current distribution, S parameters, and so on.

EMSS has also developed antenna aided design tool Antenna Magus, which integrates more than 200 functions such as standard antenna design, array design, waveguide conversion design, and antenna design knowledge management (Fig. 4.21). The designed antenna model can be seamlessly exported to FEKO for optimization and detailed design as shown in Figs. 4.22 and 4.23.

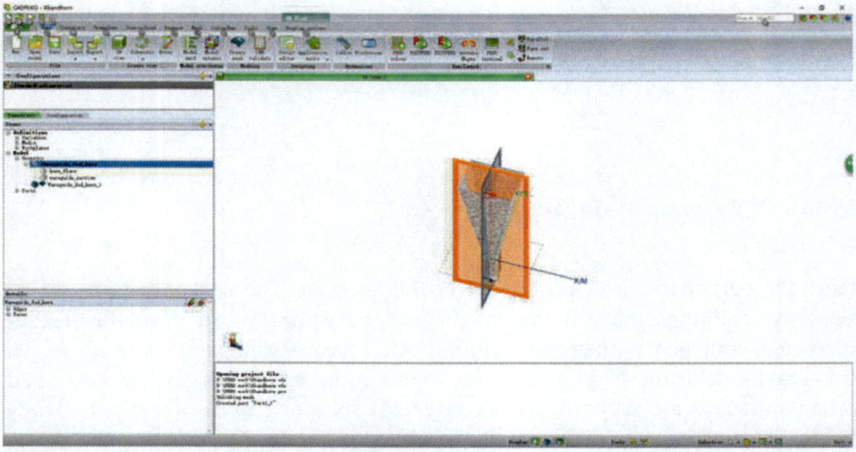

Fig. 4.19 Interface of FEKO solving module CADFEKO

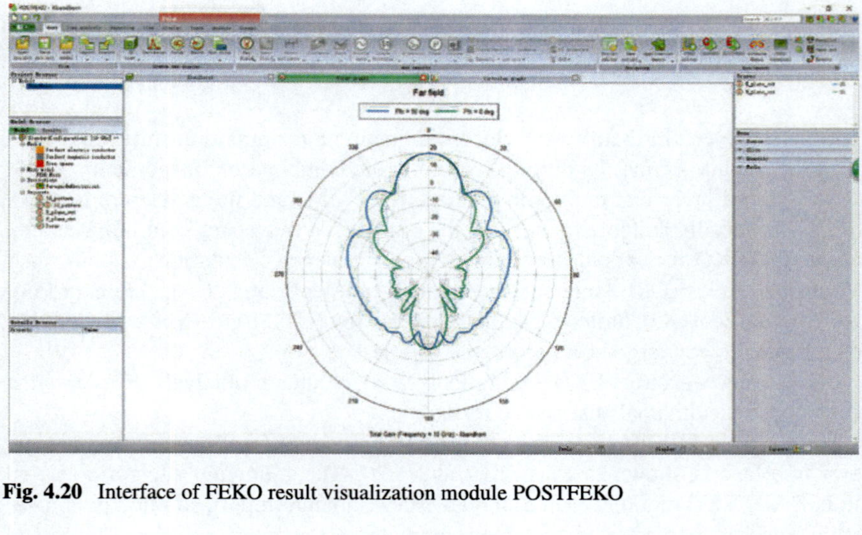

Fig. 4.20 Interface of FEKO result visualization module POSTFEKO

Fig. 4.21 Antenna Magus has more than 200 standard antenna template libraries

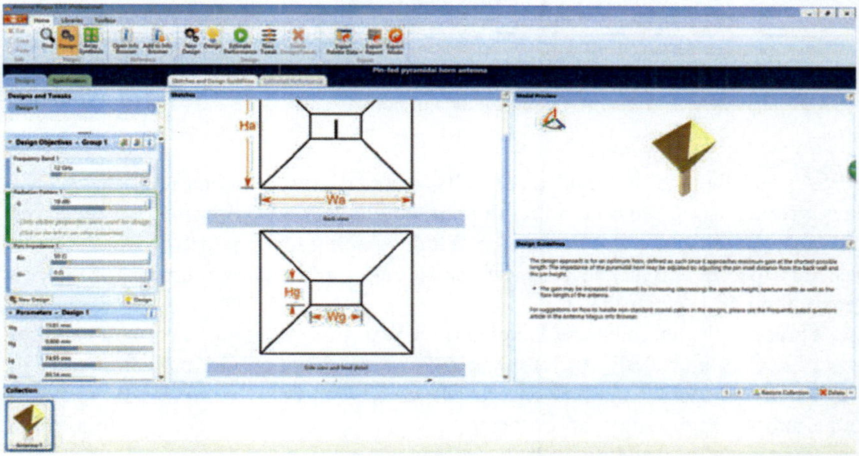

Fig. 4.22 Rapid parametric design of a horn antenna (center frequency 12 GHz, gain 18dBi)

4.5 Development Trend of Antenna of Jamming System

4.5.1 AESA and DBF

The development of radar antenna has experienced the development process from single antenna of the whole machine to passive electronically scanned array antenna

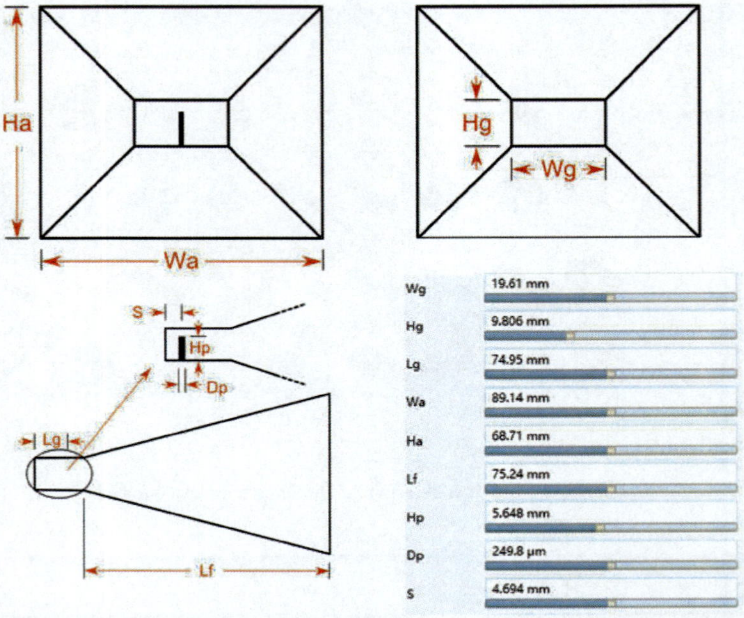

Fig. 4.23 Parameterized results of rapid solution of horn antenna

(PESA) and then to active electronically scanned array antenna (AESA). Nowadays, ground-based radar systems, shipborne radar systems, airborne radar systems, and even spaceborne radar systems are all developing toward AESA, which is now basically mature and has become the mainstream standard configuration of the new generation of equipment [12].

At present, the antennas used in electronic warfare equipment are also developing toward AESA and DBF. Active phased array antennas are adapted in electronic warfare active countermeasure equipment, which can flexibly control the scanning range and scanning angle of the beam, and can achieve simultaneous jamming of multiple targets by dividing multi-beam subarrays [13, 14].

The antenna with active phased array system can achieve higher power capacity in effective space. The traditional jammer usually has a transmitting power of tens of watts or even hundreds of watts from a single tube, which puts forward great demands on the power capacity of the antenna and the heat dissipation of the transmitter. Especially in the high frequency band, only vacuum tube can be used to realize high-power transmission. However, using active phased array technology, the power of each array element is only a few watts, and the power capacity requirements and the heat dissipation requirements of the antenna are greatly reduced, at the same time, improving the reliability of the transmitting system. The Jammer with phased array antenna can flexibly allocate jamming resources and realize the effective simultaneous multi-beam jamming to multiple targets, which is the first choice of

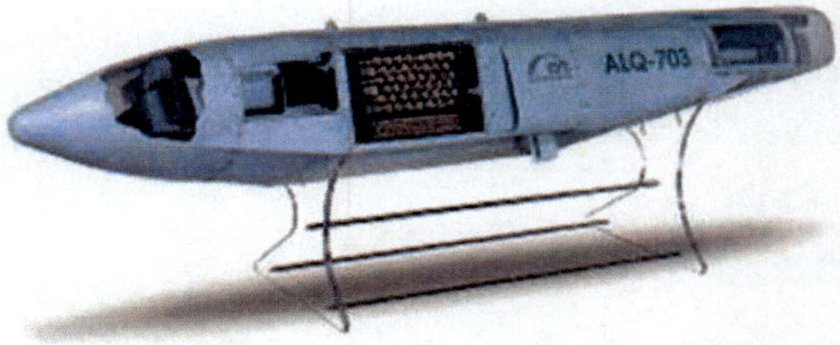

Fig. 4.24 ALQ-703 pod with a fully phased array system from ELT

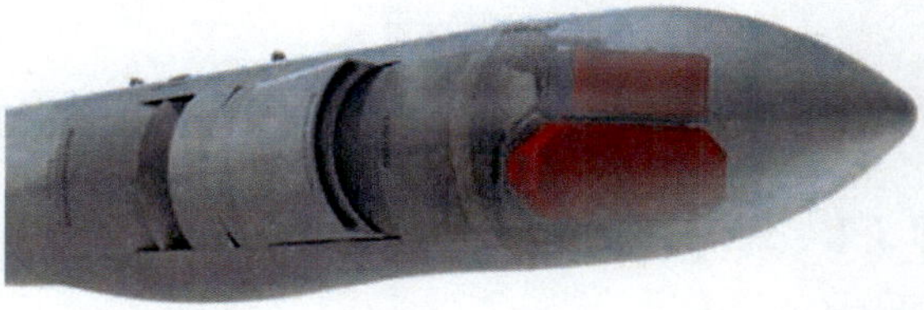

Fig. 4.25 Raytheon's next-generation jamming pod with phased array (NGJ)

the next-generation airborne jamming device. The above advantages of phased array antenna are at the expenses of high cost, so it is necessary to evaluate the cost and performance when selecting the antenna system (Figs. 4.24 and 4.25).

DBF technology is a technology that uses digital method to realize beamforming. DBF technology makes full use of the spatial information obtained by the array antenna, makes the beam obtain super-resolution and low sidelobe performance through complete digital signal processing technology, and realizes the scanning and adaptive zeroing of multiple independent controlled beams at the same time. Compared with the existing phased array technology, the digital array antenna programming is more flexible and can use the whole array aperture when realizing simultaneous multi-beams without losing aperture gain. It is becoming increasingly popular in radar systems and electronic countermeasures systems.

The traditional phased array antenna is a kind of analog beamforming, which is realized by controlling the phase shifter by the wave-control unit. DBF technology utilizes digital technology to process the sampled array metadata to realize beam control, as shown in Fig. 4.26.

The principle of receiving DBF is as follows. The equidistant array receiving antennas independently perform low-noise amplification and down-conversion to the required intermediate frequency. Each ADC channel independently collects the IF signals and enters the beamforming network in parallel. The essence of beamforming network is to use the digital processing method to compensate the phase of the incident signal in a certain direction, so as to realize the maximum energy reception and complete beamforming in that direction.

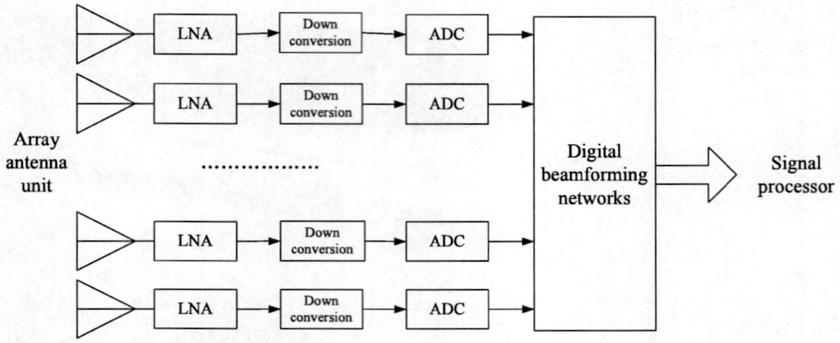

Fig. 4.26 Hardware implementation block diagram of receiving DBF

4.5.2 Conformal

Conformal antenna refers to an antenna or antenna array that is consistent with the shape of the platform.

The use of conformal antenna in airborne platform device is mainly considered from the perspective of aerodynamic structure and RCS reduction of the whole body, which has nothing to do with the electrical performance of the antenna itself. In the past, airborne ECM devices were usually in the form of external pods, such as the ALQ-99 pod adopted by the EA-6B and EA-18G. The advantage of these traditional pods is that they can achieve high jamming ERP. But the drawback of the external pod is also obvious; that is, it will damage the aerodynamic shape of the body at high speed and seriously affect the maneuverability and flexibility of the fighter at high speed. Since the pod is exposed outside the body, it may form strong radar reflection and destroy the stealth characteristics of the fighter. From the perspective of the design ideas of electronic warfare devices of the fifth-generation stealth fighters such as the F-22 and F-35 of the US army, almost all abandon the design of the external pod, which is considered from the aspect of ensuring the stealth characteristics of the whole body.

In the future, airborne emitter (including radar antennas, electronic warfare antennas, communication antennas, navigation antennas, among other antennas) will be combined with the fuselage skin to form conformal arrays. Applying conformal arrays can increase the effective aperture of the antenna and broaden the visual field of view, unlike traditional radar and electronic warfare antennas, which are limited to the front and rear. By using conformal array, the antenna can be arranged at the leading edge of the wing, the trailing edge of the wing, the side of the fuselage, the side of the vertical tail, etc., so as to realize the omnidirectional detection requirements of the airframe. The U.S. Air Force Research Institute, Raytheon, Northrop Grumman, and European Aeronautic Defense and Space Company (EADS) all carried out preliminary research on conformal radar antenna in the late 1990s and have made phased progress. It is foreseeable that conformal antennas will be the development trend of electronic warfare antennas in the future.

References

1. The ARRL Antenna Book (Chinese version) [M]. Posts and Telecommunications Press, Beijing (2009)
2. Yongnian T (2012) Radar countermeasures engineering [M]. Beihang University Press, Beijing
3. Adamy DL (2001) EW101: a first course in electronic warfare [M]. Artech House Inc., Norwood
4. Jianqing Z (2000) Electromagnetic wave engineering [M]. National University of Defense Technology Press, Changsha
5. Poisel RA (2012) Antenna systems and electronic warfare application [M]. Artech House Inc, Norwood
6. Chen ZN, Chia MYW (2015) Design and application of broadband planar antenna [M]. Hu Laizhao Trans. National Defense Industry Press, Beijing
7. Xiaozhuan L, Fei Y (2009) A small planar [C]. National Antenna Annual Conference Proceedings (I), Chengdu, pp. 553~556
8. Zhaohui S, Hongmei L, Hanying Y et al (2009) Research on miniaturization of Archimedes with zigzag arm [J]. J Microw 25(2):53–57
9. Tingwei X (2010) Research on broadband miniaturized four-arm [D]. Xidian University, Xi'an
10. Kang Y (2015) Design method of UWB double ridge horn antenna [D]. Nanjing Ship Radar Institute, Nanjing
11. Mingyang L, Min L (2014) HFFSS antenna design [M], 2nd edn. Publishing House of Electronics Industry, Beijing
12. Jiegui W, Jinqin L (2007) A wideband digital beamforming method with constant beam-width [J]. J Astronaut 06:1458–1461
13. Rui Y (2009) Broadband DBF and digital channelization technology [D]. Xidian University, Xi'an
14. Xiaojing D (2011) Research on broadband DBF technology in electronic reconnaissance [D]. Xidian University, Xi'an

Chapter 5
Receiver and Processor of Jamming System

In order to degrade the detection quality of radar, radar jamming needs to know the change of radar and therefore produce the precise jamming signal in the dimensions of time, space, and frequency [1]. If we compare an active radar jamming system to a warrior, the receiver was the eyes and ears and the transmitter was the fists and feet. Without the receiver, the jamming system is just like a blind person and cannot perform jamming correctly. The main functions of the receiver include interception, sorting, and identification. Interception is to acquire the surrounding microwave signal and measure its parameters. Sorting is to cluster the signal according to the signal parameter. After the sorting, the signal from the same kind of emitter is expected to be classified into one group, and the signal that we do not care would be excluded. Recognition is to confirm the radar type, the platform type, or even the platform unit that the cluster of signals belongs to, and then to estimate the threat level of the signal. For the jamming system, the warning information of the receiver is the precondition for launching jamming. The parameters of the threat radar outputted by the receiver are the basis of generating the jamming signal.

In this section, we will introduce the requirements, types, design methods, and development trends of receivers. The first subsection will introduce the requirement of a receiver, where the functions and performance indices requirement of the receiver from the aspect of the jamming system and the relations between the indices and the system functions are discussed as well. In the second subsection, the types of the receiver are presented. We will summarize the common radar jamming system receiver types and emphatically describe their principles. In the third subsection, the design methods of a receiver are introduced. In the fourth subsection, we will give the most commonly used processing methods of signal and data. In the fifth subsection, the latest development of receivers is introduced.

G. Tang et al., *Techniques and System Design of Radar Active Jamming*,
https://doi.org/10.1007/978-981-19-9944-4_5

5.1 Requirements Analysis of Receiver and Processor

Two questions should be answered before performing radar jamming. The first one is to determine whether the radar to jam exists, and the second question is to find the radar's location and signal feature. The answer to the first question is to intercept, sort, and identify the target to generate the warning information through the receiver. To answer the second question, we need to use the receiver to transmit the location and parameters of the threat to the jamming system. In this subsection, we will demonstrate the receiver's requirement from the aspects of interception capability, sorting capability, and recognition capability.

5.1.1 Interception Requirements

The interception capability of a receiver is measured from the aspects of the frequency range, instantaneous bandwidth, sensitivity, dynamic range, simultaneous multiple signals adaption, and parameter measurement.

5.1.1.1 Frequency Range

The frequency range of a receiver is the frequency range of the radar signal that it can process. Frequency range is crucial for intercepting certain radars. If the working frequency of the radar is out of the frequency range of the receiver, the receiver does not have the capability of intercepting this radar, as shown in Fig. 5.1.

Significantly, the term frequency range here is not the instantaneous frequency range of a receiver but the overall frequency range the receiver can achieve. For the receivers that are unable to adjust their receiving frequency range, the frequency range is the same as the instantaneous frequency range. For the receivers that can adjust their receiving frequency range, they reside within a small range in frequency at a particular time, but achieve a larger range in frequency via a certain frequency sweep strategy.

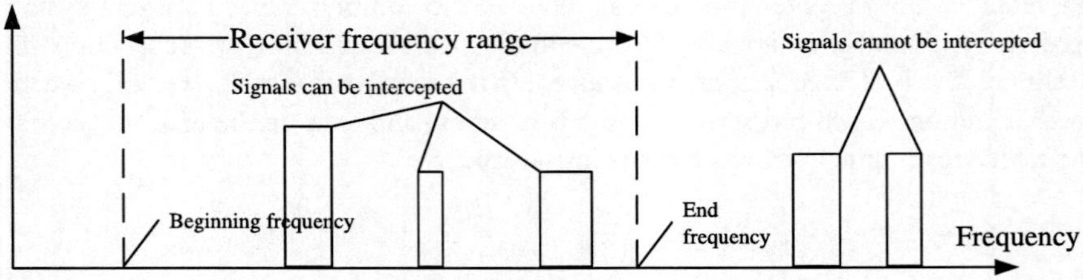

Fig. 5.1 Frequency range of receiver

In the history of radar jamming, the original radar worked from tens of MHz to hundreds of MHz. The receiver of a radar jamming system only needs to cover the frequency range under 1 GHz. However, with the development of technology, the radar frequency was raised to L, S, C, X, Ku, and even millimeter wave band. As a result, the frequency range of a receiver was extended to the same frequency range rapidly [2].

Different radar jamming systems require different working frequency ranges of receivers. The working frequency range of a receiver is determined according to the working frequencies of the main jamming targets. For example, a jamming system for airborne fire control radar requires a receiver whose working frequency range from 8 to 12 GHz, since the working frequency of most airborne fire control radar is in this domain. For the jamming systems for early warning detection radar, the receiver frequency range should be able to cover the bands of P, L, and S, since early warning detection radar usually works in these frequency bands [3].

5.1.1.2 Instantaneous Bandwidth

Instantaneous bandwidth is the difference between the maximal frequency and the minimal frequency of the signal the receiver is able to process simultaneously. It is not only required to satisfy the requirement of covering the bandwidth of radar pulse signal but also required to satisfy the requirement of radar frequency agility in order to design a suitable instantaneous bandwidth. For example, the pulse signal bandwidth of SAR/ISAR radar might be over 1 GHz. To completely receive the signal of SAR/ISAR radar, the instantaneous bandwidth of the receiver should be over 1 GHz [4–7]. The frequency agility range of regular radar varies from hundreds of MHz to several GHz. In order to satisfy the complete reception of a single radar signal, the instantaneous bandwidth of the receiver should achieve hundreds of MHz, or up to several GHz. The specific instantaneous bandwidth requirement should be determined according to the specific jamming object.

It is difficult for most types of receivers to cover all the working frequency ranges. Instead, the receivers use smaller instantaneous bandwidths and adjust the working frequency range by switching the center frequencies. Besides, the receivers can exploit multiple receivers with small instantaneous bandwidth to form a large instantaneous frequency range to achieve the same goal, as shown in Fig. 5.2.

The wider the instantaneous bandwidth is, the better the receiver's interception capability is from the point of view of the frequency domain. However, the wider the instantaneous bandwidth is, the stronger the receiver noise is. If channelized or frequency detection methods were not adopted, the sensitivity of the receiver would be decreased. Another influence brought by large instantaneous bandwidth is the number of signals entering the receiver simultaneously would increase, which demands more from the processing capability of the receiver. The choice of receiver instantaneous bandwidth needs to compromise among the interception probability, the sensitivity, and the processing capability [8–11].

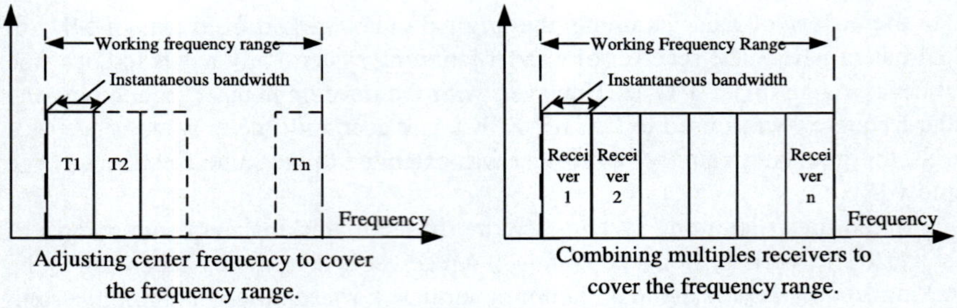

Fig. 5.2 Relationship between the frequency range and the instantaneous bandwidth

Assume the instantaneous bandwidth is B_w, the working bandwidth of the receiver is B. If the receiver uniformly scans within the working bandwidth, the dwell time for the frequency would be about $\frac{B_w}{B}$. Assume the instantaneous bandwidth of a receiver is 500 MHz, and its working bandwidth requirement is 5 GHz. If the receiver uniformly scans within the working bandwidth, the dwell time ratio for certain frequency would be 10%. The dwell time ratio will affect the interception probability of radar, which is described in the book *Introduction to Modern EW System* by Andrea De Martino in detail and would not be discussed in this book.

5.1.1.3 Sensitivity

The sensitivity of a receiver determines the minimal power of the intercepted signal, and therefore, the sensitivity is a necessary index to be considered while designing a receiver. The requirement of receiver sensitivity relates to the jamming target and the reconnaissance capability of the jamming system, and radar peak power, radar antenna gain, the distance between radar and jammer, and the receiving antenna gain should be taken into account. The sensitivity requirement of receiver can be calculated through Formula 5.1,

$$S_e = \frac{P_t G_t}{4\pi R^2 L_t} \cdot \frac{G_e \lambda^2}{4\pi L_e} \cdot L_a \tag{5.1}$$

where S_e is the signal power received by the receiver, R is the distance between the radar and the receiving antenna, G_t is the radar transmission antenna gain in the direction of the receiving antenna, L_t is the radar loss factor from the radar transmitter to its antenna, G_e is the receiving antenna gain, λ is the wavelength, L_e is the loss factor from the receiving antenna to the receiver, and the L_a is the atmospheric loss which relates to the weather and the transmission distance.

From Formula 5.1, we can find that the requirement of receiver sensitivity is in proportion to the radar peak power. The higher the radar peak power is, the higher the power entering the receiver is, and the lower the required receiver sensitivity is. For example, under certain conditions, the receiver sensitivity requirement is −60 dBm in order to intercept radar whose peak power is 100 kW. If the radar peak power decreased to 10 kW, it would require the receiver sensitivity to rise to –70 dBm while all the other conditions remain unchanged.

The requirement of receiver sensitivity is in proportion to the radar transmission gain in the direction of the receiver's antenna. For self-protection systems, they only receive the main lobe signal of radar, and therefore, the radar transmission gain in the direction of the receiver's antenna is the main lobe gain of the radar. For standoff jamming systems, they are required to have the capabilities of side lobe reconnaissance and jamming, and therefore, the radar transmission gain in the direction of the receiver's antenna is the side lobe gain of the radar. Since the side lobe of radar is rather fluctuant, we should carefully estimate the average level and fluctuation of the radar side lobe gain and make sufficient margin, while analyzing the sensitivity requirement of side lobe receiving systems.

The requirement of receiver sensitivity is in proportion to the receiver's antenna gain. If the receiver gain was high, then the requirement of receiver sensitivity would be low, and vice versa. For example, if the receiver antenna gain was 0 dB and the requirement of receiver sensitivity was −60 dBm, the receiver sensitivity requirement could decline to −50 dBm when the receiver antenna gain rises to 10 dB with all the other conditions unchanged.

The requirement of receiver sensitivity is in proportion to the square of R which is the distance between the receiver and the radar. The further the distance is, the lower the requirement of receiver sensitivity is. For example, if the sensitivity requirement is −60 dBm of the receiver when the reconnaissance distance is 100 km, the requirement of the receiver would decline to −66 dBm when the reconnaissance distance is required to increase to 200 km.

5.1.1.4 Dynamic Range

The dynamic range of a receiver means the input signal range that the receiver could detect the signal without distortion. The dynamic range is determined by the minimal and the maximal signal power that can be detected and processed. Hence, the dynamic range requirement is closely related to the application scenario [12].

The estimation of receiver dynamic range usually starts from the potential application scenarios of the radar jamming system and analyzes the threat radar possibly encountering, the radar's radiation level, and the maximal and minimal distance from the radar to the receiving antenna. With the parameters above, the maximal and minimal power entering the receiver can be calculated by the reconnaissance formula and the receiver dynamic range requirement also can be determined.

Sometimes, the calculated receiver dynamic range requirement exceeds the design limit of the receiver. In this case, a compromise is necessary. We could divide the scenario into multiple scenes, and use different sensitivities in different scenes in order to satisfy the optimal detection capability.

5.1.1.5 Simultaneous Multiple Signals Adaption Capability

When multiple signals in the environment enter the bandwidth range of the radar receiver, the capability of the receiver to separately intercept and measure the signals is called simultaneous multiple signals adaption capability [13].

The higher the simultaneous multiple signals adaption capability is, the stronger the environment adaption capability is. Since the high simultaneous multiple signals adaption capability is at the expense of hardware and computation resources, compromise should be considered when asking for high simultaneous multiple signals adaption capability for specific receivers. Receivers of different systems have different simultaneous multiple signals adaption capabilities. For example, an analog instantaneous frequency measurement receiver only can adapt to single signal detection, while a digital receiver is able to detect multiple signals simultaneously, and channelized receiver can adapt to simultaneous signals in different channels [14, 15].

5.1.2 Parameter Measurement Capability Requirements

The parameter measurement capability of a receiver is determined based on the requirements of sorting, identification, and jamming guide. In general, the parameters a receiver needs to measure include carrier frequency, pulse width, time of pulse arrival (TOA), angle of pulse arrival (AOA), and pulse amplitude (PA).

5.1.2.1 Measurement of Carrier Frequency

Carrier frequency is the center frequency of the electromagnetic wave emitted by the radar. Carrier frequency is important for signal sorting and radiation source recognition and is one of the key parameters for the jamming guide. Hence, all kinds of jamming system receivers are capable of carrier frequency measurement capability. In the early age, radar's carrier frequency is fixed, and therefore, it was easy to differentiate them and convenient to guide jamming. As the development of radar technology, frequency agility and temporary frequency switching capability have been widely used in modern radar, and it becomes difficult to distinguish and identify only depending on the frequency information in which time other signal parameters should be used.

One of the important indicators of carrier frequency measurement is frequency-measuring precision. Different systems require different frequency-measuring precisions.

5.1.2.2 Measurement of Pulse Width

Pulse-based target detection is used by most radars, and therefore all the receivers are required to be able to measure pulse width. Different radar uses different pulse widths. The level of radar pulse width is associated with the system of radar signal processing. For example, the pulse width of most airborne fire control radar that uses pulse Doppler processing is at a microsecond level, and the pulse width of ground-based radar that usually exploits pulse compression processing is much larger and could be tens or hundreds of microseconds. Continuous wave radar does not have the parameter of the pulse width. However, the receiver would define a special value for this kind of radar to unite the expressions.

Adaptive pulse width and pulse width measuring precision are the performance indicators of receiver pulse width measurement capability.

5.1.2.3 Measurement of Time of Pulse Arrival

The time of pulse arrival is the time when the jamming system receiver receives the radar pulse. The measurement of TOA arrival is fundamental to the pulse sorting, pulse repetition interval (PRI), and time difference-based localization. TOA is usually measured according to the leading edge of a pulse.

Pulse repetition interval can be obtained by calculating the TOA difference between every pulse and its neighbor. Radar of early age used fixed PRI, and PRI reflected the features of different radars. Most modern radar exploits multiple sets of PRI. For example, some radar uses stagger PRI, which makes the PRI of radar even more complicated. Analyzing methods also emerged at the same time. PRI is gotten through the analysis of the pulse sequence of radar, and hence, the measurement of PRI is finished after the sorting procedure or at the same time as the analysis.

The indicator for the measurement of TOA is measurement precision. For different applications, the requirements of TOA measurement precision are different. For the systems that do not have the duty of time difference-based localization, they do not require high precision for measuring TOA. For time difference-based localization systems, the TOA measurement precision directly affects the precision of localization. The specific requirement can be calculated by the localization precision decomposition.

5.1.2.4 Measurement of Angle of Pulse Arrival

The angle of pulse arrival indicates the specific direction the pulse comes from, and it is important for the sorting of pulses. What's more, AOA is also used to guide the direction of the jamming beam. The measurement precision of AOA will satisfy the requirement if the precision is better than the half of jamming beam width, while the system delay and influence of dynamic change can be neglected. For the jamming system whose spatial relation with radar greatly varies, there should be enough room for the jamming beam direction error caused by spatial change.

5.1.2.5 Measurement of Pulse Amplitude

Pulse amplitude reflects the signal power received by the receiver. The higher the PA is, the stronger the signal power is and the higher the signal-to-noise ratio (SNR) is. When the system is illuminated by the main lobe of radar, the PA change is mainly related to the distance and is a slow variable. Hence, PA is sometimes used as the evidence of sorting, and used to analyze the beam scan feature of radar.

5.1.2.6 Measurement of Other Parameters

Besides the parameters mentioned above, the measurements of radar location, signal modulation method, and fingerprint feature might be proposed as the requirements of the jamming system receiver too. Certain jamming system requires to be able to measure the location of radar to guide jamming better and support other applications. In order to measure a radar's location, a jamming system needs to use corresponding architecture and processing methods which will be discussed in the following sections. Modern radar wave shape includes linear frequency modulation, nonlinear frequency modulation, and phase encoding, and the modulation parameters are different for the same type of wave shape, which benefits sorting and identification. Hence, some jamming systems require their receivers be capable of measuring the modulation method of the radar signal. Fingerprinting feature is a new technique developed in recent years. The radar individual can be identified by measuring the radar signal's fine features of multiple dimensions.

5.1.3 Sorting Capacity Requirements

There are a great number of electromagnetic signals in space, from aspects of both type and amount. Jamming systems will receive all kinds of signals in the environment which are deeply interleaving and have to separate them. This demands the signal sorting capability from receivers [16]. The distribution of the environmental signals is shown in Fig. 5.3, where signals of different frequencies, pulse widths, modulations, and amplitudes are illustrated. Some of the signals are separable in the frequency domain while others are overlapped in the time domain and are not separable. In order to identify the radar signal in the environment and evaluate the threat, signal sorting is necessary.

It is difficult to propose quantitative indicators for the sorting capability, and it would be difficult to examine too even if quantitative indicators were proposed. Intuitively, there are several aspects to be considered for the sorting capability which are

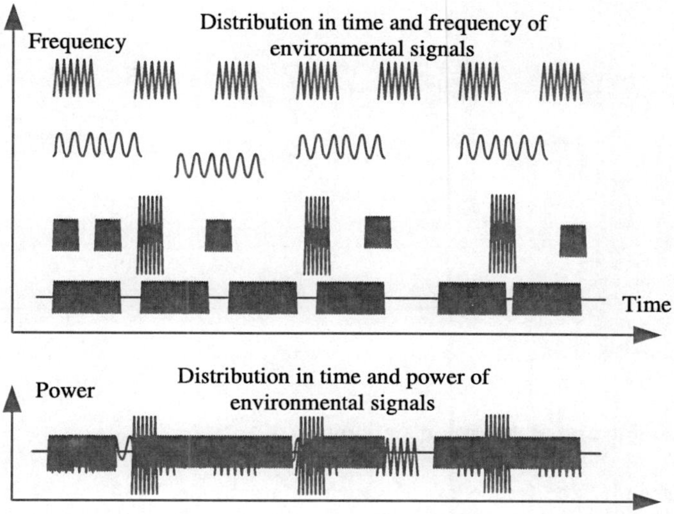

Fig. 5.3 Distribution of environmental signals

environmental adaptation capability, radar signal adaptation capability, and sorting accuracy.

5.1.3.1 Environment Adaption Capability

Signal sorting is to cluster the environmental signals according to their features and then classify the clusters into different radiation sources. Other microwave signals in the environment might affect the sorting processing, causing increasing batch and incorrect classification results. The signal features used in the sorting process include amplitude, DOA, carrier frequency, pulse width, intra-pulse modulation, TOA, etc. It will be easy to make mistakes if the environmental microwave signals are similar to the radar signal from the aspects mentioned above.

Besides, the existence of environmental signals might lead to errors in the procedure of signal detection. For instance, pulse width measurement error, frequency measurement error, pulse loss, TOA measurement error, and modulation measurement error. As shown in Fig. 5.4, the erroneous measurement might make the sorting more difficult or lead to sorting errors.

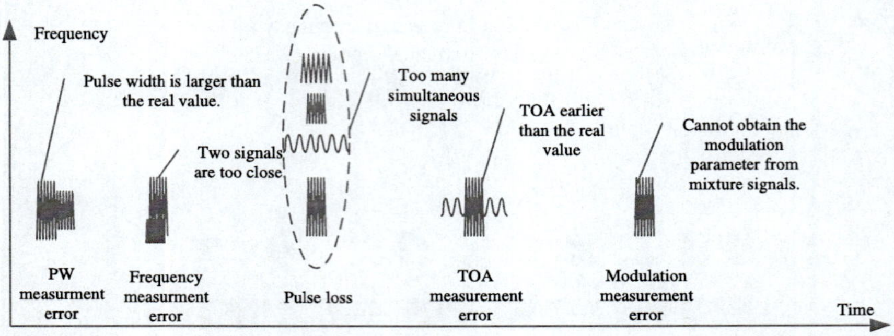

Fig. 5.4 Illustration of parameter measurement mistakes caused by environmental signals

5.1.3.2 Radar Signal Adaption Capability

For a radar with fixed frequency, pulse width, amplitude, DOA, modulation, PRI, and polarization, the sorting of its signals is simple and effective. It only needs to set the filter bank and classify signals according to the parameters.

However, most modern radar has a changeable frequency, pulse width, modulation, PRI, and even changeable amplitude. It requires different adaption capability for different radar jamming systems since different radar has different ways of change.

Figure 5.5 gives the illustration of two typical ways of radar frequency change. Batch-to-batch frequency agility refers to the capability of radar to change its transmitting carrier frequency on a batch-to-batch basis, in which the frequencies are the same within a batch and the PRI are usually the same as well. From the perspective of reconnaissance, it is easy to classify within one batch according to pulse width, frequency, and PRI, while the classification across batches needs other information like angle, position, or known radar knowledge. Pulse-to-pulse frequency agility implies that the pulse frequencies are different for two neighboring pulses. If the frequency agility range is large, it is difficult to cluster according to the frequency information.

Figure 5.6 gives the illustration of several typical PRI variations. Pulse-to-pulse PRI agility means the PRI of radar changes irregularly and can be treated as a random number from the aspect of reconnaissance. PRI stagger indicates that the PRI of radar alternatively changes within several fixed values, whose sum is a constant. That is, there is a fixed overall period. Batch-to-batch PRI agility means the radar signals come in batches, and the PRI value is fixed within one batch. Bursting pulse train refers to the radar pulses radiated by batches. Each batch has a fixed PRI value, but there is a long time duration between two batches. The received phased array radar signals received usually show this feature.

In order to adapt to the variation of frequency, pulse width, and PRI, the receiver has to adopt customized sorting algorithms and needs to integrate several algorithms to adapt to multiple changes.

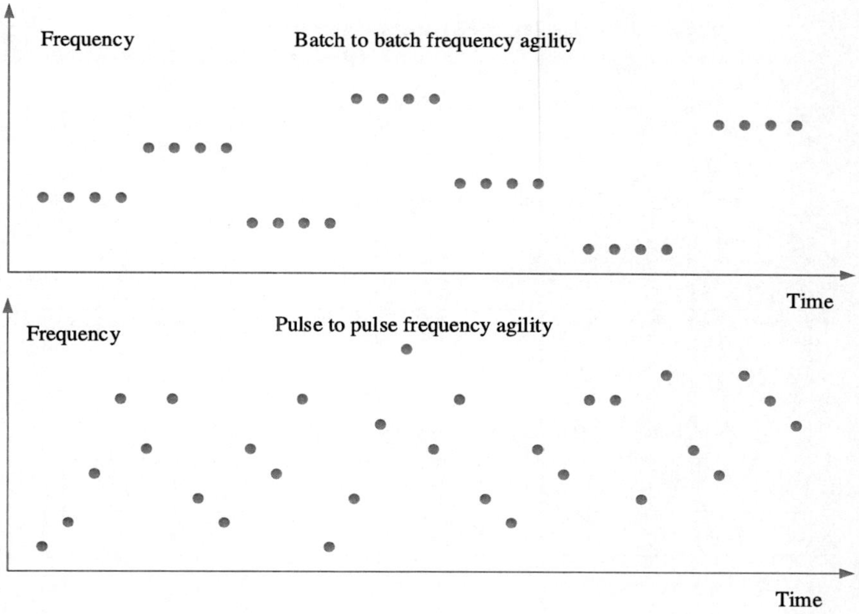

Fig. 5.5 Illustration of the change of radar pulse frequency

However, most modern radar has a changeable frequency, pulse width, modulation, PRI, and even changeable amplitude. It requires different adaption capability for different radar jamming systems since different radar has different ways of change.

5.1.3.3 Sorting Accuracy

Sorting accuracy refers to the ratio between the number of accurately sorted PDW and the overall PDW number. Since the performance of sorting is affected by the environmental complexity, radar signal type, radar number, and the distribution of radars, the requirement of sorting accuracy is for a specific electromagnetic environment and radar signal type. The clearer the electromagnetic environment and radar signal type description is, the easier it is to examine the sorting accuracy. However, it might not be generally applicable while it is strongly pertinence.

There are a number of sorting algorithms now, such as dynamic association, histogram-based sorting, cluster-based sorting, and support vector machine-based sorting. The adaption features of the algorithms are different, and therefore, the electromagnetic environment and target radar type should be clearly described when analyzing the requirement of sorting requirement.

Increasing batch and missing batch are common mistakes in sorting results. Increasing batch means two or more targets are outputted while sorting the signal

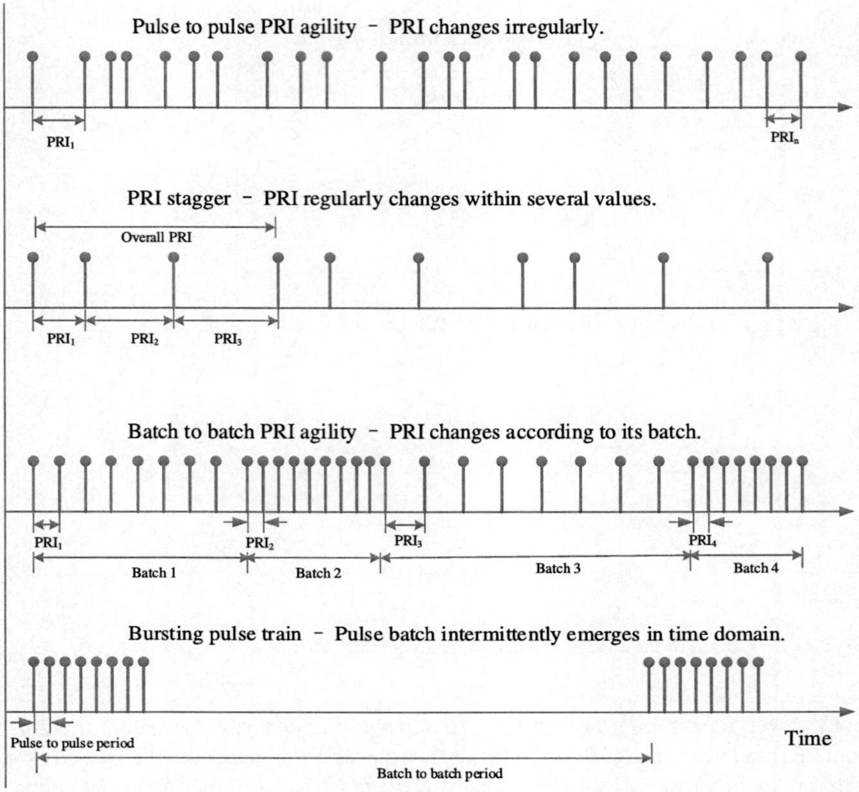

Fig. 5.6 Illustration of the typical PRI variations

from a single radar. Increasing batch usually happens when processing agile radar signals when the sorting algorithm cannot merge the signals before and after the agility to one batch. Missing batch means the algorithm cannot successfully cluster the radar signal and this radar is missing from the sorting result.

Increasing batch makes the radar number after processing greater than the real radar amount which makes the jamming decision more difficult and degrades the jamming effect. Missing batch reduces the capability of threat recognition and might bring recognition delay or mistake which affects the performance of jamming.

5.1.4 Recognition Capability Requirements

Threat recognition is the precondition of performing jamming. A jammer emits jamming signals only when it finds the threat radar. Hence, the requirement of recognition capability is an important factor that should be considered by a receiving system. Threat recognition capability is described from the aspects of recognition processing time, the feature items of recognition result, recognition number, and recognition accuracy rate.

5.1.4.1 Recognition Processing Time

Recognition processing time is the time consumed on finishing all the computation for identifying the threat radar according to the sorting result. Since the radar-guided weapon systems can launch their attack very fast, it requires the radar jamming system to perform effective jamming before the end of the attack. Threat recognition is a key step in launching jamming, and therefore, the radar jamming systems have a high standard for the recognition processing time. The exact demand of processing time is related to the jamming system. For example, an airborne self-defense system requires the recognition to be finished within hundreds of microseconds, and for support jamming systems, the required recognition time could be longer and can even be finished manually.

5.1.4.2 Feature Items of Recognition Result

Information describing the threat would be given if the receiving system finished the threat identification. The information above might contain radar name, radar type, carrier type, related weapon, threat level, credibility, and so on. The exact information that should be outputted is determined by the system designers. In general, the threat description would be more comprehensive for the radar with more priori information.

5.1.4.3 Recognition Number

The recognition of threat radar relies on the priori information database. The size of the database and recognition processing capability determine the maximal number of threat radars that can be identified. The size of the database can be very large by the design. The priori information is the determining factor of recognition amount, and it requires the constant improvement of the database. From the aspect of requirement, the recognition number should be determined based on the threat kind and amount in the worst environment.

5.1.4.4 Recognition Accuracy Rate

The recognition accuracy rate is to measure how accurate the recognition result is and is defined as the ratio of the radar number of correctly identified and the overall identified radar number. For example, there are totally 100 targets to be identified. If 90 of the targets were correctly identified, the recognition accuracy rate would be 90%, or say, the recognition credibility was 90%. Accurate threat recognition is the premise of performing correct jamming. Inaccurate recognition would lead to the omission of threat radar or jam radars of our side. Hence, the receiver's processor should try its best to increase the target recognition rate.

5.2 Typical Receiver

EW receiver developed from analog receiver to digital receiver. Types of the receiver include crystal video receiver (CVR), instantaneous frequency measurement receiver (IFMR), superheterodyne receiver (SHR), channelized receiver, conversion receiver, digital receiver, and so on. Receivers of different types have their own features and applicable situations, which will be introduced in this subsection [17]. Conversion receiver exploits Bragg cell or microscan compression technique to detect radar signal, which is equivalent to Fourier transform. However, this kind of receiver is already outdated due to its poor platform adaption. Hence, the conversion receiver will no longer be further introduced in this book.

5.2.1 *Crystal Video Receiver*

Crystal video receiver was the most widely used previous generation EW receiver. It uses video detection for radar signal and obtains the information about radar signal envelope. CVR consists of prefilter, pre-amplifier, video detector, video logarithmic amplifier, and processor, as shown in Fig. 5.7.

The function of prefilter is to filter the microwave signals outside the concerned frequency band. Especially in complex electromagnetic environment, the prefilter can effectively filter the strong interference signals, such as cell phone base station

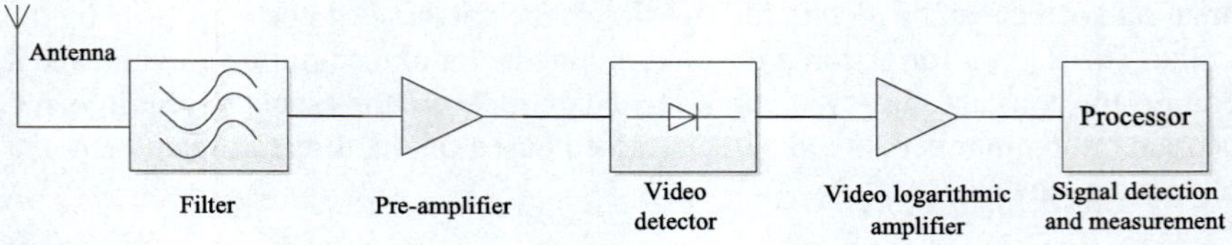

Fig. 5.7 Schematic diagram of CVR

signal. Some receiving would add a limiter in order to enhance the anti-burn capacity against alien electromagnetic signals.

The function of pre-amplifier is to reduce the noise and amplify the signal. It is close to the receiving antenna to reduce the signal loss before amplification since the loss before the first amplification would directly affect the noise factor of the receiver and the sensitivity of the receiver.

The function of video amplifier is to extract the envelope of the RF signal and form the video signal of the envelope. The sensitivity of CVR is largely determined by the sensitivity of video detection and its sensitivity is low. The dynamic range of traditional detection is not large. In order to enlarge the dynamic range, the newly exploited Successive Detection Logarithmic Video Amplifier (SDLVA) extends its dynamic range through cascading amplitude-limited amplification detection.

The features of CVR contain large instantaneous bandwidth, low sensitivity, and large dynamic range. Radar PRI of early ages had audio band, and it is able to identify or confirm the radiation source utilizing its PRI. Today's CRV, combined with instantaneous frequency measurement technique, is able to measure pulse amplitude, pulse width, frequency, direction of arrival, and time of arrival.

5.2.2 *Instantaneous Frequency-Measuring Receiver*

Instantaneous frequency measurement receiver measures the phases of the original and the delayed signals and calculates the radar's frequency according to the difference between the two phases. The principle of instantaneous frequency measurement receiver is given in Fig. 5.8.

The transmission line difference L is the key point of obtaining the phase information. For a fixed frequency f, the wavelength difference after transmitting along the cables is $\frac{L}{\lambda} = \frac{Lf}{c}$, and the caused phase difference is

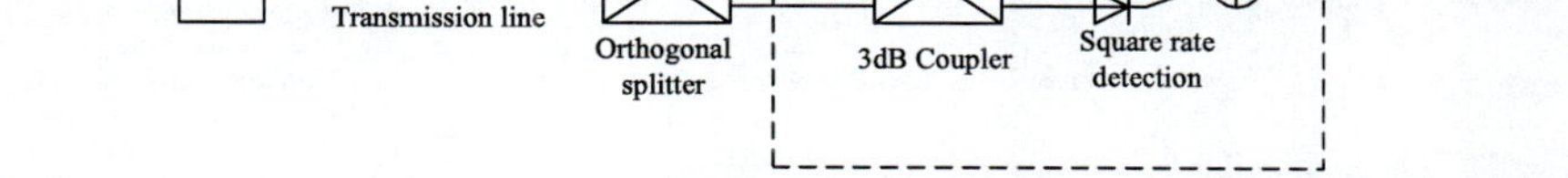

Fig. 5.8 Schematic diagram of instantaneous frequency measurement receiver

$$\Delta\varphi = \frac{2\pi L}{c} f. \tag{5.2}$$

In theory, this value is in proportion to the frequency. Through the phase detector mentioned above, it is able to obtain the information of sin θ and cos θ. The frequency information is mapped by the phase and can be measured based on a reasonable encoding circuit design.

Instantaneous frequency measurement receiver usually has large instantaneous bandwidth, small dynamic range, and lower sensitivity. Besides, the frequency measurement precision is not high for this kind of receiver, which is about 2–10 MHz and is inadaptable to multiple signals situations.

5.2.3 Superheterodyne Receiver

Superheterodyne receiver mixes the local oscillator produced signal and the input signal to transform the frequency of the input signal to a predefined range, detects the signal and measures the parameters, as shown in Figs. 5.9 and 5.10. Superheterodyne receiver processes the signal within a related narrowband frequency range and realizes wide band scan by changing the local oscillator's frequency.

The input signal first goes through a prefilter amplifier after being received by the antenna. This procedure is to filter out the signals outside the working band and reduce the potential intermodulation during the frequency changing period followed up. In order to better restrain the out-band signal, some prefilters adopt a series of switched filters to adapt to different frequency ranges. The filtered signal is transformed into intermediate frequency. The frequency of the local oscillator determines the frequency range selected. For most receivers, AD sample, signal detection, and parameter measurement are performed after frequency mix, and video detection is also used by some receivers.

Superheterodyne receiver is good at frequency selection and has high sensitivity, which makes it the most commonly used receiver.

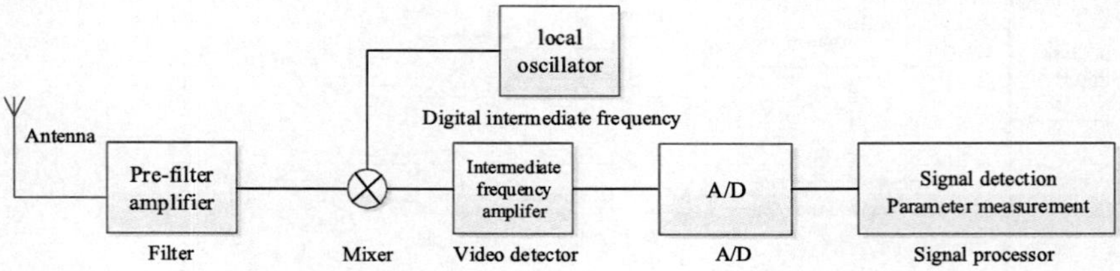

Fig. 5.9 Schematic diagram of superheterodyne receiver

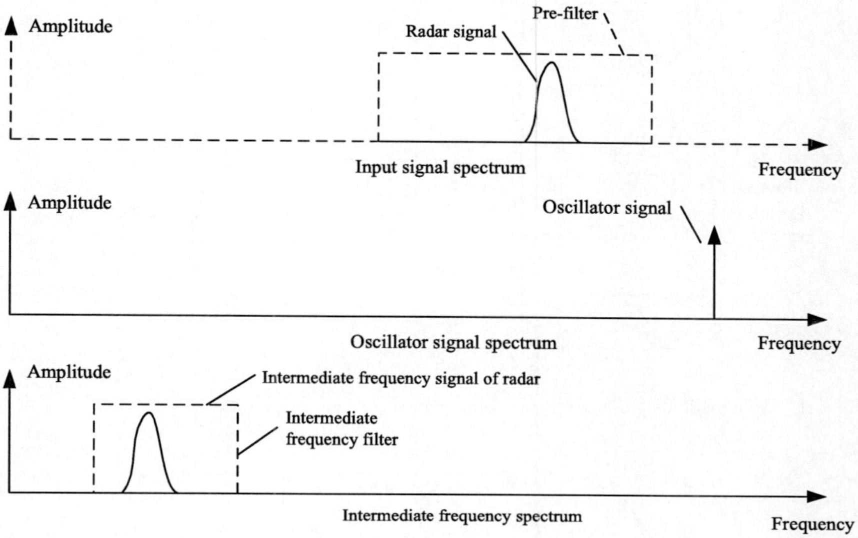

Fig. 5.10 Illustration of the frequency change of superheterodyne receiver

5.2.4 Channelized Receiver

Channelized receiver exploits filter banks to divide signals of different frequencies and finishes signal detection and parameter measurement inside each filter. The principle of channelized receiver is shown in Fig. 5.11. The received signal is divided into multiple signals with equal amplitude after going through a power divider, and the divided signals enter different filters in order to realize frequency division. Finally, the pulse description word (PDW) can be formed by performing signal detection and parameter measurement for each divided signal.

The most significant advantage of channelized receiver is the high sensitivity obtained under large bandwidth scenario. Since the parameter measurement is performed within narrow bandwidth in which the thermal noise base is lower, higher signal-to-noise ratio (SNR) can be achieved under the same conditions.

The method of channelized was realized by analog filter at the very beginning. As the development of digital techniques, channelized receiving process is also realized in digital methods. Analog channelized receivers exploit dielectric filter banks, electromodulated filter banks, surface acoustic wave filter banks, etc. The disadvantage of this kind of filter is the large volume, bad consistency, shift in high and low temperature, and poor restrain capability outside the band. In the digital era, as the rise of DSP and FPGA performance, channelized receiver is usually realized by digital filter banks, and this kind of receiver is named digital channelized receiver.

Digital filters have excellent channel consistency, and the order can be extremely large. However, if each bank utilized an individual digital filter, it would require lots

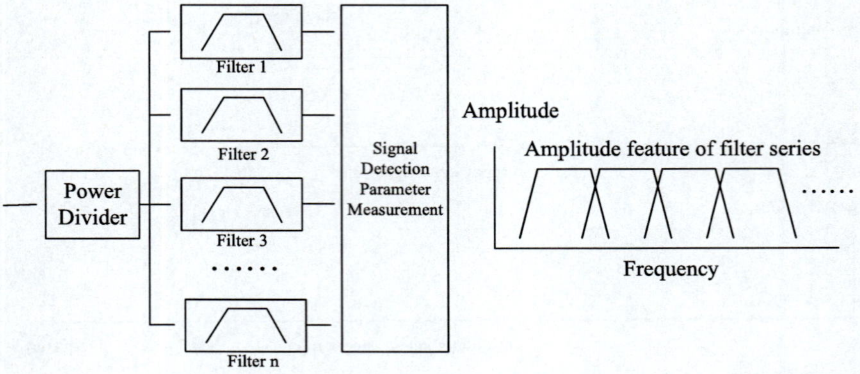

Fig. 5.11 Schematic diagram of channelized receiver

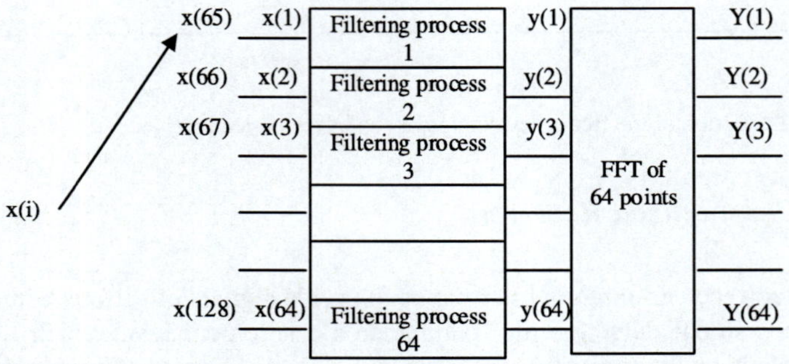

Fig. 5.12 Principle of channelized filtering

of logical resources to realize channelized receiver since a large number of filters need to be used each of which has good restrain capability outside its working bank. In engineering, the multiple-phase filter method is used. In each bank, filtering is performed after data sampling, and FFT is finally calculated. In this way, the order of each filter bank is small, but the equivalent filter order is the sum of each filter bank's order. Hence, the requirement of channelized filter can be achieved with fewer logical resources. Channelized filter with 64 channels is illustrated in Fig. 5.12.

5.2.5 *Digital Receiver*

Digital receiver utilizes sampling and retaining circuit to digitize radar signal, then detects signal, and finally measures parameter with digital signal process techniques.

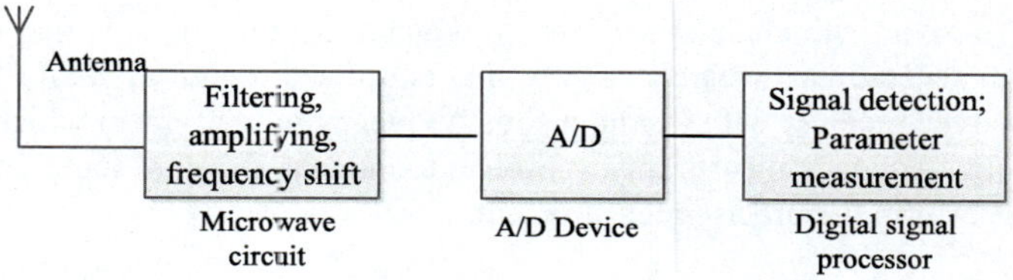

Fig. 5.13 Schematic diagram of digital receiver

The components of a typical digital receiver are shown in Fig. 5.13, which contain microwave circuit, A/D device, and digital processor. For signals of lower frequency (under 1 GHz), direct sampling can be used, and therefore, microwave circuit only needs to filter and amplify the signal. For high-frequency signal (above 1 GHz), it requires transforming the signal to the baseband or intermediate band and performing sampling. In this case, the microwave circuit includes signal frequency shift.

The signal processing flow of a digital receiver is shown in Fig. 5.14. First, the baseband complex signal is formed by transforming the real signal of intermediate frequency. Next, fast Fourier transformation (FFT) is performed to obtain the frequency expression of the signal. Finally, signal detection and measurement in the frequency domain are executed. The processing time length of FFT determines the frequency resolution and frequency measurement precision as well as the precision of TOA. The longer the FFT processing time is, the higher the frequency resolution and frequency measurement precision are, while the TOA and pulse width measure precision would be worse.

Digital receiver has high sensitivity, good parameter measurement precision, and strong multiple signal adaption capability and is widely used in radar reconnaissance receivers. Since the processing bandwidth cannot satisfy the requirement of wide frequency reconnaissance of 10 GHz level, it is usually combined used with other receiving systems to achieve wideband, high sensitivity, and precise measurement. Commonly used combinations include: (1) digital receiver with superheterodyne system which realizes wide band frequency coverage by particular search strategy;

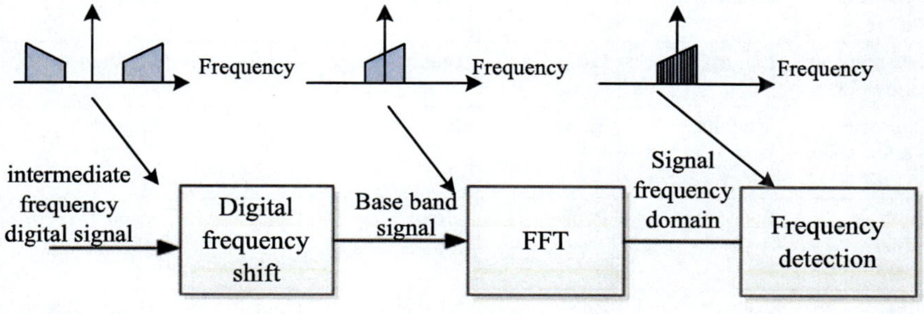

Fig. 5.14 Signal processing diagram of digital receiver

(2) digital receiver with channelized system which the input signal is first channelized to form several narrowbands each of that is utilized digital system; (3) digital superheterodyne receiver with instantaneous frequency measurement which utilizes instantaneous frequency measurement receiver to guide and digital superheterodyne receiver to perform the precise measurement.

5.2.6 Performance Comparison of Receivers

The performance comparison of several regular receivers is given in Table 5.1.

Since the detection bandwidth of crystal video receiver is large and its noise base is also high, its sensitivity is low. For direct detection receivers, their dynamic ranges are not large; for crystal video receivers with SDLVA, their dynamic ranges are large. Since crystal video receiver exploits direct analog detection, it has poor multiple signal capability. Crystal video receiver's cost and power are low because its processor has a simple structure. Hence, crystal video receiver is suitable for scenarios where strong power radars exist and the electromagnetic environment is not complex. In early radar countermeasure systems, crystal video receivers were widely used since the environment was not complex and the techniques were limited. Besides, crystal video receiver is also used in self-protection jamming systems since this kind of system only needs to receive the signal from the main lobe of radar which has strong power.

Instantaneous frequency measurement receiver has large detection bandwidth, and as a result, its sensitivity is low. IFMR performs amplitude limit before delayed phase detection, and therefore, its dynamic range is small. Since IFMR utilizes phase-based detection, it is naturally maladaptive to multiple signals. Its cost and power are

Table 5.1 Performance comparison among EW mainstream receivers

Index	CVR	IFMR	SHR	Channelized	Digital
Sensitivity	Medium low	Medium low	High	High	High
Dynamic Range	Medium Large	small	Large	Large	Large
Instantaneous bandwidth	Large	Large	Medium large	Large	Medium small
Parameter measurement precision	Medium	Medium	High	High	High
Multiple signal capability	Poor	Poor	Good	Good	Good
Cost	Low	Low	Medium	High	Medium high
Power	Low	Low	Medium	High	Medium high

low due to its simple processing. For the features mentioned above, IFMR is suitable for self-protection systems where the weight, volume, and power cost are strictly limited.

Superheterodyne receiver has small detection bandwidth and high sensitivity. Its detection precision is high and has strong multiple signal processing capability. Superheterodyne receivers are widely used in radar countermeasure systems, especially in repeater jamming systems whose instantaneous bandwidth is small.

Channelized receiver has high sensitivity and measurement precision since its detection bandwidth is small. It achieves large instantaneous bandwidth by combining multiple detection channels. Since the number of devices is large, this kind of receiver has large volume and cost. Channelized receivers fit the systems that do not have high requirements in volume, weight, and cost, e.g., special-purpose EW aircraft, special-purpose ground jamming stations, and shipborne jamming systems.

Digital receiver has high sensitivity, good measurement precision, large dynamic range, and strong multiple signal adaption capability. However, its instantaneous bandwidth only can achieve up to GHz level due to its AD and digital processing devices. Digital receivers need high-end AD devices and digital signal processors, and therefore cost and power are both high.

5.3 Receiver Performance Design

Receiver sensitivity, dynamic range, instantaneous bandwidth, and interception probability are the key indexes, which will be discussed in this subsection about how to design them.

5.3.1 Sensitivity Design

Receiver sensitivity is determined by the detection bandwidth, noise factor, and the minimal SNR (detection factor) necessary to extract information from the received signal.

EW engineers usually estimate the sensitivity of a receiver by the following formula:

$$P_{\min} = -114 + 10 \log B + \mathrm{NF} + M \tag{5.3}$$

The unit of the result calculated above is dBm, while -114 is a parameter calculated by Boltzmann constant k, and working temperature $T = 290$ K, whose unit is dBm/MHz.

In the formula, B is the bandwidth of the receiver, whose unit is MHz. NF is the noise factor of receiver. M is detection factor, that is, the minimal SNR when the receiver works properly. The detection factor is related to the receiver's detection method and can take 15 dB under certain conditions.

Assume there is a receiver whose bandwidth $B = 100$ MHz, and noise factor NF $= 8$ dB. The sensitivity of this receiver can be calculated by the above formula:

$$P_{\min} = -114 + 10 \times \log(100) + 8 + 12 = -74\ \text{dBm}$$

(1) Noise factor

Noise factor is the value that expresses the magnitude of the noise inside the receiver. Noise is the source reason that constrains receiver sensitivity. Noise factor is defined as the ratio between the SNR of the output side of the receiver's linear section and the SNR of the input side.

$$\text{NF} = \text{SNR}_o/\text{SNR}_i \tag{5.4}$$

where SNR_o is the output SNR, and SNR_i is the input SNR. Receivers usually exploit multistage amplifier, frequency mix and filter, and the general noise factor formula can be given by

$$\text{NF} = F_1 + \frac{F_2 - 1}{G_1} + \frac{F_3 - 1}{G_1 G_2} + \ldots + \frac{F_n - 1}{G_1 G_2 \ldots G_{n-1}} \tag{5.5}$$

where $F_i (i = 1 \sim n)$ is the noise factors of each stage, and $G_i (i = 1 \sim n)$ is the signal gains of each stage.

According to the formula, the noise factors and gains of the preceding stages contribute more to the overall noise factor. In order to reduce the total noise factor, the following reasons should be considered while the designing procedure:

(a) Since the first stage noise factor directly increases the receiver noise factor, it should be reduced by all means. The insert loss before the first stage amplifier should be minimized and the device number before the first stage amplifier should be decreased as well. Try to shorten the distance between the antenna and the first stage. If necessary, put the pre-amplifier directly behind the antenna, separating from the main body of the receiver.

(b) The noise factors of the back stages will be reduced by the gains of the front stages. The contribution to the whole receiver's noise factor of the back stages would be amplified if the front stage gain is smaller than the noise factor of the back stage. Hence, designers should try their best to make sure the gain of the front stage is greater than the noise factor of the back stage.

(2) Detection bandwidth

Detection bandwidth is the noise bandwidth when the receiver is detecting radar signals. It is different from the instantaneous bandwidth of receiver. The following example will demonstrate the difference between them.

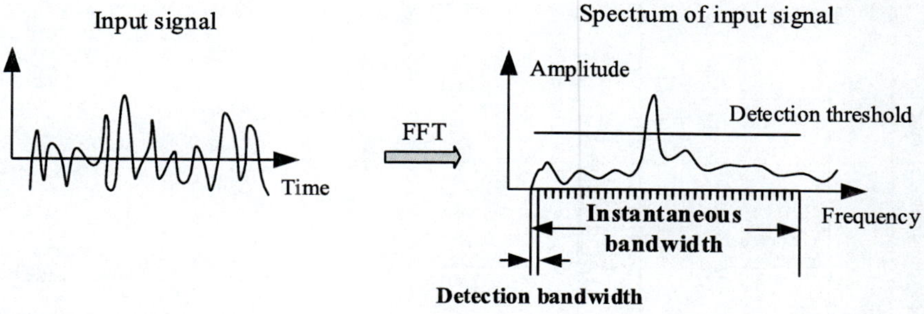

Fig. 5.15 Relationship between digital receiver's detection bandwidth and instantaneous bandwidth

For a digital receiver whose instantaneous bandwidth is 1000 MHz and overall noise bandwidth is 1000 MHz. There are two potential processing methods for the digital receiver. The first one is to perform FFT, such as 128 points FFT with each frequency unit as 7.8125 MHz. In this case, the noise base of each frequency resolution cell should be the sum of all the noise within the range of 7.8125 MHz, if frequency domain target detection is performed. Hence, the detection bandwidth should be 7.8125 MHz other than 1000 MHz, as shown in Fig. 5.15.

If digital channelized technique was exploited by the receiver, as shown in Fig. 5.16, the overall bandwidth 1000 MHz would be divided into 128 channels, each bandwidth of which was 7.8125 MHz. The time domain signal detection would be performed in every channel, and the detection bandwidth should be 7.8125 MHz since the noise bandwidth of each channel was 7.8125 MHz.

(1) Detection factor

Detection factor is designed according to the requirements of the detection probability and the false alarm rate and is determined based on a given false alarm rate in general. The detection theory can be described by the following equations:

$$\begin{aligned} &H_0 : x[n] = t[n] + b[n] \quad n = 0, 1, \ldots, N-1 \\ &H_1 : x[n] = b[n] \qquad\qquad n = 0, 1, \ldots, N-1 \end{aligned} \tag{5.6}$$

where H_0 denotes the sum of radar signal and noise, H_1 denotes the background noise, $x[n]$ is the data for detection, $t[n]$ is radar signal, and $b[n]$ is noise. The target detection task in fact is a binary test problem (Fig. 5.17).

Assume the noise followed Gaussian distribution, the detection factor under a given false alarm rate could be derived based on maximum likelihood detection method. In general, let the detection factor be 15 dB is able to meet the requirement.

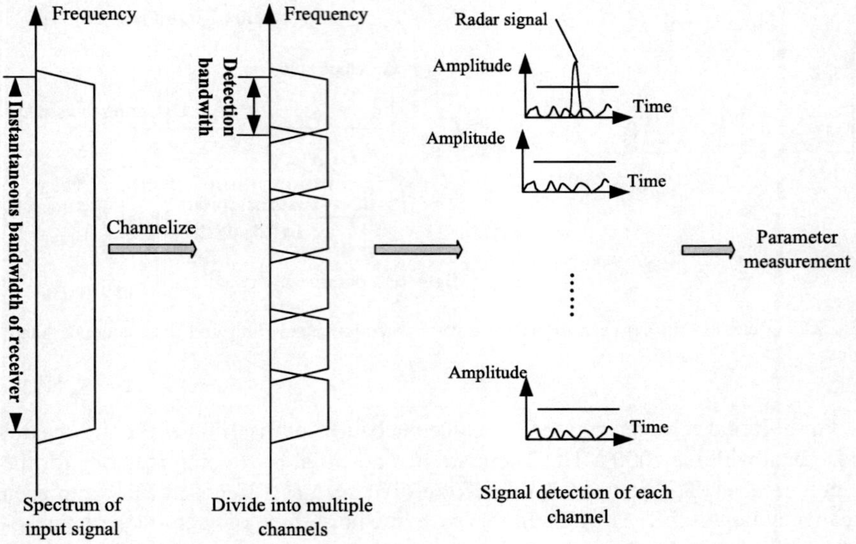

Fig. 5.16 Relationship between channelized receiver's detection bandwidth and instantaneous bandwidth

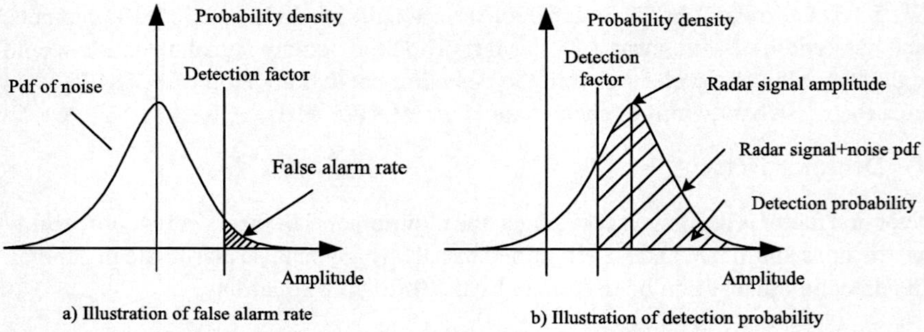

Fig. 5.17 Relationship between detection probability and false alarm rate

5.3.2 *Instantaneous Dynamic Range Design*

Instantaneous dynamic range is the ratio between the minimal radar signal power that can be properly received (sensitivity) and the maximal power and is usually denoted by decibel. For example, if the minimal power for the properly receiving signal of a receiver was −70 dBm, and the maximal power was −30 dBm, the dynamic range of this receiver would be 40 dB.

For digital receivers, their dynamic range is affected by the dynamic ranges of microwave channel and ADC. The dynamic range of microwave channels should be carefully designed based on the requirement of the receiver's dynamic range. The dynamic range of ADC is determined by the level of its devices, which should be selected according to the requirement of dynamic range. The following factors should be considered to design a receiver's dynamic range.

(1) The gain of a receiver's microwave channel is determined when the receiver's sensitivity and ADC devices are given. The gain of a microwave channel is equal to the ratio between the minimal input signal power of the ADC device and the receiver's sensitivity. As mentioned above, in order to decrease the noise factor of the receiver, the front gain should be as large as possible. However, constrained by the maximal input signal level of the mixer, the dynamic range would be reduced since too large a front gain makes the input signal exceed the maximum of the receiver which causes the receiver out of work. Hence, the requirements of the receiver noise factor and dynamic range should be taken into consideration while designing to assign channel gain more properly.
(2) Design or use devices with a large dynamic range.
(3) Exploit logarithmic amplifier to enlarge dynamic range. A logarithmic amplifier transforms linear power into logarithmic power, which can increase the dynamic range of a receiver significantly.
(4) Use devices with large bit number and large maximal power while choosing ADC devices.
(5) When the distance between the antenna and the receiver is far, the signal loss difference is also large for different frequencies, especially for EW receivers with large working bandwidth. For example, between 2 and 18 GHz, the loss difference would several decibels for 10 m. Hence, it is able to conclude frequency separated transmission and compensator.
(6) It is better to leave enough margin while designing the dynamic ranges of microwave channel and ADC, e.g., a margin of 5–10 dB, to satisfy the dynamic range requirement of the whole receiver.

5.3.3 Instantaneous Bandwidth Design

The instantaneous bandwidth requirement of a radar jamming system's receiver usually achieves hundreds of MHz or even several GHz. In order to realize broadband receiving, the following methods could be used.

(1) Exploit an instantaneous frequency measurement receiver to directly achieve large instantaneous bandwidth. Since instantaneous frequency measurement receiver does not fit multiple signals, and its sensitivity is not high. Other receivers should be taken into consideration if the two indexes above cannot meet the requirement.
(2) Use a digital receiver. Owing to the development of ADC devices and digital processing devices, the digital receiver's instantaneous bandwidth is able to achieve 1 GHz or higher.

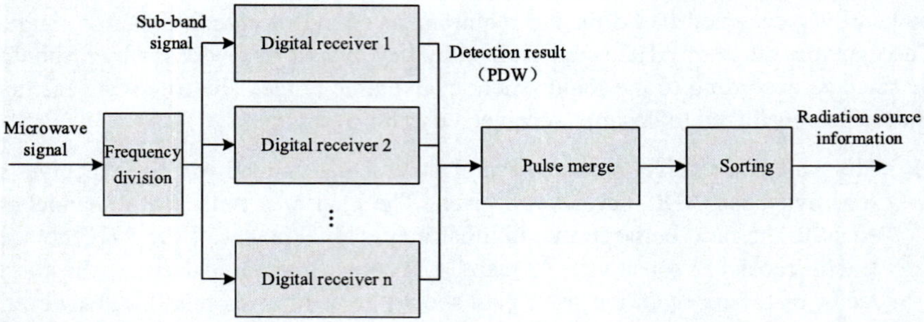

Fig. 5.18 Relationship between detection probability and false alarm rate

(3) If a digital receiver still failed to satisfy the requirement, channelized receiver could be considered. Channelized receiver divides the signal into several subchannels, each of which will process the signal within its own bandwidth. Finally, the processing results of each channel will be merged, as shown in Fig. 5.18.

5.3.4 *Intercept Probability Design*

The interception probability of a receiver for a given radar is defined as the probability that the receiver finds the target within a given time range. Besides the influence of signal detection probability, the discontinuous interception caused by frequency scan, space scan, and time window also should be considered. The interception probability can be given by:

$$P = P_d \cdot P_t \cdot P_f \cdot P_s \tag{5.7}$$

where P is the overall interception probability; P_d is signal detection probability which is related to the amount that the radar signal exceeds the threshold, receiver noise distribution, and noise power; P_t is time window interception probability. If the receiver is in full-time reconnaissance mode, the time window interception probability will be 100% since there is no time window. Since a jamming system usually does not jam while reconnoitering, and does not reconnoiter while jamming, there is a collision probability problem while jamming. The time window interception probability within a specific time range can be calculated according to the reconnaissance time sequence and radar working time sequence. Figure 5.19 depicts the relationship between the time of the radar signal's arrival and receiving time sequence. According to the figure, sometimes the radar signal can be intercepted and sometimes not.

P_f is the interception probability of the frequency domain. If the receiver covered the entire reconnaissance range instantaneously, the interception probability of the

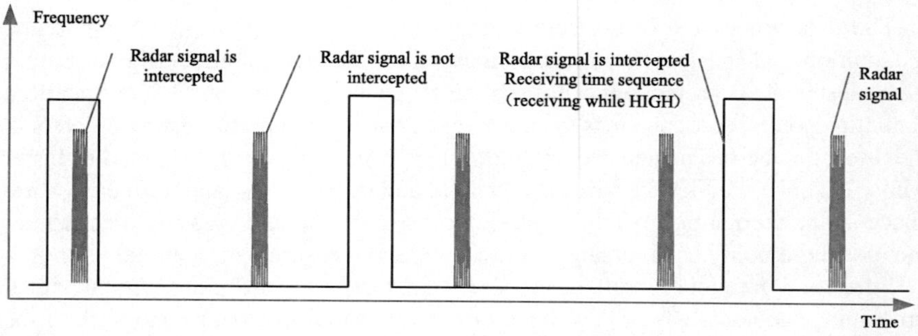

Fig. 5.19 Relationship between radar signal and receiving time sequence

frequency domain would be 100% for the radar signals within the bandwidth. In case the receiver does not cover the whole reconnaissance band, the receiver would apply a superheterodyne-based method to realize full-band interception by a specific receiving frequency switch program.

P_s is the interception probability of the space domain. If the receiver covered the entire spatial range to reconnoiter, the interception probability of the space domain would be 100%. If the receiving system instantaneously receives the signal only within a small angle range and fulfills space search by beam scanning, its interception probability of space domain will be less than 100% in most cases. For the systems that can only receive the signal within the radar's main lobe, this probability will be the possibility that the beams of two radars lock each other.

5.4 Signal and Data Processing Methods

The reception processing of a radar active jamming system's receiver includes signal processing and data processing. Signal processing contains filtering and frequency shift, signal transformation, signal detection, and parameter measurement. Data processing contains signal sorting and radiation source identification. The signal filtering methods of a receiver include analog filter and digital filter, the effect of which is to let the interested signal pass and restrain the signal outside the band. Frequency shift contains analog shift and digital shift. Analog shift usually converts the received signal to the intermediate frequency or baseband, realized by the combination of mixer and filter. Digital shift usually converts the intermediate digital signal sampled by ADC to baseband, realized by digital devices. Signal transformation indicates the transformation of signal from the time domain to frequency or time–frequency domain, realized by digital processing devices. Signal detection is to perform constant false alarm rate (CFAR) detection to fulfill interception. Parameter measurement is to measure the frequency, angle, pulse width, and amplitude of the signal.

The data processing of a receiver includes signal sorting and radiation source recognition and the location measurement of the radiation source. Signal sorting is to cluster the detected signal for the subsequent radiation source identification. Radiation source recognition is to compare the sorting result with the parameters in the threat database. The nearest neighbor rule is commonly used to select the cluster whose distance satisfies the specific threshold and is the nearest one from the feature space as the recognition result. Time difference location and crossing location are the most commonly used radiation source position measurement methods.

Filter and frequency shift and signal transformation are conventional signal processing methods, which have been discussed in multiple books and will not be introduced in this book anymore. In the rest part of this subsection, signal detection, parameter measurement, signal sorting, and radiation source recognition will be introduced in detail.

5.4.1 Signal Detection

Signal detection is to detect whether the radar signal exists based on the predefined or adapted threshold. Time domain detection and frequency domain detection are the most commonly used methods [18].

(1) Time domain detection

Time domain detection performs signal detection based on the time–amplitude relationship of the signal, as shown in Fig. 5.20. The detection procedure is to exact the envelope of the radio frequency or intermediate frequency signal and then compare the signal amplitude with a specific threshold to output the radar signal detection result.

Signal envelope exaction can be realized by detection or SDLVA in the analog domain and can be realized by calculating the absolute value of amplitude and then passing a digital low pass filter in the digital domain.

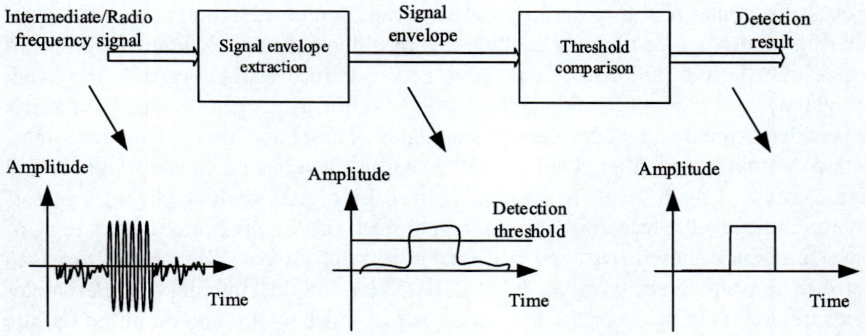

Fig. 5.20 Schematic diagram of time domain signal detection

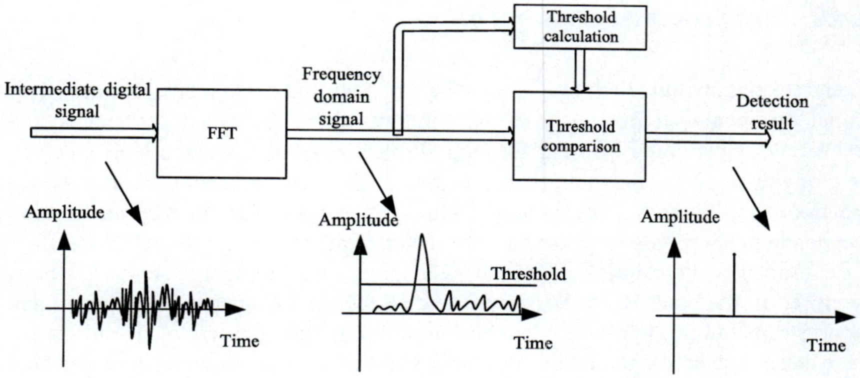

Fig. 5.21 Schematic diagram of frequency domain signal detection

The threshold comparison contains a fixed threshold and a CFAR threshold. The fixed threshold is a predefined value, which keeps constant while the noise base changes. CFAR threshold is calculated by multiplying the threshold factor by the mean value of the reference unit. The value of the threshold factor determines the amount of false alarm rate.

(2) Frequency domain detection

Frequency domain detection is to transform the signal to the frequency domain and perform signal detection in the frequency dimension, as shown in Fig. 5.21. First, perform the time–frequency transformation on the input digital signal, which is usually fulfilled by FFT [19]. Next, calculate the detection threshold based on the threshold factor. Similar to time domain detection, there are fixed thresholds and dynamic thresholds in frequency domain detection. The computation method of threshold is similar with that of time domain detection too [20, 21]. Finally, the detection result is obtained.

Frequency domain detection is to perform the time–frequency transformation on the signal of the time slice. Hence, the longer the time slice is, the higher the frequency resolution is, and the lower the time resolution is. [22] For example, if a time slice of 100 ns was used, the frequency resolution would be 10 MHz, and the time resolution would be 100 ns. Limited by this case, the average error of frequency measurement would be 5 MHz, and the average error of time measurement would be 50 ns in the general case. If a time slice of 1000 ns was used, the frequency resolution would be 1 MHz, and the time resolution would be 1000 ns. The average error of frequency measurement would be 0.5 MHz, and the average error of time measurement would be 500 ns in the general case. If particular frequency and time measurement technique were applied, the measurement error could be improved to about 1/5 of the resolution. Hence, the time slice length is required to be a compromise between the frequency measurement precision and the time measurement precision.

5.4.2 Frequency Measurement

There are two methods for frequency measurement: the first method is to receiver signal by opening a frequency strobe window, whose frequency is equal to the frequency of the signal passing through. Typical examples include YIG receiver, heterodyne receiver, channelized receiver, etc. The second method is to measure the frequency by signal processing, including the measurement by transforming frequency to amplitude or phase, and the measurement by direct digital processing.

The detailed description of various frequency measurement systems has been proposed in the book Radar Reconnaissance Receiver Design by Hu Laizhao, and readers can read this book for related information. In this subsection, we only introduce these measurement methods in short, except the digital frequency measurement which will be discussed in detail.

(1) YIG frequency measurement

YIG filter can be used to perform selective filtering for the signal. The pass band of a YIG filter can be controlled by current driver. The frequency of output signal could be identical to the frequency of the pass band, as shown in Fig. 5.22. The advantage of YIG frequency measurement is simple to realize, and the disadvantages include the frequency measurement precision is constrained by the filter and poor, the adjustment speed of the YIG filter is slow, the interception probability is low and the YIG frequency selections lack stability. Nowadays, YIG frequency measurement has been already out of date.

(2) Heterodyne frequency measurement

Using a mixer, heterodyne frequency measurement transforms the radio frequency signal to the intermediate frequency signal, performs video detection, and outputs the signal frequency which is the radio frequency corresponding to the intermediate frequency. Assume the local oscillator's frequency is f_0, the central frequency of intermediate filters is f_1, and the intermediate filter's bandwidth is B, the input signal's frequency will be

$$f_s = \begin{cases} f_0 + f_1, & \text{for high local oscillator} \\ f_0 - f_1, & \text{for low local oscillator} \end{cases} \tag{5.8}$$

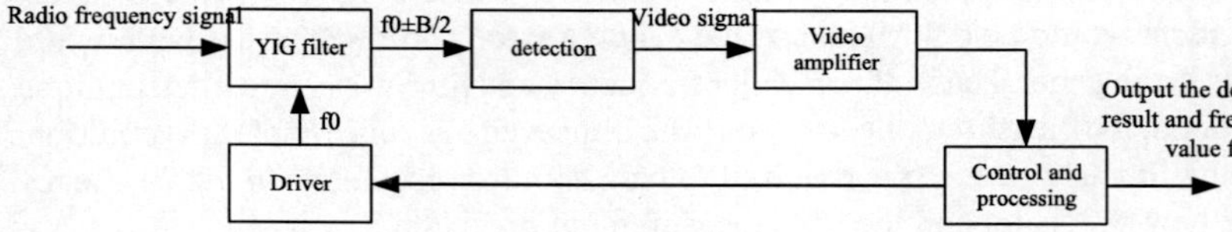

Fig. 5.22 Schematic diagram of YIG frequency measurement

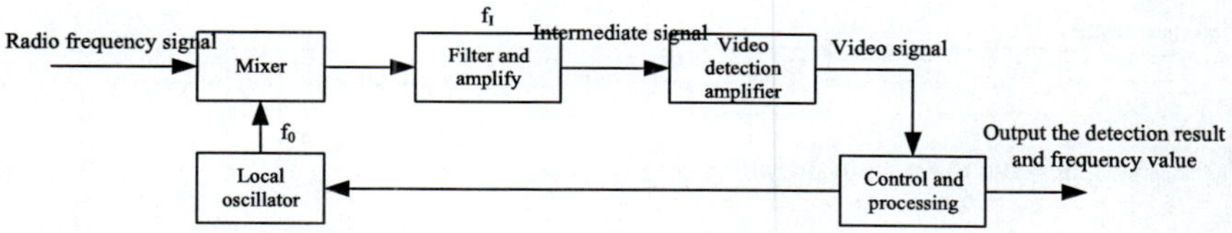

Fig. 5.23 Schematic diagram of heterodyne frequency measurement

The frequency measurement error is *B*/2.

Since heterodyne frequency measurement transforms the signal to the intermediate frequency, the consistency is good, and the performance is more stable than YIG frequency measurement. However, due to the constraint of the filter's pass bandwidth for the measurement precision, the frequency measurement precision is poor. The use of heterodyne frequency measurement is getting less since only one signal could be processed at the same time, the pass bandwidth (i.e., the instantaneous bandwidth) is narrow, and the interception probability is low (Fig. 5.23).

(3) Channelized frequency measurement

Channelized frequency measurement divides the receiving bandwidth into several subfrequency bands. Video detection is performed in every subfrequency band, and the central frequency of each subfrequency band is the detected radar signal's frequency, as shown in Fig. 5.24. Channelized processing method has larger instantaneous coverage than heterodyne frequency measurement. However, the disadvantage is the volume of the device is large. That is, the increase of frequency measurement instantaneous bandwidth is at the cost of adding more devices.

(4) Frequency transforming to amplitude or phase frequency measurement

This method transforms the frequency to amplitude or phase and then measures the amplitude or the phase. Amplitude or phase has a relationship to radar frequency, and the signal frequency can be derived based on the relationship. The receiver measuring frequency by transforming to phase is also called instantaneous frequency measurement receiver , whose working principle can be found in Sect. 5.2.2.

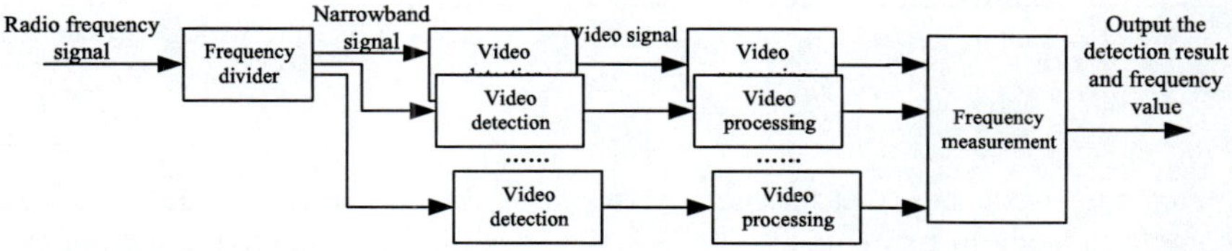

Fig. 5.24 Schematic diagram of channelized frequency measurement

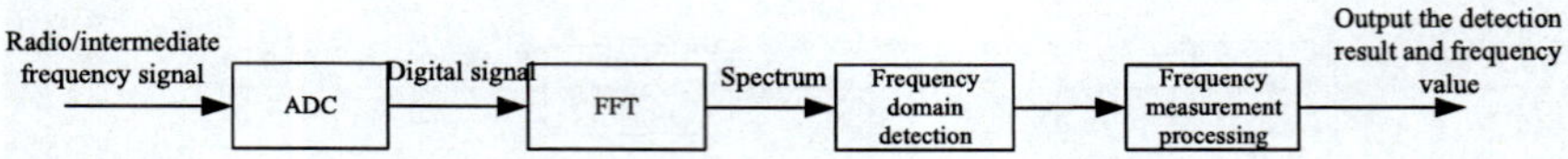

Fig. 5.25 Schematic diagram of digital frequency measurement

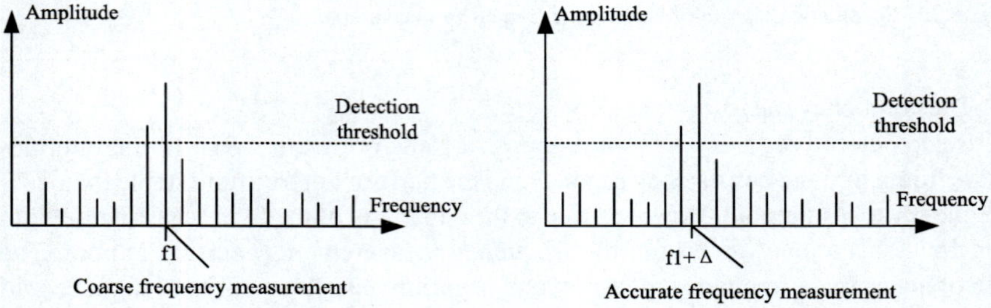

Fig. 5.26 Illustration of the procedure of digital frequency measurement

(5) Digital frequency measurement

Digital frequency measurement is used to perform frequency measurement based on the digital domain signal detection, as shown in Fig. 5.25.

Digital frequency measurement includes accurate frequency measurement and coarse frequency measurement. Coarse frequency measurement is to use the frequency marker, where the signal with maximal detected amplitude dwells, as the frequency of the signal. Accurate frequency measurement is to use the frequency corresponding to the energy center of a number of units where the maximal signal and its neighbors exist, as shown in Fig. 5.26.

Assume the sample rate of digital signal is f_s, the FFT point number is N, and the unit number where the maximal signal value is detected is k, the frequency measured by coarse frequency measurement is:

$$f_c = \frac{k}{N} \cdot f_s \tag{5.9}$$

And the maximal error of coarse frequency measurement is

$$\sigma = \frac{1}{2N} \cdot f_s \tag{5.10}$$

Accurate frequency measurement performs correction based on the result of coarse frequency measurement. First, it calculates the deviation value δ from the energy centroid to the channel where the signal with maximal amplitude exists. Then, it adds the deviation value to the corresponding frequency and gets the accurate frequency value which is:

$$f_j = \frac{k+\delta}{N} \cdot f_s \tag{5.11}$$

The channel deviation can be calculated by

$$\delta = \frac{\sum_{m=-M}^{m=M} \left[m \cdot |X(k+m)|^2 \right]}{\sum_{m=-M}^{m=M} |X(k+m)|^2} \tag{5.12}$$

where $X(k)$ is the amplitude of the k'th frequency unit. M determines the date number required to finish the deviation calculation. In general case, let $M = 1$ can satisfy the requirement, and in this case the equation will be simplified to

$$\delta = \frac{|X(k+1)|^2 + |X(k)|^2 - |X(k-1)|^2}{|X(k+1)|^2 + |X(k)|^2 + |X(k-1)|^2} \tag{5.13}$$

5.4.3 *Time Measurement*

Time measurement includes the measurement of the radar signal's arrival time, pulse width, and pulse repetition interval which is calculated based on the measured arrival time in the following sorting.

For receivers designed for the time domain detection, their measurement principle is shown in Fig. 5.27. The precision of time measurement depends on the steepness of rising and falling edges of the signal processed by the receiver. In general, the steepness of the video signal is determined by the video bandwidth and actually is inversely proportional to the video bandwidth.

For receivers designed for the frequency domain detection, their measurement principle is shown in Fig. 5.28. The arrival time is the central time of the data frame where the signal is first detected, and the pulse width is the time difference between

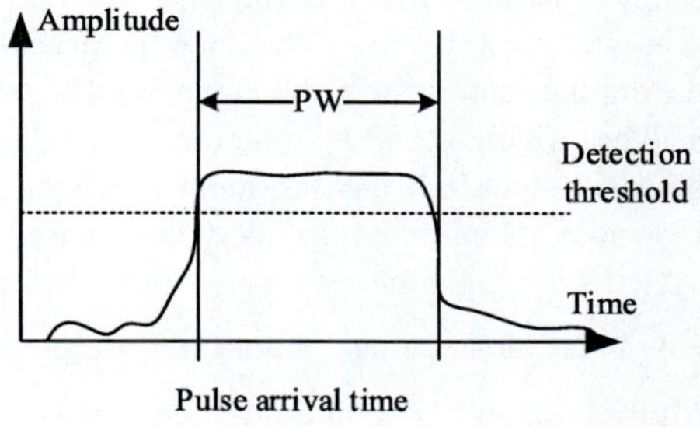

Fig. 5.27 Illustration of the time measurement in time domain

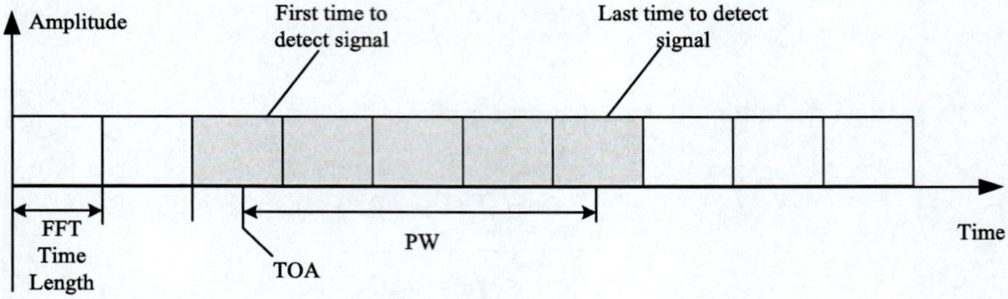

Fig. 5.28 Illustration of the time measurement in frequency domain

the time when the signal is lastly detected and the time when the signal is first detected.

The precision of time measurement depends on the time length of FFT. The longer the data frame is, the poorer the time measurement precision is. For example, if the FFT data's time length was 100 ns, the maximal time measurement error would be 50 ns.

5.4.4 Angle Measurement

Angle measurement is to measure the radar signal's direction of arrival (DOA). Angle measurement has two classes. The first class is the single antenna's maximum-based method, and the second class is the multiple antennas' comparison-based method. Single antenna maximum-based method uses the rotation of a single antenna and compares the received radar signal's amplitudes of different antennas' directions to derive the DOA value. Multiple antennas' comparison-based method is to receiver radar signal with multiple antennas simultaneously, and compare the amplitudes, phases, or the TOAs from the antennas to get the DOA, including amplitude comparison-based direction finding, phase comparison-based direction finding, and time difference-based direction finding.

(1) Single antenna's maximum-based method

Single antenna's maximum-based method is to use an antenna to perform a special scan with uniform rate. The antenna direction when the strongest signal is received during the scan is the radar signal's DOA, as shown in Fig. 5.29.

The principle and calculation of single antenna's maximum-based method are both simple. High measurement precision can be achieved by high antenna gain. However, its interception rate is low due to the antenna scan. As a result, receiving systems of single antenna's maximum-based method are not common nowadays.

(2) Multiple antennas' amplitude comparison-based direction finding

Multiple antennas' amplitude comparison-based method is to receive radar signal with multiple receiving antennas whose beams are overlapping, and the radar signal's

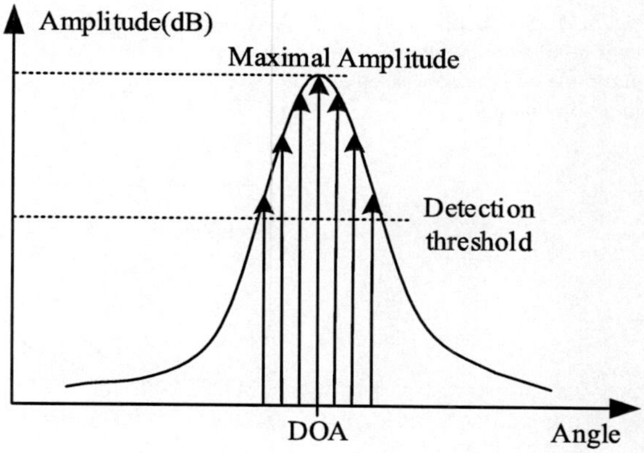

Fig. 5.29 Illustration of Single antenna's maximum-based direction finding

TOA can be derived by the comparison of the received amplitudes. The principle of bi-antenna amplitude comparison-based direction finding is given in Fig. 5.30. Assume the radar signal's amplitude received by antenna A was A1, and the radar signal's amplitude received by antenna B was A2, the radar signal's direction of arrival could be calculated by the following formula:

$$\text{DOA} = \eta \cdot [\arctan(A1/A2) - \pi/4] \tag{5.14}$$

where η is angular factor which is related to the antenna pattern and can be 1.5 in general. The angular factor's value can be adjusted according to actual situation.

The precision of amplitude comparison-based direction finding is related to the shapes of the antenna's beams. If there was enough SNR, the angle measurement precision would be higher within the range where the antenna gain changed fast. Limited by the beam width, bi-antenna amplitude comparison-based direction finding

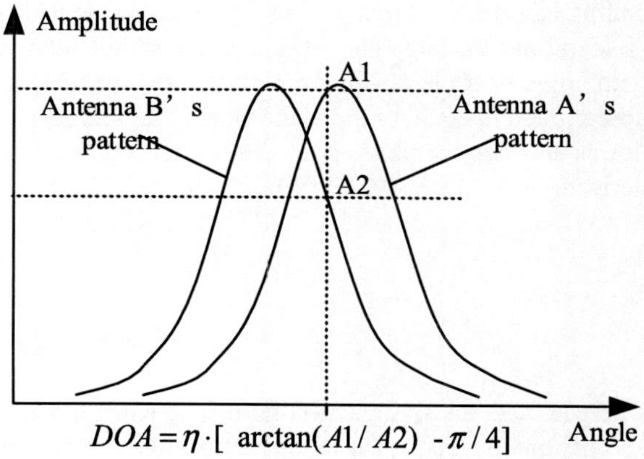

Fig. 5.30 Schematic diagram of bi-antenna amplitude comparison-based direction finding

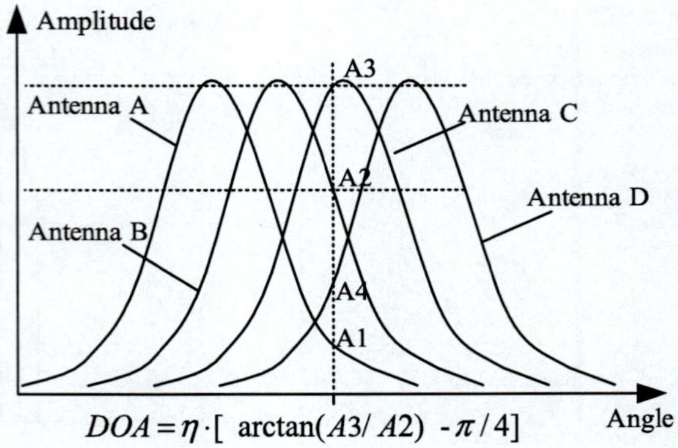

Fig. 5.31 Schematic diagram of four antennas amplitude comparison-based direction finding

is only able to cover a region smaller than its beam width. In order to realize the coverage of 360 degrees, multiple overlapping antennas can be exploited, as shown in Fig. 5.31.

The advantage of multiple antennas' amplitude comparison-based method is the large simultaneous space coverage and the high interception probability. The disadvantage is the low system sensitivity caused by the low antenna gain.

(3) Phase comparison-based direction finding

The radar signals received by different antennas have the phase difference, which is caused by the range difference arriving at different unit antennas, and the phase differences in different directions are not equal. Phase comparison-based direction finding is also called interferometer-based direction finding. It builds the relationship between the phase difference and the DOA and calculates the radar signal's DOA by measuring the phase difference. The antenna arrangement of phase comparison-based direction finding includes linear arrangement, circle arrangement, grid arrangement, non-uniform arrangement, etc.

We take linear arrangement as an example to illustrate the principle of phase comparison-based direction finding for convenience, and Fig. 5.32 depicts the principle of linear arrangement. The incident signal enters from the upper right direction with angle θ. If the far-field condition was considered, the incident wave would be approximately equal to the plane wave. The line perpendicular to the incident direction is equal to the phase line. Under this condition, the following formulas can be derived:

$$\Delta R_n = d_n \sin\theta \tag{5.15}$$

$$\psi_n = 2\pi \cdot \Delta R_n/\lambda = 2\pi \cdot d_n \sin\theta/\lambda \tag{5.16}$$

where ΔR_n is the signal traveling distance difference between the 0'th antenna and another antenna. d_n is the distance from the 0'th antenna to another antenna, that is

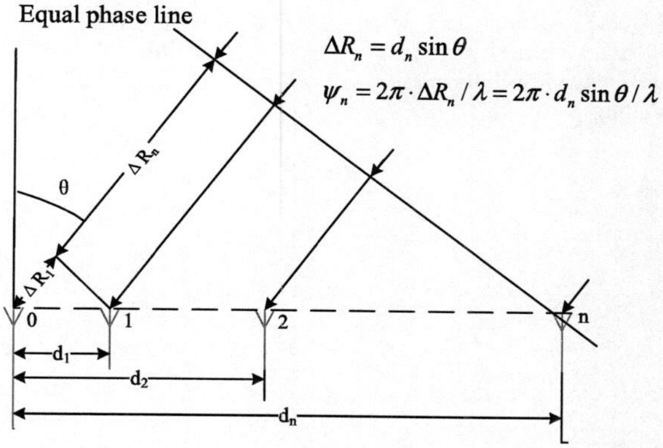

Fig. 5.32 Schematic diagram of phase comparison-based direction finding

the baseline length. ψ_n is the phase difference between the 0'th antenna and another antenna. λ is the wavelength of the electromagnetic wave.

The direction measurement precision can be expressed by

$$\sigma_\theta = \frac{c\sigma_\psi}{2\pi d_i f \cos\theta} = \frac{c}{2\pi d_i f \cos\theta \sqrt{2\text{SNR}}} \tag{5.17}$$

where σ_θ is the root mean square error of direction measurement, σ_ψ is the root mean square error of phase measurement, c is the speed of light, f is the frequency of the electromagnetic wave, and SNR is signal-to-noise ratio.

According to the direction measurement precision formula, the direction measurement precision is related to SNR. The higher the SNR is, the smaller the direction measurement error is. The baseline distance also affects the direction measurement precision. The longer the baseline distance is, the smaller the direction measurement error is. However, angular ambiguity is introduced due to the 360 degrees flip caused by the large baseline distance. In order to achieve high precision without angular ambiguity, baseline combination is usually exploited, to obtain high angle measurement precision with a long baseline, and solve angular ambiguity with a short baseline. Besides, direction measurement precision is also related to the signal frequency and incident angle value. The higher the frequency is, the smaller the direction measurement error is. The smaller the incident angle is, the smaller the direction measurement error is. Considering the baseline length, frequency, and the speed of light together, the ratio between the baseline length and the wavelength is obtained, that is the direction measurement precision is affected by the multiple of the baseline length over the wavelength.

(4) Time difference-based direction finding

For two receiving antennas that have certain distance, radar signals from different directions have different TOAs. If the relationship between the time difference of arrival (TDOA) and the direction of arrival is built up, the radar signal DOA can be obtained by measuring the TDOA.

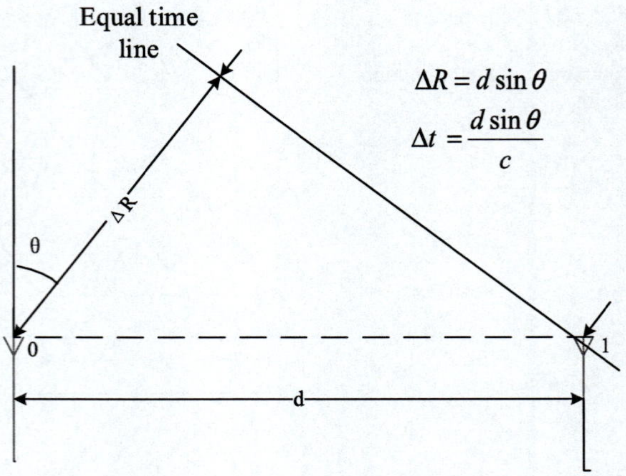

Fig. 5.33 Schematic diagram of time difference-based direction finding

The principle of time difference-based direction finding is illustrated in Fig. 5.33. If far-field condition is satisfied, the incident wave can be nearly equal to the plane wave. Assume the incident angle is θ and the plane perpendicular to the incident direction is equal time plane, the antenna receiving signal's transmission distance difference and the TDOA are

$$\Delta R = d\sin\theta$$
$$\Delta t = \frac{\Delta R}{c} = \frac{d\sin\theta}{c} \tag{5.18}$$

where ΔR is the transmission distance difference, and Δt is TDOA.

The expression of time difference-based direction finding's precision can be derived by:

$$\sigma_\theta = \frac{c\sigma_{\Delta t}}{d\cos\theta} \tag{5.19}$$

where σ_θ is angle measurement error, and $\sigma_{\Delta t}$ is time difference measurement error. According to the equation above, time difference-based direction finding's precision is mainly related to the time difference measurement precision, baseline length and the DOA. The higher the time difference measurement precision is, the higher the direction measurement precision is. The longer the baseline length is, the higher the direction measurement precision is. The smaller the incident angle is, the higher the direction measurement precision is. Different from phase comparison-based direction finding, time difference-based direction finding has nothing to do with the single's frequency. Based on the equation above, if the time difference measurement precision is 1 ns and baseline length is 10 m, the direction measurement precision is about 1.7°at the incident angle around 0°.

5.4.5 Position Measurement

The information of the target's location is able to support jamming. With the development of technology, a number of target positioning methods were invented and applied in projects. Typical location measurement algorithms include cross-localization, time difference localization, movable target frequency difference localization, single platform localization for fixed radiation source, etc., while this book focused on cross-localization and time difference localization.

(1) Cross-localization

Cross-localization uses two or more separately distributed stations to find the directions of the same radiation source and localize the radiation source based on the intersection of the bearing lines. The principle of cross-localization of two stations is shown in Fig. 5.34, where the direction of the radar signal received by the first antenna array is θ_1, and the direction of the radar signal received by the second antenna array is θ_2. The radiation source's location can be calculated according to the geometric relation in a two-dimensional plane. While designing a practical cross-localization system, the following two problems should be solved.

(a) The problem of matching the received signal of two stations. Fake location information would be obtained, if the cross-localization calculation used the signals that were not from the same radiation source. Since there are plenty of signals of radar, communication, or guidance in the environment, a number of fake location points would be produced by wrong associations and make the location results unavailable. Hence, the essential precondition cross-location is to identify and match the signals.
(b) The precision problem of cross-localization. For identical receiving systems and deployment, the localization precisions for different areas are different. The further the distance is, the larger the localization error is. The shapes of the error zone of different areas are different as well. For the analysis of localization

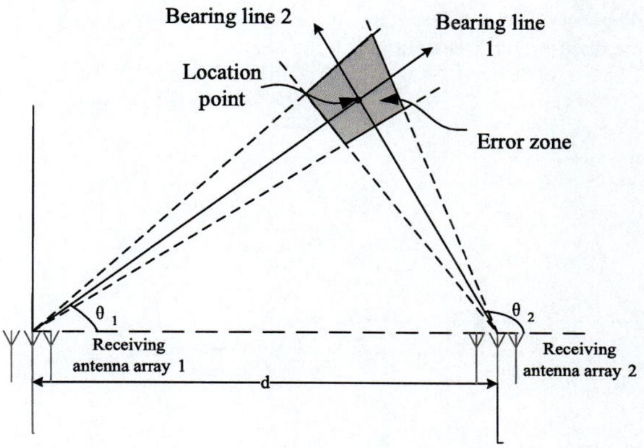

Fig. 5.34 Schematic diagram of cross-localization of two stations

error, see *Introduction to Modern EW System.* We would not further introduce it in this book.

(2) Time difference localization

Time difference localization uses the arrival time difference of radar signals received by multiple antennas to perform location calculations. The time difference of the three receiving stations can be used to derive the position of the radiation source in a two-dimensional plane. The schematic diagram of time difference localization based on three receiving stations in a two-dimensional plane is illustrated in Fig. 5.35. In Fig. 5.35, the arrival time of the signal received by antenna 0 is denoted by TOA_0, the arrival time of the signal received by antenna 1 is denoted by TOA_1, and the arrival time of the signal received by antenna 2 is denoted by TOA_2. The TOA difference between antenna 1 and antenna 0 is shown by time difference line 1, and the TOA difference between antenna 2 and antenna 0 is shown by time difference line 2. The intersection of the two curves is the localization result. The detailed localization method can be found in related data or references. The following factors should be considered in practice.

(a) Signal match problem. The measurement of arrival time and the time difference is of pulse level. It requires performing the time difference calculation for one or a series of pulses generated by the same radiation source, to obtain the correct time difference value. Pulse level signal match is necessary in this case. For radars with fixed radiation frequency, pulse, and PRI, the incorrect pulse match would introduce the time difference ambiguity problem, the ambiguity value of which is an integral multiple of the PRI value.
(b) Time synchronization problem among stations. In time difference localization, even time difference measurement error at 10 ns level will significantly affect the localization. The time measurement precision of commercial GPS receiver is around 100 ns and does not satisfy the requirement. Hence, plenty of special

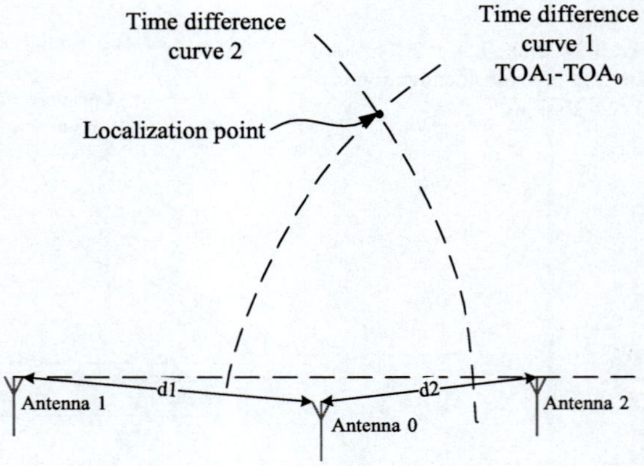

Fig. 5.35 Schematic diagram of time difference localization of three stations

technical means are applied in time difference localization systems to maintain time synchronization among stations.

(c) Precision problems related with locations. Time error of 10 ns is equal to the location error of 3 m, and therefore, time difference localization system requires higher reconnaissance station's localization precision than that of a cross-localization system. Besides, the precisions of time difference-based localizations are different for different geometry configurations, which should be also considered in system design and application.

5.4.6 Signal Sorting

Signal sorting is a kind of classification based on the pulse description information obtained by the radar receiver and extracts the parameter information of the radiation source. The procedure of signal sorting contains classification processing, parameter extraction, and similar items merging, as shown in Fig. 5.36.

(1) Classification processing

Classification processing is to classify the pulses whose DOA, frequency, and PW are similar and time-related, in order to form a set of pulses of the same kind. The methods of classification processing are various, such as histogram-based method, box-based method, clustering-based method, etc.

Histogram-based method builds up a statistical histogram of certain parameter of the radar signal, such as frequency histogram, PW histogram, and DOA histogram. The relatively concentrated pulses in the histogram are outputted as a class, as shown in Fig. 5.37.

The principle of box-based sorting method is shown in Fig. 5.38. Build a box with a certain tolerance range, whose center is the PDW. The PDWs that subsequently enter this box will be classified into this class, and the box's size and position will be adjusted at the same time. If the new signal failed to fall into any of existing boxes, a new box would be created. A number of boxes can be made in this way, and each box is a class of signals. From the perspective of processing method and procedure, box-based sorting is equal to cluster naturally.

(2) Parameter extraction

Parameter extraction should be performed for each class of signal after classification. Since each pulse has the information of frequency, PW, PA, and DOA, the parameter

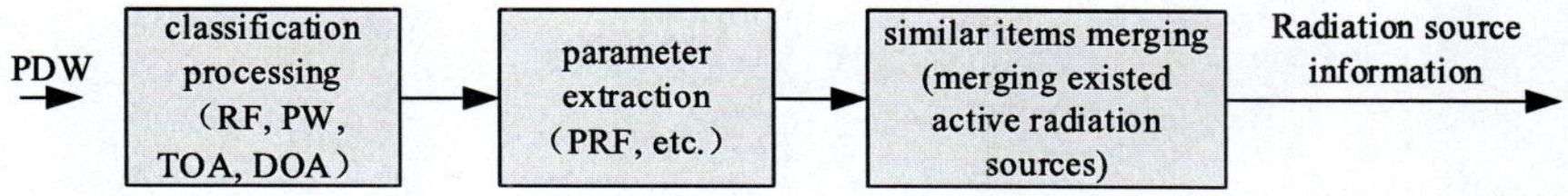

Fig. 5.36 Block diagram of signal sorting procedure

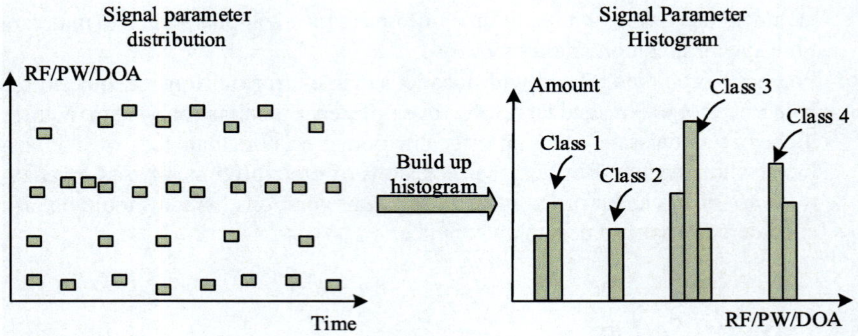

Fig. 5.37 Schematic diagram of histogram-based classification processing

Fig. 5.38 Illustration of box-based classification

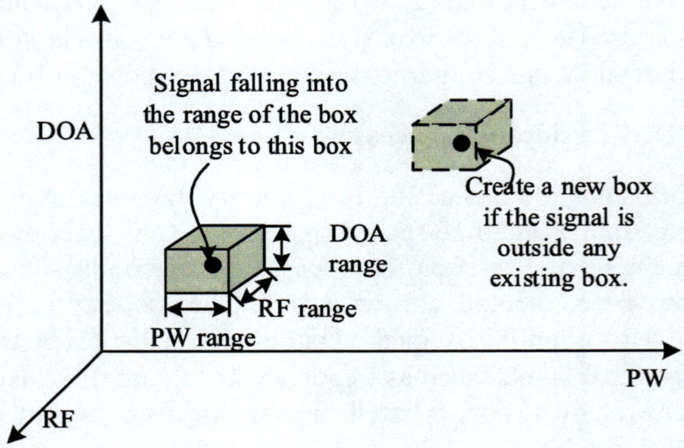

estimations for the above items are based on the values and the distribution of these parameters in each class. Fox radiation sources with fixed parameters usually need a simple calculation for the mean value, which is shown in the following equation.

$$
\begin{aligned}
\overline{\mathrm{RF}} &= \frac{1}{N}\sum_{n=1}^{N} RF(n); \\
\overline{\mathrm{PW}} &= \frac{1}{N}\sum_{n=1}^{N} PW(n); \\
\overline{\mathrm{DOA}} &= \frac{1}{N}\sum_{n=1}^{N} DOA(n); \\
\overline{A} &= \frac{1}{N}\sum_{n=1}^{N} A(n);
\end{aligned}
\tag{5.20}
$$

For radiation sources with dynamic parameters, they require the calculation of the parameter changes and the corresponding parameters. For example, the change range needs to be computed for pulse-to-pulse frequency agility radar, and the batch number and frequency values of each batch need to be computed for batch-to-batch frequency agility radar.

The extraction of PRI is more complicated. For fixed PRI, it only needs to calculate the difference of neighboring pulses' TOAs. For jittered PRI, we need to confirm it is this kind of PRI and then compute the frequency amount and every frequency value.

(3) Similar items merging

The description of classified radiation sources should be compared with the radiation sources intercepted before. If they were determined to be the same kind, they should be classified into the same class. If it was a new radiation source, a new item should be added to the radiation source list and radiation source recognition would be performed next.

5.4.7 Radiation Source Recognition

Radiation source recognition is to determine the category and model of the radiation source based on priori-knowledge.

The input for radiation source recognition includes:

(a) Frequency: the category, value, or value range of frequency.
(b) PW: the category, value, or value range of PW.
(c) PRI: the category, value, or value range of PRI.
(d) Inner Pulse Modulation: modulation category and modulation parameter.
(e) Amplitude.

Since amplitude does not reflect the inherent feature of the radiation source, radiation source recognition usually does not use amplitude information.

The current main way of identifying radiation sources is to compare the result with the threat target database. A threat target database is an information list of known radar stored inside the jamming system. The accuracy of the database-based radiation source recognition method strongly relies on the accuracy of the database as well as the integrity of the intercepted radar signal.

Radiation source recognition is a classification procedure from the perspective of mathematics, and plenty of classification algorithms can be adopted. During the procedure of comparing with the database, a tolerance range should be set, which is determined based on the parameter measurement precision of the receiving system.

5.5 New Development of Receiver Technology

Along with the developments of signal processing technique, digital processing device, optical processing technique, and microsystem technique, receiver technique keeps developing as well. New receiver techniques at present mainly include compressed sampling (sensing), digital array, optical processing, and correlation processing technique, which will be briefly introduced in this subsection.

5.5.1 Compression Sampling Technique

Compressed sampling technique utilizes the sparse nature of signal to perform sampling with the sample rate which is lower than what Nyquist's law requires. It recovers, detects, and measures the signal with signal processing technique [23].

Compressed sampling technique is able to solve the contradiction between the requirement of wide instantaneous bandwidth for EW receiver and the sampling rate of ADC device [24]. For example, for a receiver requiring 2 GHz instantaneous bandwidth, it at least demands ADC devices with a bichannel and 2.4 GHz sampling rate. Even if the above requirement can be satisfied with high-end devices, it would be difficult to satisfy the demand for larger instantaneous bandwidth. For instance, an instantaneous bandwidth covering the whole 2–18 GHz range cannot be realized directly by digital sampling processing. To deal with this case, instantaneous frequency measurement receivers or superheterodyne receivers are applied at present. However, it would be possible to solve this problem, if compressed sampling technique was applied.

Based on the observation of the real electromagnetic signal environment, the signal does not always exist for the whole reconnaissance window, and multiple signals do not occupy all the bandwidth. For a fixed observation time, there is very little information that is useful. That is, the signal is *sparse* if a professional term is used. In 2004, Compressed sampling (CS) theory was proposed by D. L. Donoho, E. Candes, and T. Tao. In this theory, if the signal to be detected was assumed to be sparse, its observation sample could be obtained with a sampling rate much lower than what Nyquist's law requires. The original signal can be reconstructed with a specific algorithm, and the contradiction between the sampling rate and system instantaneous bandwidth can be resolved. It builds up a nonlinear measure with samples of a relative small number to obtain all the sparse signals or information of the compressible signal in mathematics (Fig. 5.39).

Current researches mainly focus on how to precisely reconstruct the original signal with a specific algorithm. The success of compressed sampling depends on two conditions: sparsity and non-correlation. If the signal or its transformation in some domain has sparse nature, the original signal could be significantly compressed and sampled with an observation matrix (measurement function) that is non-correlated

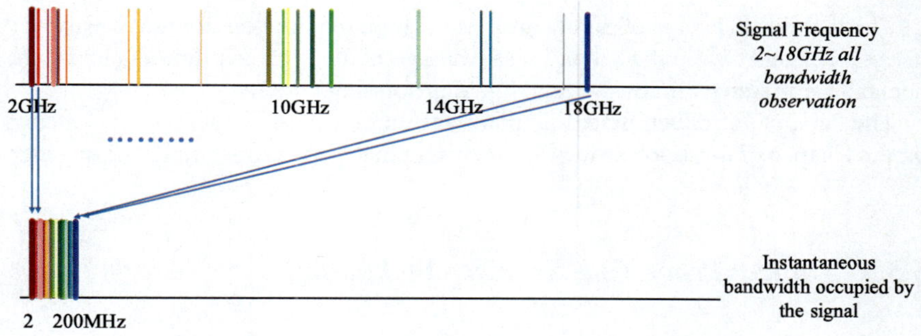

Fig. 5.39 Illustration of the sparse nature of signal

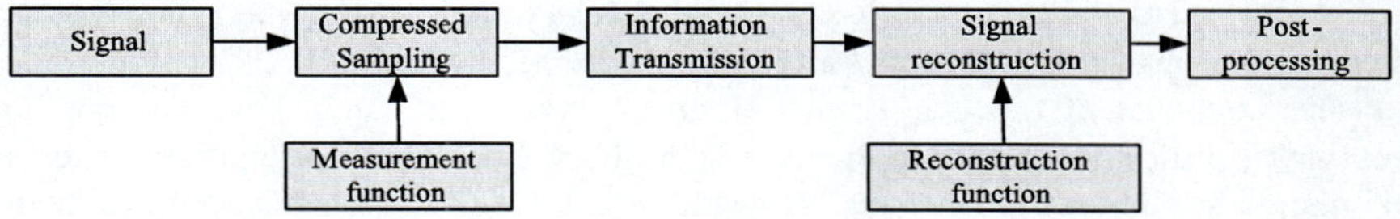

Fig. 5.40 Flowchart of compressed sampling signal processing

with the transformation base and accurately reconstruct the original signal with a very high probability by a reconstruction algorithm (Fig. 5.40).

Nowadays, the research of compressed sampling still stays in the theoretical stage, and there are plenty of problems that need to be studied in the process of engineering. The problems include observation function design and realization, stable, efficient, and low computation complexity reconstruction function design, high-efficient expression of the original signal, and so on. Compressed sampling technique has high potential in remote sensing imaging, medical imaging, wireless communication, biological sensing, astronomy, radar detection, lossless compression, optical spectrum analysis, image super-resolution reconstruction, and geological prospecting. We believe it will bring revolutionary change to the EW domain in the future.

5.5.2 Digital Array Technology

Digital array technique utilizes digital techniques to realize beam control of antenna array. A digital array receiver samples the signal received by a single antenna or an antenna group in the antenna array and performs array signal processing to form wave beams in the digital domain for signal detection and parameter measurement. Digital array is able to realize beam agility capability with a speed similar to phased array radar, as well as simultaneous reception via multiple beams.

The first disadvantage of digital array is the high requirement for the consistency of each channel, while the second disadvantage is the high cost introduced by the demand for massive microwave and digital processing circuits.

The detailed introduction to digital array techniques can be found in the antenna part in Chap. 4 of this book as well as other reference books about digital array radar.

5.5.3 Optical Processing Receiver Technology

An optical processing receiver is a kind of new receiver which is designed based on the microwave photon technique. It combines microwave radio frequency technique and optical fiber technique and takes the advantage of the wide processing bandwidth, low energy cost, and small volume nature of optical processing [25–27]. The typical flow of microwave optical processing is shown in Fig. 5.41.

The institutes of the US, Britain, France, Israel, and Japan have started the research on the theory and application in military electronic information systems of microwave photon processing. Nowadays, a lot of research results have been outputted and published. Defense Advanced Research Projects Agency (DARPA) of the US proposed the research plans named Analog Optical Signal Processing (AOSP) and Ultra-Wideband Multifunction Photonic Transmit/Receive Module (ULTRA-T/R). The participating corporations include Raytheon Company, Lockheed Martin Space System Company, Rockwell Company, Air Force Research lab of the US and JPL lab of the US. Great progress has been made by them in plenty of optical processing fundamental techniques.

The current research interests in microwave photon processing mainly focus on microwave photon frequency transformation, microwave photon measurement, microwave photon filter, arbitrary waveform optical formation, and photon analog–digital conversion.

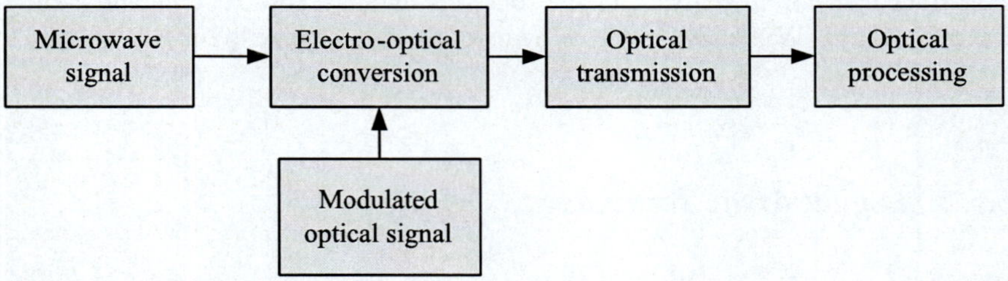

Fig. 5.41 Flowchart of microwave optical processing

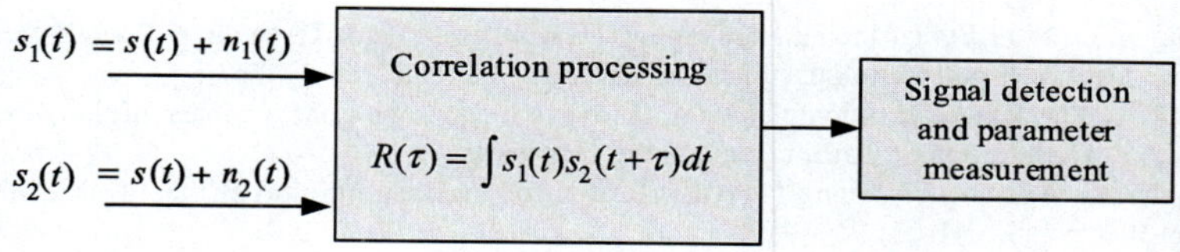

Fig. 5.42 Schematic diagram of correlation processing

5.5.4 Correlation Receiver Technology

Correlation receiving technique utilizes the autocorrelation nature of radar signal and performs cross-correlation processing on the signals received by two channels to improve the SNR of the received signal and the receiver's sensitivity. The principle of cross-correlation is given in Fig. 5.42. $s_1(t)$ is the received signal of the first channel, $s_2(t)$ is the received signal of the second channel, and $s(t)$ is radar signal. Assume the two channels are consistent, and the received signals should be equal. $n_1(t)$ and $n_2(t)$ are the noises of the two channels, which are unrelated since the two channels are independent. Hence, we have

$$R(\tau) = R_{ss}(\tau) = \int s(t)s(t+\tau)dt \tag{5.21}$$

For radar signals using linear frequency shift or phase encoding techniques, the autocorrelation function has a correlation peak, and autocorrelation algorithms can be used to improve SNR, which is benefit to signal detection under low SNR situations.

Although correlation processing benefits signal detection, the parameter measurement cannot be performed since the information of radar signal's frequency, TOA, and PW is lost after the correlation processing. Hence, corresponding parameter measurement methods should be introduced to satisfy the requirements of EW receiver.

References

1. Adamy DL (2001) EW101: A first course in electronic warfare. Artech House Inc., Norwood
2. Laizhao H (2002) Instantaneous frequency measurement. National Defense Industry Press, Beijing
3. Laizhao H (2010) Design of radar reconnaissance receiver. National Defense Industry Press, Beijing
4. Hong L, Yingke Y, Baomin X (2010) Introduction to synthetic aperture radar countermeasure. National Defense Industry Press, Beijing
5. Xiaokang Y (2003) Introduction to spaceborne synthetic aperture radar. National Defense Industry Press, Beijing
6. Chengbo Z (1989) Principle, system analysis and application of synthetic aperture radar. Science Press, Beijing

7. Yongtan L (1999) Radar imaging technology. Harbin Institute of Technology Press, Harbin
8. Tsui J (1995) Digital techniques for wideband receivers. SciTech Publishing
9. Xin W (2008) Research and implementation of key technologies of broadband digital receiver. Doctoral Dissertation of Harbin Engineering University
10. Shixian G, Xizhang W, Xiang L (2013) Review of wideband digital channelized receivers. Acta Electron Sin 41(05):949–959
11. Guangzu L, Jianxin W, Dalong X (2012) Design and implementation of efficient structure for digital channelized receivers. Syst Eng Elect 34(2):391–395
12. Yong C (2008) Realization of receiver in EW comprehensive simulation system. Modern Radar 30(3):86–88
13. Wei H, Ziyue T, Zhenbo Z (2013) Target tracking method using bearing information of ESM in doppler blind zone. Syst Eng Elect 35(8):1650–1656
14. Xudong W, Maozhong S (2010) Wideband digital ESM receiving technique based on STFT. Syst Eng Elect 32(9):1811–1814
15. Lin L, Hongbing J (2009) Estimation of multipath ESM signals based on L-Wigner distribution. Syst Eng Elect 31(11):2618–2622
16. Jiegui W, Xueming J, Jingqing L (2006) Airborne emitter recognition based on multi-sensor data fusion of ESM and ELINT. Acta Electron Sin 34(3):424–428
17. Lufei D, Jianchun C (2009) Radar theory, 4th edn. Publishing House of Electronics Industry, Beijing
18. Allen DE (1982) Channelised receiver. A viable solution for EW and ESM systems. Communications Radar Sig Processing IEEE Proceedings F 129(3):172–179
19. Cheng CH, Lin DM, Liou LL et al (2012) Electronic warfare receiver with multiple FFT frame sizes. IEEE Trans Aerosp Electron Syst 48(4):3318–3330
20. Khalighi MA, Nayebi MM (2000) CFAR processor for ESM systems applications. IEEE Proceedings Radar Sonar Navigation 147(2):86–92
21. Jeffries, Russell F, et al (2001) Further development of a future ESM channeliser with high temperature superconducting filters. IEEE Trans Appl Superconductivity 11(1):410–410
22. Whittall NJ (2008) Signal sorting in ESM systems. IEEE Proceedings F-Communications, Radar Sign Process 132(4):226–228
23. Yijiu Z (2012) Research on sparse analog signal compression sampling and reconstruction algorithm. Doctoral Dissertation of University of Electronic Science and Technology, Chengdu
24. Zuohao Z (2014) Research on compression observation technology and its application in radar signal. Doctoral Dissertation of Xi'an University of Electronic Science and Technology, Xi'an
25. Li Y (2013) Research on microwave photon multi-frequency measurement technology based on compression sampling. Doctoral Dissertation of Beijing University of Posts and Telecommunications, Beijing
26. Yuyang G (2015) Research on broadband RF photonic receiver based on under sampling. Doctoral Dissertation of Beijing University of Posts and Telecommunications, Beijing
27. Ze L (2012) Research on key issues in microwave photonic signal processing. Doctoral Dissertation of Zhejiang University, Hangzhou

Chapter 6
Jamming Source

The core function of the radar active jamming system is to generate radar jamming signals to form suppression or deception jamming effect on radar, and destroy or prevent radar from detecting, tracking, and identifying targets. Among them, the jamming source, also known as the technology generator (TG), is a key component responsible for generating jamming signals in radar active jamming system, which can best reflect the essence of "active" jamming in the system. The design of the jamming source is closely related to the requirements of radar active jamming system for jamming style, and the quality of the jamming signal produced by the jamming source often determines the jamming effect on the radar to a great extent.

This chapter will introduce the design requirements, main types, design methods, and development trends of jamming sources. Among them, the first section introduces the design requirements of the jamming source, mainly centering on the final application of jamming to comb the functional requirements and performance index requirements of the jamming source. The second section introduces the typical types of jamming sources. The third and fourth sections, respectively, take the two most common jamming source systems as examples to introduce the principle, composition, characteristics, and main design methods of jamming sources. The fifth section analyzes the development trend of technology and capability of jamming sources.

6.1 Requirements Analysis of Jamming Sources

6.1.1 Functional Requirements

The most ideal jamming source is an arbitrary waveform generator; that is, it can generate jamming signals that fully meet the requirements and change rapidly and arbitrarily according to the requirements of radar active jamming system for jamming waveform and jamming style. However, the implementation process of this kind of system will be very complex and often unnecessary.

G. Tang et al., *Techniques and System Design of Radar Active Jamming*,
https://doi.org/10.1007/978-981-19-9944-4_6

In practice, jamming sources with corresponding functional characteristics can be selected according to the jamming mode requirements of radar active jamming system. The common two types are suppression jamming and deception jamming. Different jamming modes will correspond to different types of jamming sources. The long-distance support radar jamming system generally adopts the suppression jamming mode. The design of jamming source will focus on the generation of jamming signals such as noise, dense false target suppression, and signal modulation with different suppression bandwidth. The self-defense jamming system may include not only the suppression jamming, but also the deception jamming at range, velocity, or angle. However, the design of general jamming source often needs to meet the requirements of multifunction, integration, and rapid response.

The most basic function of the jamming source is to generate jamming signal. Usually, the jamming signal should include two aspects, one is the basic waveform of the jamming signal, and the other is the different jamming styles that can be formed by the jamming modulation. Therefore, the function of the jamming source mainly includes the generation of jamming signal waveform and the modulation of jamming signal.

(1) Waveform generation function

The function of jamming source waveform generation mainly refers to the generation of some basic jamming signal waveforms, such as noise, sine wave signals with different frequencies, and pulse signals with different intra-pulse modulations. They can form different styles of jamming signals after being modulated by parameters such as amplitude, frequency, phase, and polarization or controlled by time sequence.

The type requirements of jamming signal basic waveform in jamming source often come from the signal waveform style of radar. Among them, noise signal is widely existed in the environment and radar receiving system, which has an important jamming effect on radar processing. Therefore, it is the most common and basic jamming waveform in radar active jamming design. Radar always works within a certain frequency range. Sine waves with different frequencies can be combined into jamming styles such as comb spectrum jamming and can also be used as the input of Doppler modulation signal in jamming modulation. In order to achieve coherent jamming to radar and obtain the intra-pulse matched filter gain of radar, the jamming source often needs to generate jamming pulse signals consistent with radar pulse modulation types, such as linear frequency modulation and phase coded modulation.

(2) Signal modulation function

The most important function of the jamming source is to form jamming signals with different styles by means of different signal parameter modulation or timing control modes according to different application requirements. The radar active jamming system may adopt a variety of jamming signal styles, which will also change in the process of countermeasures. Therefore, in order to adapt to the change of radar signal and jamming requirements, and obtain better jamming effect, it is necessary for the jamming source to have the ability of real-time modulation and fast switching of the basic jamming waveform.

Tamming signal modulation is mainly carried out in time domain, frequency domain, and energy domain. The parameters to be modulated usually include signal frequency, bandwidth, Doppler, amplitude, phase, delay time, and jamming quantity. The modulation in spatial domain and polarization domain generally requires multiple jamming signal channels and the cooperation of antennas. Table 6.1 presents several common radar active jamming styles and their characteristics, which have different main modulation parameters.

Different modulation modes of jamming signals can be illustrated by the following mathematical model. Supposed that the transmitting signal of the radar is $s(t)$, and the radar target echo signal is:

$$s_T(t) = \sigma s(t - \tau_{\mathrm{T}}) \exp(j2\pi f_T) \tag{6.1}$$

In the formula, σ is target RCS, τ_T and f_T are time delay of distance and Doppler shift respectively.

Radar active jamming system receives radar transmitting signal, and false target signal can be generated by using repeater jamming.

$$J(t) = \sum_{i=1}^{N} A_i s(t - \tau_T - \tau_i) \exp[j2\pi(f_T + f_i)] \tag{6.2}$$

In the formula, N is the number of false targets, A_i, τ_i and f_i are the amplitude, relative time delay, and relative Doppler shift of each false target signal, respectively.

By setting different parameters such as target number, relative time delay, Doppler shift and signal amplitude, radar active jamming system can generate many different types of false target signals, which have different characteristics and jamming effects respectively. In Eq. (6.2), when τ_i is small and N is large, it is dense false target suppression jamming. When τ_i is large and N is small, it is multiple false target deception jamming. By adjusting the change of τ_i or f_i with time t, the range gate pull-off jamming and velocity gate pull-off jamming can be realized respectively. The timing relationship between them is shown in Fig. 6.1.

Table 6.1 Common radar active jamming modulation styles

Number	Jamming style	Main modulation style	Typical modulation parameter
1	Noise suppression jamming	Direct RF noise (no modulation), Noise amplitude modulation, Noise phase modulation, Noise frequency modulation; Spot jamming, blocking jamming Frequency-swept jamming Intermittent noise jamming	Center frequency, Signal bandwidth, Jamming power and duration
2	Comb spectral jamming	Comb spectrum structure Fixed frequency jamming Random frequency jamming	Number of jamming frequency points The frequency and amplitude of jamming signal, etc.
3	False target deception jamming	False point target and false track; Range false target and Doppler false target; Single false target jamming and multiple false targets jamming	Number of false targets The frequency, delay time and amplitude of jamming signal, etc.
4	Dense false target suppression jamming	The suppression effect is similar to noise jamming Range dimension dense false target suppression jamming and Doppler dimension dense false target suppression jamming	Jamming signal frequency, initial delay time, jamming replication interval, jamming replication number and signal amplitude
5	Range gate pull-off jamming	Generally, the amplitude of jamming signal is much larger than that of the real target echo signal Single false target's wave gate dragging jamming and multi-false target's simultaneous dragging jamming	It is divided into four stages: capture, pull-off, dwell and stop Jamming signal amplitude and continuous change delay time, etc.
6	Velocity gate pull-off jamming	It is the same as the principle and working process of the range gate pull-off jamming	Jamming signal amplitude, Continuously changing Doppler frequency shift parameters, etc.
7	Angle deception jamming	Blinking jamming, cross-eye jamming, cross-polarization jamming Multiple antennas or polarized antennas are generally required	Signal frequency, amplitude and phase of interference signal, amplitude ratio and phase difference among multiple channels, etc.

(continued)

Table 6.1 (continued)

Number	Jamming style	Main modulation style	Typical modulation parameter
8	Combined jamming	It can generate a variety of jamming signals at the same time and to quickly realize different modulations according to different timing requirements	It may include all dimensions in time domain, spatial domain, frequency domain, polarization domain, and energy domain

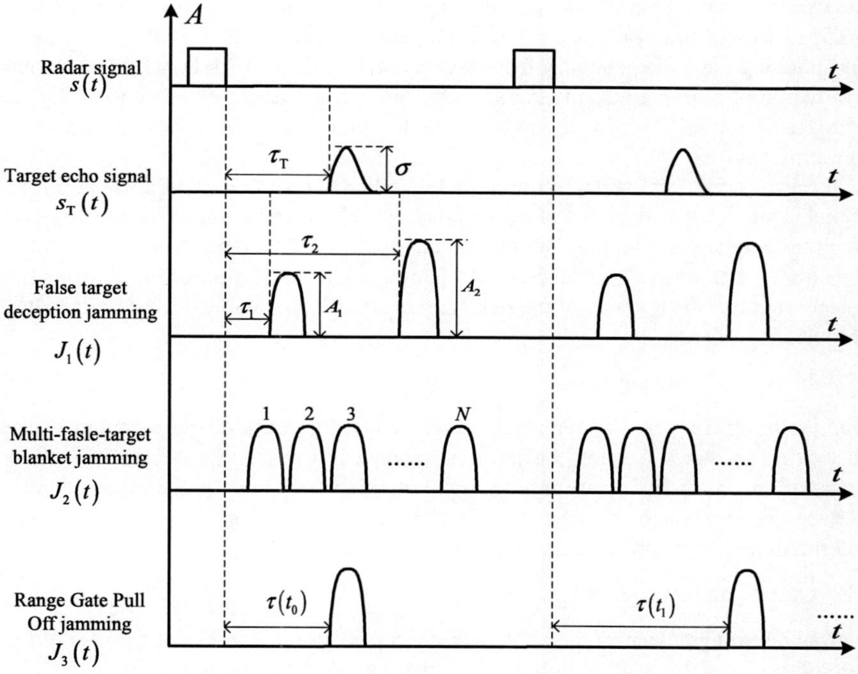

Fig. 6.1 Schematic diagram of radar target echo and repeater jamming signals

6.1.2 Indicators Requirements

Different types of jamming sources generally have different technical index requirements. For example, noise jamming sources are mainly used to generate noise jamming signals to suppress radars and require high technical indicators such as the frequency range and instantaneous bandwidth of the jamming signals. The jamming sources used for deceptive jamming pays more attention to the frequency error, signal coherence, and waveform fidelity between the generated jamming signal and radar signal. In Sects. 6.3 and 6.4, the specific design methods of DDS and DRFM jamming sources will be introduced in detail respectively. Here, only a few major functional and technical indicators that are generally considered in the design process of jamming sources are given.

(1) Maximum instantaneous bandwidth

Maximum instantaneous bandwidth refers to the maximum frequency width of the jamming signal that can be generated when the jamming source is working normally. The maximum instantaneous bandwidth of the jamming source generally needs to be greater than or equal to the maximum signal bandwidth of the threat radar. It can be the working bandwidth of one jamming signal output channel, or a large bandwidth grouped by multiple output channels. The signal bandwidth of traditional search and tracking radars is generally from several MHz to tens of MHz, while the signal bandwidth of high-resolution radars, such as imaging radar, is often up to hundreds of MHz to several GHz, which puts forward higher requirements for the design of jamming sources.

Typically, for noise suppression jamming, when the bandwidth of noise jamming signal is much larger than that of radar signal, it is blocking jamming. When the noise jamming signal bandwidth cannot significantly exceed the radar signal bandwidth, the aiming jamming or frequency-swept jamming is generally adopted, in which the former requires the receiver of the radar active jamming system to be able to conduct frequency guidance.

(2) Jamming frequency range

The jamming frequency range refers to the frequency range of the jamming signal that can be generated when the jamming source is working normally. This index generally needs to be corresponding to the working frequency range index of radar active jamming system. It can be realized by up-conversion and down-conversion of intermediate frequency jamming signals in the jamming source.

(3) Output jamming power

Output jamming power refers to the maximum peak power and average power of the jamming signal that can be output when the jamming source is working normally. The output signal power of the jamming source is generally small, which needs to be amplified by the transmitter to form a high-power jamming signal. Therefore, the output jamming power needs to be designed according to the input signal power index of the transmitter.

(4) Output jamming style

The output jamming style refers to the type of jamming waveform and jamming style that can be generated by the jamming source. It is usually required that the jamming source can flexibly control and output a variety of jamming waveforms and form different jamming modulation styles according to the system needs. Common jamming styles are shown in Table 6.1.

(5) Jamming signal spurious

Jamming signal spurious refers to the part of the output jamming signal of the jamming source that is not required but has not been fully suppressed. The stray energy not only reduces the effective jamming power of radar active jamming system but also is easier to be recognized by the enemy as a feature of jamming. This index represents the purity of the output jamming signal of the jamming source. Generally, the smaller the stray of the jamming source is, the better. The stray part of the output signal of the jamming source will appear more serious after being amplified by the subsequent transmitter.

As for the noise jamming signal, the spurious mainly refers to the noise unintentionally generated by the jamming source within the radar instantaneous working bandwidth and outside the noise jamming signal bandwidth. For other jamming signals, stray signals include noise and harmonics caused by the jamming source. It mainly comes from the channel noise inside the jamming source, device nonlinear output and mixer intermodulation signal, etc., especially in the digital jamming source system, the finite word length effect will lead to the quantization error of ADC and DAC and will inevitably introduce the noise and harmonic components.

(6) Modulation accuracy of jamming signals

The modulation accuracy of jamming signal refers to the error between the parameters of the jamming signal output by the jamming source and the set parameters, which can represent the fine modulation ability of the jamming source to the jamming signal. It mainly includes aiming frequency accuracy, amplitude modulation precision, phase modulation accuracy, and Doppler modulation accuracy.

(7) Processing delay time

It refers to the minimum time required for the jamming source to output the jamming signal from receiving the jamming command and jamming guidance parameters or input signals. In general, in order to realize the rapid response of radar active jamming system, the processing delay time of jamming source should be shortened as far as possible.

For some particular jamming sources, there are generally some other typical index requirements. For example, the repeater jamming also includes the minimum pulse replication interval, the maximum time delay, and other indicators, which will not be repeated here.

6.2 Main Types of Jamming Sources

Different signal waveforms generation methods are the main basis to distinguish different jamming source systems. According to the different sources of jamming signal waveform components, jamming sources can be divided into internal waveform generation and external waveform generation. According to the different circuit structures of jamming signal waveform generation, jamming sources can be divided into analog jamming sources and digital jamming sources. However, with the improvement of the design and manufacturing level of digital integrated circuit, it is an inevitable trend that digital jamming source replaces analog jamming source. The full digitization is beneficial to the integration of the jamming source system, which has the advantages of small volume, lightweight, high stability, strong programmable ability and good controllability and so forth.

In this chapter, according to different types of generated jamming signals and different jamming signal generation methods, the radar jamming sources are divided into three types: noise jamming sources, responder jamming sources, and repeater jamming sources. However, analog and digital classification methods are also used alternately. The following is a brief introduction to the different types of jamming sources.

6.2.1 Noise Jamming Source

The noise jamming source is one of the most used jamming sources. The ideal noise jamming signal should be infinitely close to the radar receiver noise. Its key advantage of noise jamming is that it can be implemented with less information about the threat radar.

Early noise jamming sources were generated in the analog way. The microwave noise generated by analog devices (resistors or triodes, etc.) when working, is subjected to band-pass filtering, modulation, and power amplification, finally forming direct RF noise or noise modulation signal with a certain central frequency and corresponding bandwidth, as shown in Fig. 6.2. The circuit diagram of a common RF noise signal generation is shown as follows. It uses the base and emitter of the triode to generate noise current, which is amplified to form noise jamming signal [1]. However, it is difficult to debug this kind of circuit. After the design is completed, it is difficult to change the probability distribution, bandwidth, and other parameters of noise signal. The quality of output signal is also easily affected by external conditions such as temperature and power supply, which leads to poor working stability and controllability.

Modern noise jamming sources are mainly realized by digital means. Figure 6.3 shows principal block diagram of noise jamming signal generation based on DDS, which directly synthesizes the noise intermediate frequency or baseband jamming signal in a digital way, and then forms noise blanket jamming signal with different frequencies, probability distribution, power gain and bandwidth after filtering, mixing, and amplification. Among them, the noise waveform library can store the noise samples obtained from the real noise signal sampling and can also store a group or more groups of pseudorandom digital sequences to modulate the DDS, such as m sequence.

Fig. 6.2 Circuit diagram of noise signal generation based on triode

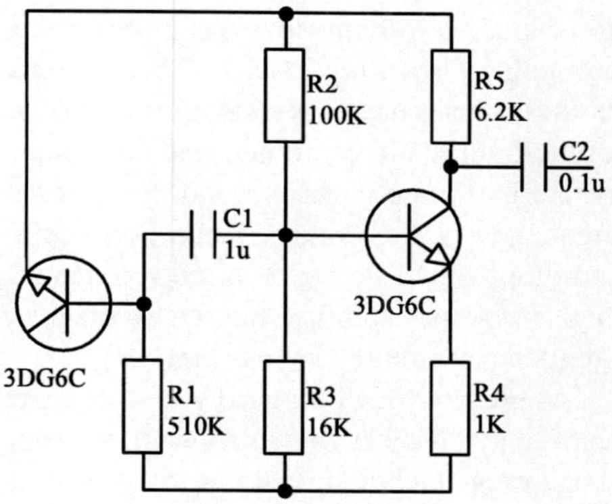

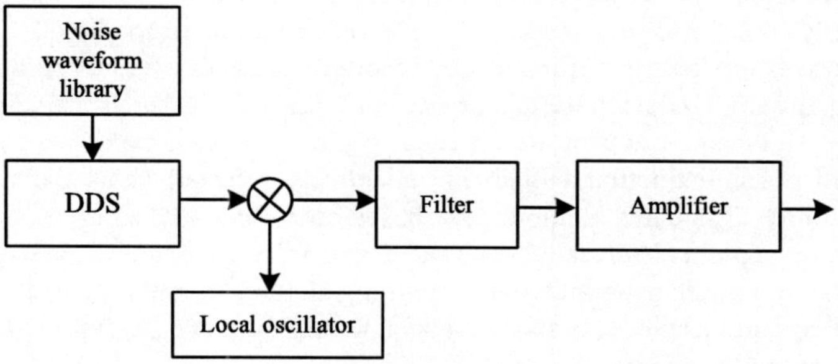

Fig. 6.3 Block diagram of noise jamming signal generation based on DDS

6.2.2 Responder Jamming Source

Response jamming is also called oscillatory response jamming. Its principle is to generate the corresponding jamming signal by frequency synthesis technology according to the carrier frequency and waveform characteristics of the threat radar signal measured and identified by the reconnaissance part of the radar active jamming system. In practical application, the fidelity between the frequency and waveform of the synthesized jamming signal and the threat radar echo will affect the jamming effect on the radar.

In the past decades, frequency synthesis technology has developed three ways: direct frequency synthesis, phase-locked frequency synthesis, and direct digital synthesis.

Direct frequency synthesis refers to using one or more different crystal oscillators as a reference signal source to directly generate the required jamming signal through frequency multiplication, frequency division, and frequency mixing. The jamming signal obtained by this method has high-frequency stability and fast frequency conversion speed, but its implementation structure is complicated and it is easy to produce too many spurious components. The phase-locked frequency synthesis technology is mainly based on phase feedback theory and analog phase-locked technology. It uses one or several reference frequency sources to generate a large number of harmonics and combined frequencies through the mixing and frequency division of the harmonic generator and then uses the phase-locked loop (PLL) to lock the frequency of the voltage controlled oscillator (VCO) and indirectly generate the jamming signal of the required frequency. This method also has the advantage of high-frequency stability, but the frequency conversion time is long, resulting in slow frequency switching speed (Fig. 6.4).

At present, DDS direct digital frequency synthesis (DDS) technology is widely used. It synthesizes frequency from the concept of phase and adopts digital sampling and storage technology. It has many advantages, such as high-frequency resolution, short frequency conversion time, continuous output phase, and programmable. The working principle and design method of DDS will be introduced in Sect. 6.3.

Generally, the noise and response mode of operation is to directly generate jamming waveform in the jamming source according to the threat radar signal parameters and jamming characteristic parameters provided by the radar active jamming system, which does not depend on the radar signal samples. Therefore, sometimes there is no strict distinction, which is collectively referred to as the responder jamming source. Since this method is not based on radar signal samples, the generated jamming signal is irrelevant to the radar transmission signal. However, except for noise jamming, the generated pulse jamming signal generally meets the characteristics of continuous phase, is autocorrelated within the pulse, and still has a certain gain for radar signal processing.

6.2.3 Repeater Jamming Source

Repeater jamming source is also called stored frequency jamming source. Its main working principle is to first receive the radar RF signal, save it for a period, and form the jamming signal through modulation processing or copying and forwarding. The jamming signal has very good coherence with the radar signal.

The analog frequency storage jamming source directly stores the incident radar RF analog signal. Generally, RF delay lines such as coaxial cable and delay fiber and RF switch cascade can be used to store and simply modulate the radar signal and form jamming signal. Among them, the design of RF delay line is pivotal, which generally requires materials with small insertion loss, large working bandwidth, small volume, and lightweight. Analog frequency storage technology has the merit of wide instantaneous frequency band, large dynamic range and high-frequency storage accuracy. But it can only output once and is difficult to form a complex modulated jamming signal.

At present, DRFM is the most extensively used. It is to sample and quantize the input radar RF analog signal to form a digital signal and store it in the digital memory. When necessary, it is read from the memory and reconverted into an analog signal

Fig. 6.4 Principal block diagrams of two frequency synthesis techniques

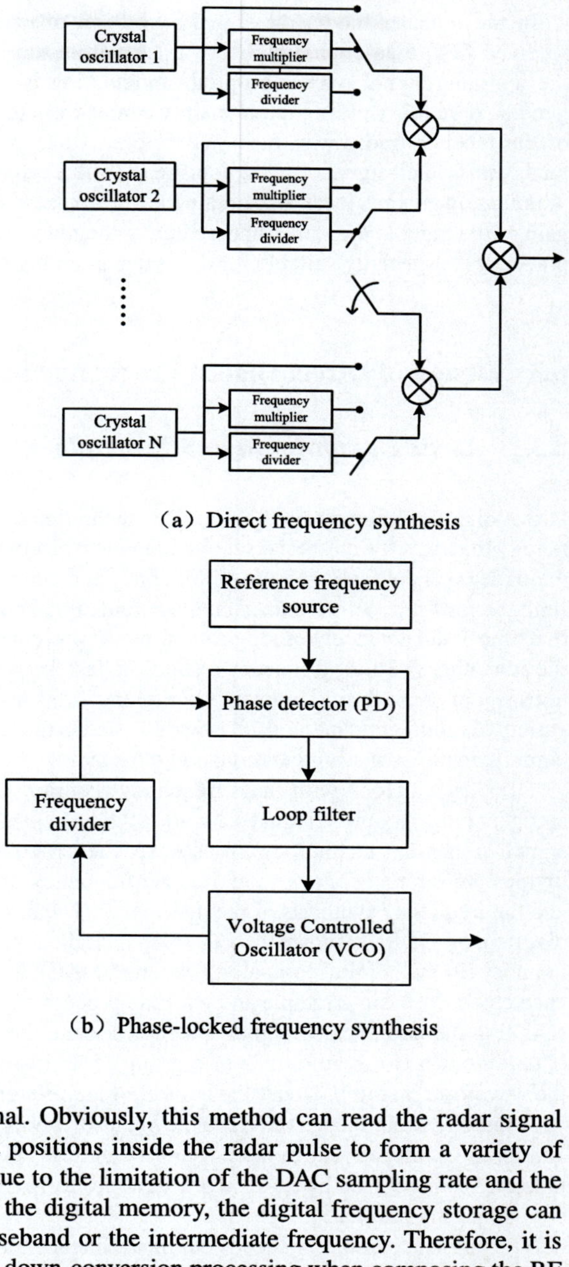

output to form jamming signal. Obviously, this method can read the radar signal many times or read different positions inside the radar pulse to form a variety of jamming signals. However, due to the limitation of the DAC sampling rate and the reading and writing speed of the digital memory, the digital frequency storage can only be performed on the baseband or the intermediate frequency. Therefore, it is necessary to perform up- and down-conversion processing when composing the RF signal memory. The working principle and design method of DRFM will be mainly introduced in Sect. 6.4.

In the repeater mode, the signal waveform of external threat radar is usually received first, then copied and forwarded within the jamming source, and finally, the jamming signal with appropriate modulation is formed. The jamming signal generated by this working mode mainly comes from the modulation and forwarding of the received radar waveform. Therefore, it has very good coherence with the radar transmitted signal, making it difficult for the radar to identify and suppress the jamming. In general, it can also obtain the intra-pulse and interpulse signal processing gain of the radar at the same time, reducing the requirements for the jamming signal power. It is especially suitable for deceptive jamming of coherent processing radar.

6.3 Design of Direct Digital Frequency Synthesize

6.3.1 Basic Composition and Principle

Direct digital frequency synthesis (DDS) technology is a frequency synthesis technology that directly synthesizes the required waveform from the concept of phase. Its main idea was put forward by Joseph Tierney and others in 1971. However, due to the limitations of microelectronic technology and digital signal processing technology at that time, it did not receive enough attention. At present, because DDS technology has the advantages of high-frequency resolution, fast frequency switching speed, continuous signal phase during frequency switching, arbitrary waveform generation, and a variety of modulation methods, it is widely used in radar, electronic countermeasures, communications, instrumentation, and other fields.

DDS uses a clock with fixed frequency accuracy as the reference clock source and generates an output signal with adjustable frequency and phase through digital signal processing technology. In essence, the programmable binary control word divides the reference clock to realize the frequency division of the reference clock and the frequency synthesis of the new signal. The basic composing principal block diagram of DDS is shown in Fig. 6.5, including phase accumulator, waveform memory (ROM), digital-to-analog converter (DAC), low-pass filter (LPF), and reference clock. Its basic principle and working process are as follows: based on Nyquist sampling theorem, analog signal is sampled, quantized or generated by software to obtain the waveform amplitude of a group of standard signals, which are stored in the waveform memory. Under the control of the reference clock, the phase accumulator linearly accumulates the frequency control words, generates a phase sequence, and looks up the waveform memory as an address. The obtained waveform amplitude data is processed by DAC and filtering to form the analog signal waveform of corresponding frequency.

The phase accumulator carries out phase linear accumulation under the action of the clock. When the accumulator reaches the full range, an overflow will be generated and the cycle will continue, forming a periodic signal at the output end. Waveform memory is to convert the phase sequence of the signal into amplitude sequence.

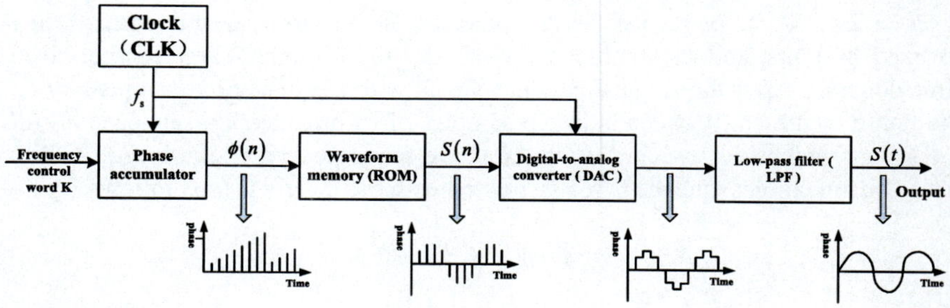

Fig. 6.5 Block diagram of basic composition principle of DDS

Theoretically, it can store any periodic waveforms, but in practical application, sine wave is the most representative.

Supposed that sine wave with frequency f is denoted as

$$S(t) = \sin(2\pi ft) \tag{6.3}$$

When sampling the signal at the sampling rate f_s, the discrete sequence of the signal is

$$S(n) = \sin(2\pi fnT_s),\ n = 0,\ 1,\ 2, \ldots \tag{6.4}$$

where $T_s = 1/f_s$ is the sampling period.

It is easy to know that the phase of a single-frequency sinusoidal signal is a linear function of time variables, so the salient feature of its phase sequence is that the phase increment between adjacent sample values is a constant and only related to the frequency of the signal, that is

$$\begin{cases} \phi(n) = 2\pi f \cdot T_s \cdot n, & n = 0,\ 1,\ 2, \ldots \\ \Delta\phi(n) = 2\pi T_s \cdot f \end{cases} \tag{6.5}$$

$\phi(n)$ is the phase sequence of $S(n)$, and $\Delta\phi(n)$ is the phase increment.

In DDS, f_s is taken as the reference clock, and assume that the frequency of the single-frequency signal is

$$f = kf_s = \frac{K}{M} f_s \tag{6.6}$$

In this expression, k is a constant. K and M are integers. The phase increment in Eq. (6.5) can be rewritten as

$$\Delta\phi(n) = \frac{2\pi}{M} K = \phi_s \cdot K \tag{6.7}$$

$\phi_s = 2\pi / M$ can be regarded as 2π phases to be uniformly quantized according to fixed M parts, and its significance is similar to T_s (signal sampling period in time domain). After the single-frequency signal with frequency f is sampled by f_s, the phase increment between adjacent samples of its quantization sequence is still a constant which is K. According to the phase sequence expression in Eq. (6.5), a phase quantization sequence is reconstructed with the invariant K as follows:

$$\phi(n) = K\phi_s \cdot n \tag{6.8}$$

Another expression of the signal with respect to ϕ_s can be obtained by continuing the backward deduction, which is equivalent to Eq. (6.4), namely

$$S(n) = \sin(Kn\phi_s) = \sin(2\pi f n T_s),\ n = 0,\ 1,\ 2, \ldots \tag{6.9}$$

Through such a process, in the case of fixed sampling rate f_s, the theoretical equivalent conversion between the conventional amplitude sampling sequence and phase sampling sequence of the signal with frequency f is completed. This is the basic principle that the corresponding discrete amplitude sequence can be generated directly from the waveform memory after the signal phase sequence is formed by the phase accumulator in DDS. $K = M \cdot f/f_s$ is referred to as the frequency control word.

In addition, according to the Nyquist sampling theorem:

$$\frac{f_s}{f} \geq 2,\ \text{ that is} \frac{M}{K} \geq 2 \tag{6.10}$$

Only when the upper expression is satisfied, $S(n)$ can only recover $S(t)$ after DAC and low-pass filtering smoothing.

6.3.2 *Jamming Signal Synthesis Method*

6.3.2.1 Complicated Waveform Modulation

In the previous section, the principle of single-frequency signal produced by DDS is detailed. However, when DDS is used as a radar jamming source, it is often necessary to generate some jamming signal waveforms with complex modulation patterns, such as linear frequency modulation (LFM) signal, nonlinear frequency modulation (NLFM) signal, and phase coded signal, which mainly involve the modulation of frequency, phase, and amplitude. Since DDS adopts all digital structure and is a phase control system, it can realize the functions of frequency modulation, phase modulation, and amplitude modulation of signals in a digital way to generate different modulation waveforms.

DDS addresses the waveform memory according to the phase sequence $\phi(n)$ formed by the frequency control word (K) and synthesizes analog signal $S(t)$ from a series of digital sampling points $S(n)$ through DAC. Its switching speed is very fast, and the phase of the output signal is continuous. Therefore, a frequency modulation module can be added before the phase accumulator. The frequency modulation module will form corresponding frequency control words respectively according to the frequency change of the signal and input them to the phase accumulator continuously. Generally, the frequency modulation module can be realized by adder. The frequency of the LFM signal increased or decreased linearly, a frequency accumulator similar to the phase accumulator can be used to realize linear frequency modulation.

Similar to frequency modulation, the output of phase accumulator in DDS is the phase sequence of the signal to be generated. Therefore, adding a phase modulation module based on adder after the phase accumulator can quickly realize instantaneous phase shift or phase modulation. The output of the waveform memory is the digital sample of the signal to be generated, which characterizes the instantaneous amplitude information of the signal. Therefore, the signal amplitude modulation can be realized by adding a multiplier and an amplitude controller after the waveform memory.

To sum up, the schematic block diagram of DDS with complex signal modulation capabilities such as frequency modulation, phase modulation, and amplitude modulation is shown in Fig. 6.6. The stepping clock is obtained by dividing the frequency of the reference clock and is used to provide timing for each parameter controller.

DDS can also complete pulse modulation because of its structural characteristics. It is realized by introducing a "reset" signal to the control end of the digital-to-analog converter or phase accumulator. Generally, there are two pulse modulation methods. One is to reset the phase accumulator and DAC at the same time. When the "reset" signal is canceled, the phase of the output signal remains continuous with that of the output signal at the time before the reset. The other is to only reset DAC, while the digital processing part including the phase accumulator continues to work. When the

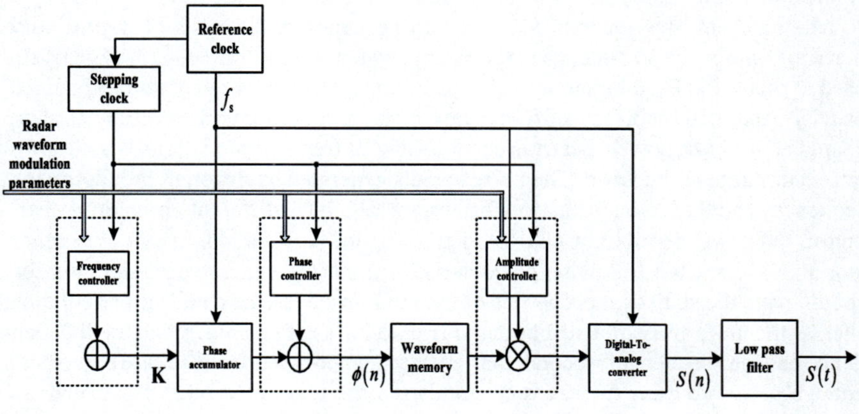

Fig. 6.6 DDS schematic diagram of complex signal modulation

"reset" signal is canceled, the phase of its output signal is synchronized with that of the output signal when the "reset" signal is not reset. This is very similar to the generation of coherent pulse train signals in modern radars, which can ensure the signal coherence of pulse modulation.

In addition, a single DDS can be used to jam several threat radars at the same time after the time division method is adopted, or multiple accumulators can be used to synthesize signals of multiple frequencies or several jamming waveforms at the same time to generate comb spectrum jamming or to deal with multiple threat radars at the same time. By using an accumulator and two waveform memories storing the sampling data of sine wave standard signal and cosine wave standard signal, respectively, the quadrature modulated jamming complex signal can also be generated.

6.3.2.2 Synthesis of Noise Signal

The generation of noise jamming signal is an important function of DDS radar jamming source. The traditional analog noise signal generation method is generally to directly amplify the microwave noise of the analog devices to form direct RF noise, or use the microwave noise signal to modulate the frequency, phase, or amplitude of the carrier frequency signal to form noise frequency modulation, noise phase modulation and noise amplitude modulation signals respectively. Similarly, the common approach for generating noise jamming signals based on DDS is to use the complex signal modulation structure shown in Fig. 6.6 in the previous section, use random sequences as the waveform modulation parameters of noise signals, and input the frequency controller, phase controller or amplitude controller of DDS to obtain noise frequency modulation, noise phase modulation, and noise amplitude modulation signals, respectively. When necessary, a group of random sequences or multiple groups of random sequences can also be input into any two or all of the controllers at the same time to form different types of noise jamming signals.

Among them, the generation of random sequence is the basis of digital noise jamming source. In practice, pseudorandom sequence with finite length is generally used to replace it. Pseudorandom sequences can be artificially generated and copied, usually generated by binary shift registers. Their autocorrelation function has sharp characteristics, the power spectrum occupies a wide frequency band, and has the properties similar to white noise. The noise signals generated by different pseudorandom sequences and their transformation forms generally have different probability distribution and power distribution characteristics. A good pseudorandom sequence generator should have the following characteristics: first, the generated random number should have the statistical properties of the random sample of a uniform population, such as the uniformity of distribution, the randomness of sampling, and the independence between sequences; second, the generated random sequence should have a long enough period to meet the needs of simulation calculation. Typical pseudorandom sequences such as m sequence.

Moreover, the generalized DDS noise jamming signal can also be generated in the following way: a noise signal waveform sample is generated by hardware such as DSP or calculated by simulation software, which is directly stored in the DDS waveform memory and then read in sequence and output by DAC and band-pass filter to form a noise jamming signal with a certain bandwidth.

6.3.3 Main Performance Analysis

Due to the full digital structure of DDS, it generally has the advantages of high-frequency resolution, fast frequency switching speed, continuous phase of signal during frequency switching and can generate arbitrary waveform, but it also has the disadvantages of large output spurious. The following focuses on the analysis of several important performance indicators of DDS for reference during jamming source design.

(1) Output jamming frequency range

According to Eq. (6.10), the frequency range of the jamming signal that DDS can generate is first determined by the system reference clock and Nyquist sampling theorem. Theoretically, the upper limit of the output frequency is $f_s/2$; that is, it is equal to half of the reference clock frequency. However, due to the non-ideal transition characteristics of low-pass filter and the limitation of high-end signal spectrum deterioration, the upper limit of DDS output frequency that can be realized in practical engineering is generally $0.4 f_s$, the lower limit is 0.

In addition, according to the structure and working principle of DDS, the upper limit of working frequency of DDS is also limited by the maximum operating speed of internal devices in the system, such as the speed of phase accumulator, waveform memory, and DAC. At present, with the progress of microelectronics materials and processing technology, the frequency of sampling clock can be as high as several hundred MHz to several GHz, and the speed of the DAC can also be as high as several GHz. Therefore, the maximum working frequency of DDS can also reach several GHz. When DDS is used as a radar jamming source, it is generally possible to use DDS to generate intermediate frequency jamming signals first and then use the up-converter to cover a larger working frequency range.

(2) Frequency resolution

According to Eq. (6.6), the frequency resolution of DDS is mainly determined by the reference clock frequency f_s and M. M is the total number of phase quantization, which is mainly determined by the number of bits of the phase accumulator. When the number of bits of binary phase quantizer is N, $M = 2^N$, the frequency resolution of DDS is:

$$\delta_f = \frac{f_s}{2^N} \tag{6.11}$$

For example, under this premise, $N = 48$ and $f_s = 2.4$ GHz, the frequency resolution of the DDS is 0.0000085265 Hz.

It can be seen that as long as the number of bits of the phase accumulator is long enough, the frequency resolution of DDS can be high enough. Using DDS as radar jamming source can effectively improve the aiming frequency accuracy of radar jamming signal.

(3) Frequency conversion time

From the digital structure and open-loop structure of DDS, it can be seen that the frequency conversion time or system response time of DDS is only related to the transmission time of frequency control word and the delay time of internal circuit. The delay time of internal circuit includes the delay time of the phase accumulator, the data reading time of the waveform memory, the delay time of the DAC, the frequency response time of the filter, etc. In high-speed DDS system, the pipelined structure can be adopted, and the frequency conversion time is very short, which can reach the order of ns.

(4) Spurious performance

Poor spurious suppression performance is one of the shortcomings of DDS, which is mainly caused by the finite word length effect and truncation in digital technology.

In addition to the main frequency spectrum, there are also many spurious components in DDS synthetic signals, which mainly come from three sources. According to the DDS processing flow, the first is the phase truncation error. In order to obtain high-frequency resolution, the number of bits of the phase accumulator is generally taken as large. In the DDS design, in order to save the capacity of the waveform memory as much as possible, on the premise of not introducing too much interference, the low-order segment data of the phase accumulator is generally cut off as much as possible, and only the high-order segment data is used as the address, so the phase truncation error is generated. The second is the amplitude quantization error. Waveform memory stores the samples of standard signal, but its storage bit width is also limited, that is, compared with its analog signal, the digitized sample signal of the standard signal has a finite word length effect in the process of waveform amplitude quantization, resulting in amplitude quantization error. The last source is the nonlinear noise of the DAC. DAC has non-ideal characteristics. At the output of the DAC, the step wave formed by digital samples contains harmonics and spurious components of the main frequency, and some of them are still difficult to be suppressed by the low-pass filter. Regarding the calculation of stray power, you can refer to the relevant literature of DDS technology research.

According to the above analysis, the spurious suppression method in DDS design is mainly realized by increasing the bits of DAC and the storage bit width of waveform memory. At present, it can also be realized by signal generation method based on waveform symmetry, signal linear interpolation, random jitter technology, or different phase-to-amplitude mapping structure.

To sum up, theoretically, DDS can digitally synthesize any jamming waveform or its modulation pattern contained in the waveform memory. However, the jamming waveforms synthesized by DDS are not completely coherent with the waveforms of a specific threat radar, which leads to the requirement for accurate frequency guidance of the radar active jamming system so that these jamming signals can first enter the radar receiver. The effect of DDS to generate signals, especially deceptive jamming signals, will also depend on the matching ability between DDS and the actual threat radar signal waveform.

6.4 Design of Digital Radio Frequency Memory

6.4.1 Basic Composition and Principle

Digital Radio Frequency Memory (DRFM) is a kind of digital jamming source widely used in the world. Especially with the rapid development of radar pulse compression technology and Doppler coherent processing technology, DRFM is a powerful means against modern advanced radar because it can well retain the waveform characteristics of radar signal, provide a variety of coherent jamming signals, and obtain the same gain as target echo in radar processing.

DRFM is based on high-speed signal sampling and digital storage technology and forms coherent jamming signals with different modulations by receiving, sampling, storing, and copying radar radio frequency signals. It can store and reproduce radio frequency and microwave signals. In the 1970s and 1980s, the basic principle of DRFM, proposed by electronic warfare designers, was considered a major revolutionary breakthrough. Nowadays, with the development of digital frequency conversion, high-speed ADC and DAC, high-speed large-scale digital storage, high-speed integration, and microcomputer control processing, DRFM has been rapidly developed and applied on a large scale and has shown good waveform jamming effect.

The basic composition and principal block diagram of DRFM are shown in Fig. 6.7, which mainly includes ADC, digital memory, controller, DAC, filter, reference clock, etc. Due to the limitation of response speed of digital devices, the digital frequency storage is usually carried out in the intermediate frequency. Therefore, taken broadly, DRFM also includes up- and down-frequency conversion and local oscillator signal and other functional modules. The main workflow of DRFM is as follows: firstly, the threat radar signal received by the radar active jamming system is down converted into intermediate frequency signal, which is then sampled, quantified, and encoded by ADC to form digital waveform information and stored in digital memory. When the jamming signal needs to be generated, the appropriate digital signal is read from the memory and modulated by the controller, such as replication, delay, and frequency shift. The signal is restored to the IF analog signal by the DAC and the filter. Finally, it is up-converted to jamming RF signal and output to the transmitter or antenna.

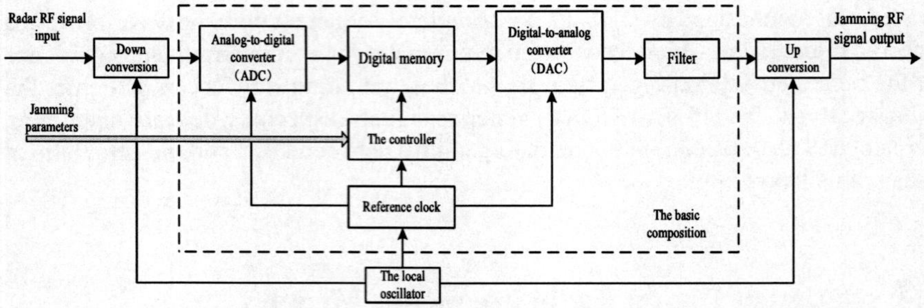

Fig. 6.7 Composition principal block diagram of DRFM

The principle of DRFM technology is relatively simple, but due to the need for high-speed digital signal processing, the engineering implementation is relatively complex. In the up- and down-conversion processing, in order to accurately receive radar RF signals and transmit jamming signals of the correct frequency, the receiver is usually required to provide frequency guidance for the local oscillator module, and the local oscillator is required to have high stability and low phase noise, so as to obtain higher quality jamming signal. The response speed of local oscillator, ADC, DAC, and digital memory should be fast in order to reduce the response time of the jamming source. The controller needs to form a control command word according to the jamming control parameters transmitted from the control unit in the radar active jamming system, quickly modulate the signals in the digital memory, and generate different jamming styles. For reasons of ensuring matching degree and coherence between the output jamming signal and the input radar signal, the DRFM system needs to use the reference clock and the local oscillator signal source from the same signal source, respectively.

DRFM adopts digital implementation structure, which has the advantages of good stability, flexibility, and versatility and is very conducive to the generation of a variety of jamming waveforms. In addition, the digital system is also convenient for high-density integration and miniaturization of jamming source modules.

6.4.2 Design of Digital Radio Frequency Memory

6.4.2.1 Number of Digital RF Channels

A channel of DRFM is defined as a signal frequency storage channel consisting of an ADC, a digital memory and a DAC. The selection of the number of channels is usually mainly related to the design requirements of the instantaneous bandwidth of the jamming signal and in some cases is also related to the jamming signal style. According to the number of channels in DRFM, it can be generally divided into single channel, orthogonal double channel, and multi-channel. They have some differences in ADC sampling rate, signal instantaneous bandwidth, and jamming signal quality.

(1) Single-channel DRFM

Single-channel DRFM is the most basic structure of DRFM system, and other types are an extension of this structure. Its basic composition is shown in Fig. 6.7. According to Nyquist sampling theorem, the sampling rate of ADC and DAC must be greater than or equal to 2 times the instantaneous bandwidth of the DRFM system or the maximum frequency of the IF signal to be processed. In application, due to the influence of the non-ideal frequency response characteristics of the filter transition band, the instantaneous bandwidth of the actual system will be less than half of the ADC sampling frequency.

The advantage of this implementation method is that it has simple structure, does not introduce too many links that may cause the deterioration of interference performance index, and has strong parasitic signal suppression ability. Its disadvantage is that the instantaneous bandwidth of the system is small when the ADC sampling rate is fixed.

(2) Orthogonal dual channel DRFM

The principal structure of orthogonal dual-channel DRFM is shown in Fig. 6.8. In the down-conversion link, the radar input signal is processed in two channels of the in-phase component (I-channel signal) and the quadrature component (Q-channel signal), respectively. After the orthogonal down-conversion, it is digitally sampling and stored, which can be regarded as the combination of two single-channel DRFM.

According to Nyquist sampling theorem, when I and Q orthogonal dual channel signals are sampled for complex signals, the original signal can be recovered without distortion only if the sampling rate of each channel is greater than or equal to the maximum frequency of the signal. Therefore, the advantage of orthogonal dual-channel DRFM over single-channel DRFM is that the sampling frequency of ADC and DAC is reduced by half, but it requires high amplitude and phase consistency of I and Q channels, which increases the complexity of the system. In the case of wideband signal processing, the orthogonal dual-channel DRFM may have mirror

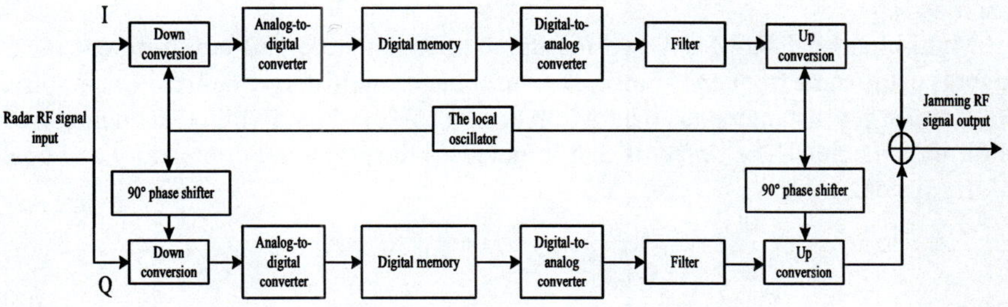

Fig. 6.8 Schematic diagram of orthogonal dual channel DRFM

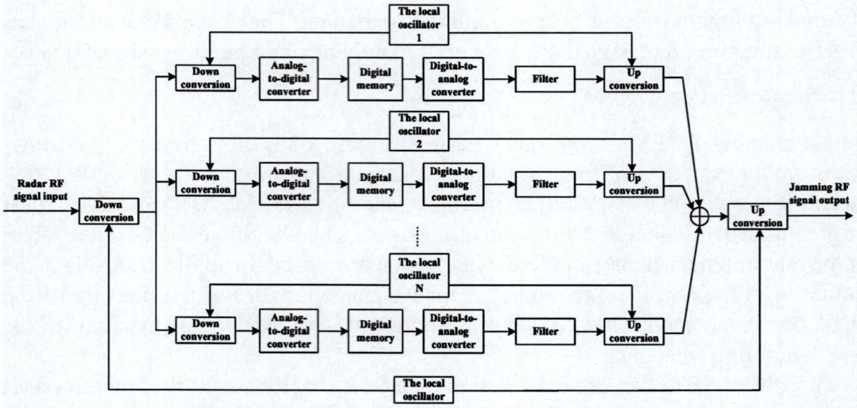

Fig. 6.9 Structure diagram of multi-channel DRFM

frequency components and cause large parasitic signals, thus affecting the spurious and coherent characteristics of jamming signals.

(3) Multi-channel DRFM

In terms of structure, multi-channel DRFM is composed of multiple single-channel DRFM, as shown in Fig. 6.9.

Multi-channel DRFM is mainly applied in radar active jamming system which has high requirement of instantaneous bandwidth. Multi-channel DRFM can be regarded as the synthesis of a wideband DRFM by using multiple narrowband DRFM to improve the instantaneous working bandwidth of the jamming source. Each DRFM can store one threat radar signal, which avoids the intermodulation between multiple radar signals and the generation of parasitic components, thus improving the quality of jamming signals and countering multiple targets simultaneously. When the operating frequency range or instantaneous bandwidth of threat radar is large, it can also be regarded as that the input radar signal is sent to multiple narrowband DRFM for processing after frequency division, which reduces the bandwidth and rate requirements for ADC, DAC, and digital memory in DRFM and is easy to implement in engineering.

Multi-channel DRFM has two merits: first, it can store and jam different radar signals in the same frequency range, so as to enhance multi-target jamming capability. Second, larger instantaneous bandwidth can be achieved by dividing frequency into different channels. The defect is that it increases the structural complexity and cost of the system.

6.4.2.2 Signal Quantization Method

DRFM quantization method refers to the method that ADC samples and quantizes the input analog signal, generally including amplitude quantization, phase quantization, and amplitude phase quantization.

(1) Amplitude quantization

Amplitude quantization is one of the most traditional quantization methods. It uses ADC to discretely sample and quantize input analog signals in the amplitude domain and convert it into digital signal. Among them, ADC quantization generally takes a fixed amplitude voltage as a reference, average quantizes to 2^N levels (N is the number of ADC bits), and then compares the instantaneous amplitude of the input signal with it to obtain the code of the closest level as its amplitude quantization value. During reconstruction, the digital signal can be restored to the analog signal through DAC and low-pass filter. The sampling rate and quantization bits N of ADC and DAC determine the fidelity of the reconstructed signal.

(2) Phase quantization

The principle of DRFM phase quantization is basically the same as that of DDS phase quantization. It means that the 2π phase cycle of the sine wave or cosine wave signal is averagely quantized into 2^N phase intervals (N is the number of bits of phase quantization) and then compared with the instantaneous phase of input signal, the code of the closest phase interval will come as its phase quantization value. During reconstruction, the digital phase quantization value can be reconverted into a stepped signal, which can be regarded as the superposition of N rectangular waves output by 1 bit phase quantization according to a certain phase relationship.

The phase quantization process is shown in Fig. 6.10. In Fig. 6.10a, taking $N = 4$ as an example, the phase is evenly quantized into 16 phase intervals, and there are a total of 16 phase codes. In Fig. 6.10b, a group of digital signals can be obtained by phase quantization of an input signal using these 16 codes. During reconstruction, a stepped signal can be reconstructed, which is similar to the characteristics of the input signal.

The phase quantization DRFM is different from the amplitude quantization DRFM in the implementation mode, mainly because the polarity quantizer replaces the ADC and the weighted addition network replaces the DAC. When $N = 4$, a typical phase quantization and sampling circuit are shown in Fig. 6.11. In order to facilitate phase sampling and quantization with analog comparator, the signal is hard limited by the limiter before that. The phase quantized DRFM has the ability to process signals with a wide dynamic range, but the adaptability of multiple signals at the same time is poor. When multiple signals with similar amplitude enter the phase quantization DRFM at the same time, the stored phase information cannot represent any signal, resulting in poor quality of output jamming signals.

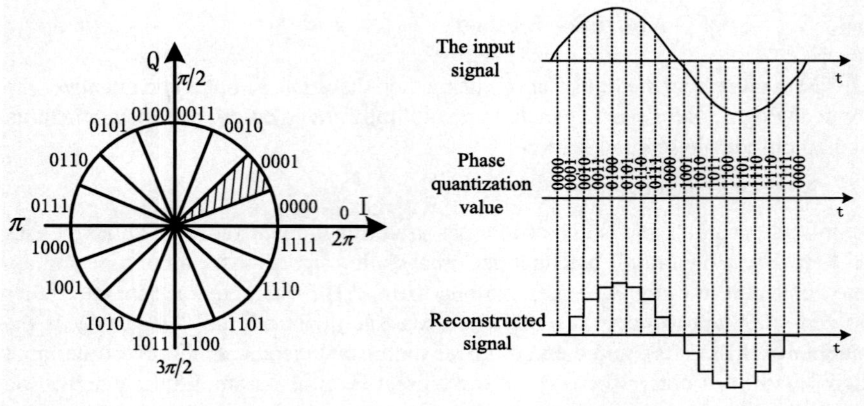

(a) Phase quantization coding (b) Phase quantization of signals

Fig. 6.10 Schematic diagram of phase quantization principle

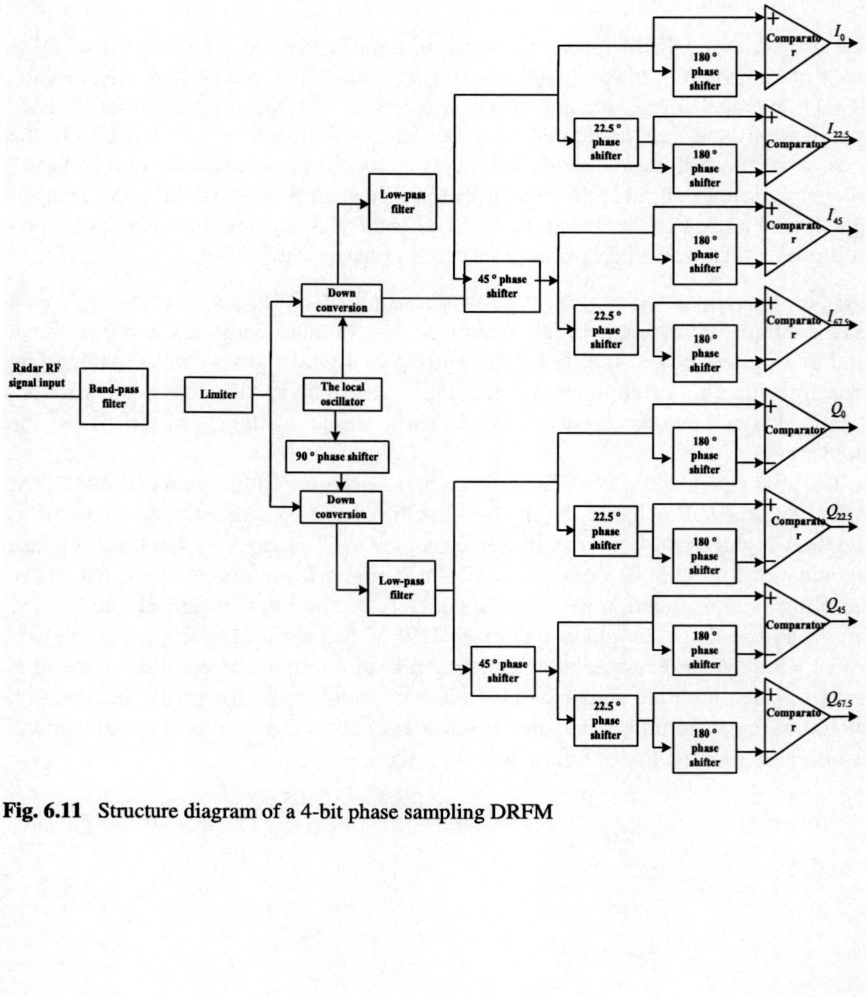

Fig. 6.11 Structure diagram of a 4-bit phase sampling DRFM

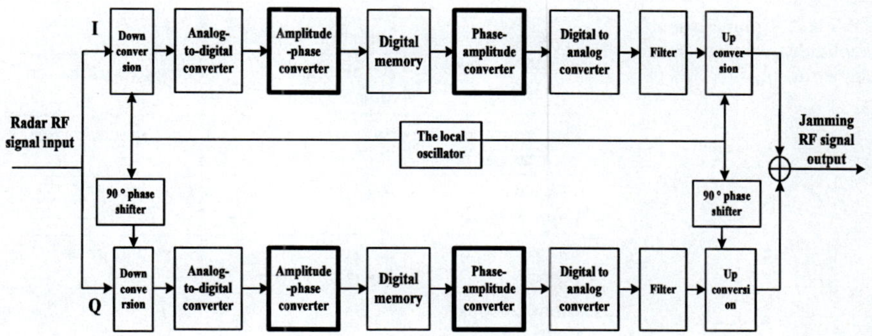

Fig. 6.12 Structure diagram of amplitude-phase quantization DRFM

(3) Phase and amplitude quantification

The DRFM of amplitude-phase quantization is an improved way based on amplitude quantization and phase quantization. Its structure is shown in Fig. 6.12. I and Q orthogonal dual channel amplitude quantization is still adopted; that is, two ADC channels are used to sample and quantize the signal amplitude, respectively, and then amplitude-phase converter and phase-amplitude converter are added between ADC and digital memory and between digital memory and DAC to realize mutual conversion between instantaneous amplitude and instantaneous phase of signal. While ensuring the high fidelity of the signal, by storing the phase quantization result of the signal, the data storage can be effectively reduced, and it is also more convenient for the controller to modulate the phase of the jamming signal, eliminate the intra-pulse amplitude change caused by the signal amplitude fluctuation, and improve the signal-to-noise ratio of the output jamming signal.

Both phase quantization and amplitude quantization have advantages and disadvantages, as shown in Table 6.2.

6.4.2.3 Signal Storage Method

The storage method of DRFM for the input radar signal is related to the jamming style and jamming time sequence of radar active jamming system. However, according to whether the input signal is continuously sampled and stored by DRFM, it can be divided into pulse sequence storage, full pulse storage, partial pulse storage, and intermittent sampling storage.

(1) Pulse sequence storage

Pulse sequence storage means that the radar pulse signal sequence input continuously for a period is stored in the digital memory, as shown in Fig. 6.13.

When it is necessary to generate jamming signals, the pulse sequence can be read out for copying, superimposing or forwarding, etc., or only part of the pulses

Table 6.2 Comparison of amplitude quantization and phase quantization

Comparative aspects	Quantitative method	
	Amplitude quantization	Phase quantization
Frequency conversion mode	ordinary	Orthogonal mixer
Resource requirements	High	Lower
Output signal power	Constant Gain, Constant Power	Constant power
Simultaneous multi-signal adaptability	YES	NO
Digital signal processing method	FFT, Amplitude Modulation	Direct phase analysis, phase modulation

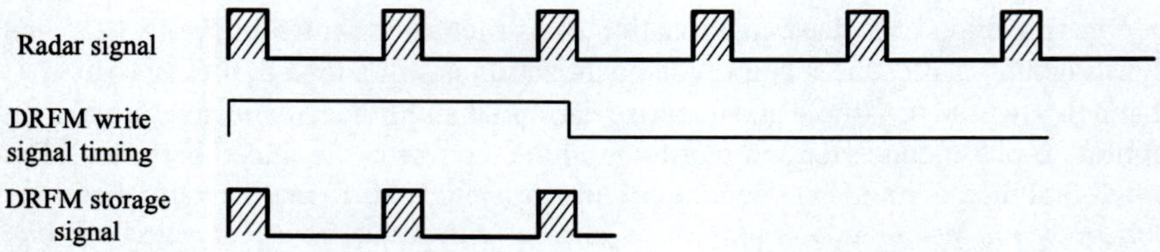

Fig. 6.13 Schematic diagram of pulse sequence storage

in the sequence can be read. Although this kind of storage method has very high requirements for memory, it can retain radar signals to the greatest extent, and it is usually more suitable for the situation where the pulse width is small under PD working mode.

(2) Full pulse storage

Full pulse storage refers to storing the input single complete radar pulse signal in the digital memory, as shown in Fig. 6.14. When the jamming signal needs to be generated, the stored data is directly read out for copying, modulation, and forwarding. The frequency of the generated output jamming signal is the same as that of the input radar signal and has very high signal coherence.

In the amplitude quantization mode, since the full pulse storage is to store all the radar signals in the whole cycle, it can completely save the subtle characteristics of the radar signal, which is more suitable for the storage of complex modulated signals or fine deception jamming.

Full pulse storage also requires high memory capacity, especially for warning radar signals with large pulse width. To solve this problem, the circular storage mode of dual-port memory can be adopted; that is, its head and tail addresses are connected; one port stores and writes data, and the other port reads data.

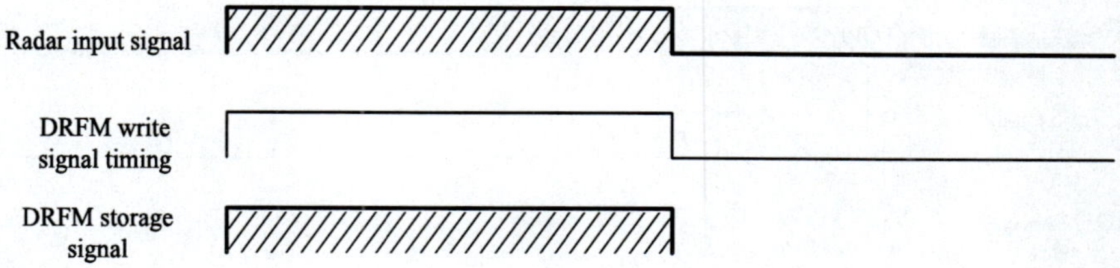

Fig. 6.14 Schematic diagram of full pulse storage

(3) Partial pulse storage

Partial pulse storage means that DRFM only stores a small segment of data at the beginning of the radar input signal, as shown in Fig. 6.15. When the jamming signal needs to be generated, the jamming signal distributed over a longer period of time can be formed by cyclic reading or copying combination. The internal phase of the jamming signal formed by data replication and combination is generally discontinuous, which may cause the frequency inconsistency with the radar signal and the signal coherence deteriorates. Generally, it is only suitable for the storage of radar signal without modulation in the pulse. However, this method reduces the requirement of memory capacity and improves the response time of the system.

(4) Intermittent Sampling Storage [2]

Intermittent pulse storage refers to intermittently storing and reading signal data by controlling the receiving and sending switches of DRFM, which can form multiple interferences within the pulse duration, as shown in Fig. 6.16. The length of each stored data and read data is relatively short, which can reduce the storage capacity of the sampling data, and realize the continuous update of the stored pulse data relative to the real radar signal, so that a certain phase continuity is still maintained between each small segment of jamming. The delay time of jamming signal generated by this method is generally shorter than that of radar signal. It is a compromise between storage capacity and signal coherence. In practical application, it can have a good jamming effect on LFM pulse compression radar.

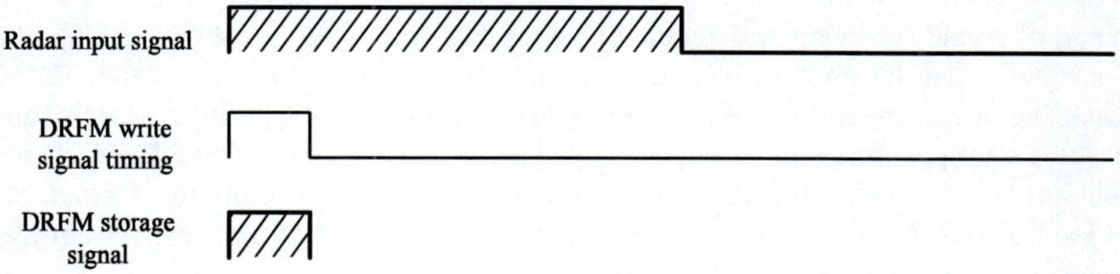

Fig. 6.15 Schematic diagram of partial pulse storage

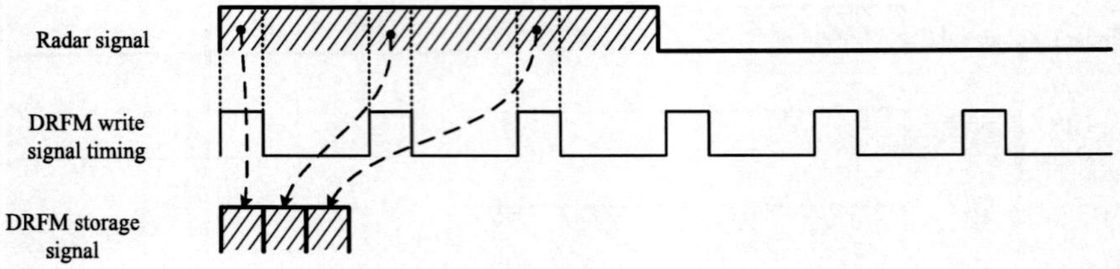

Fig. 6.16 Schematic diagram of intermittent sampling storage

6.4.3 *Main Performance Analysis*

A well-designed DRFM should have the following characteristics: it can adapt to the input radar signal with large dynamic range and large pulse width; the generated jamming signal is well matched with the radar signal in terms of frequency and coherence, etc.; low spurious component and high signal-to-noise ratio in output signal; fast system response, low processing delay, etc. Among them, the signal-to-noise ratio and coherence of jamming signal are the most important, which is the key to the success of DRFM. The following is a brief analysis of several common performance indicators in DRFM design.

(1) Instantaneous bandwidth and working bandwidth

The instantaneous bandwidth refers to the maximum bandwidth of the signal that DRFM can process. According to the structure of DRFM, its instantaneous bandwidth mainly depends on the sampling rate of ADC and DAC. For single-channel DRFM, the instantaneous bandwidth is slightly less than half of the sampling rate. For orthogonal dual-channel DRFM, the instantaneous bandwidth is slightly less than the sampling rate.

The working bandwidth refers to the frequency range of RF signals that DRFM can receive and process. The working bandwidth can be extended by up- and down-frequency conversion.

(2) Input dynamic range

The input dynamic range refers to the maximum amplitude change range of input radar signal that DRFM can adapt to, which is mainly determined by the dynamic range of signal receiving and acquisition part. For amplitude quantization DRFM, the input dynamic range mainly depends on the quantization bits N of the ADC. Each bit of quantization increases the dynamic range by about 6 dB, which can process multiple simultaneous arrival signals. For phase quantization DRFM, there is a limiter before signal quantization, so it is not affected by the amplitude fluctuation of input signal. Although the input dynamic range is large, it does not have the ability to adapt to multiple signals at the same time.

(3) Storage capacity

Storage capacity refers to the size of DRFM digital memory. The larger the capacity, the more signals can be stored and processed or the longer the signal duration. The design of storage capacity is also related to the quantization method, sampling rate, and quantization bits. When the sampling rate and quantization method are fixed, the storage capacity can often be replaced by the maximum storage depth index, which can be expressed by time, or the size of storage space. The time representation is more intuitive.

(4) Minimum propagation delay

The minimum propagation delay refers to the minimum time interval from the time when the radar signal is intercepted to the time when the DRFM outputs the jamming signal. It is mainly related to the response speed of the system, including the time delay of frequency conversion, ADC, digital memory reading and writing, controller jamming modulation, DAC, and filtering.

(5) Delay resolution and Doppler resolution

Delay resolution refers to the minimum interval of range time delay during DRFM jamming signal modulation. Doppler resolution refers to the minimum interval of Doppler frequency during DRFM jamming signal modulation. Delay resolution, also known as delay increment, is generally no less than the time to read the digital memory once and is usually related to the clock frequency of DRFM. When the controller performs jamming modulation on the stored radar signal, the signal forwarding delay in range dimension and the frequency shift in Doppler dimension are generally integer multiples of their respective resolutions.

(6) Parasitic signal analysis

The parasitic signals of DRFM mainly come from the process of frequency conversion and quantization, including local oscillator leakage, mirror response, cross-modulation, and quantization of parasitic signals.

The LO leakage is caused by the insufficient isolation between the local oscillator signal and the output signal in the frequency converter, resulting in the continuous wave signal of the local oscillator frequency in the output signal. The mirror response is caused by the inconsistency of amplitude and phase of channel I and channel Q in the orthogonally modulated DRFM, resulting in an additional mirror sideband in the jamming output. Cross-modulation refers to the new frequency component generated by the mutual modulation of different frequency and harmonic components between the signal and the local oscillator during up- and down-conversion. The quantization parasitic signal is generated by the quantization error during the amplitude quantization or phase quantization of DRFM, so the quantization parasitic signal of DRFM can be reduced by designing appropriate quantization bits.

(7) Coherence analysis

As an important coherent jamming source, the coherence index DRFM signal is particularly essential. From the perspective of radar signal processing technology, the coherence of jamming signal is mainly manifested in two aspects: one is the coherence between the output jamming signal and the input radar signal, the other is the coherence within or between the output jamming signal pulses.

From the structure of DRFM, the input radar RF signal has gone through down-conversion, sampling, quantization, reconstruction, and up-conversion, which may have an impact on the coherence of the jamming signal, mainly including the local oscillator, sampling clock, and quantization error.

DRFM needs to continuously store and forward the input radar signal. Therefore, in the up- and down-conversion processing link, the frequency stability of the local oscillator will affect the frequency accuracy of the output signal in the up- and down-conversion processing link; that is, frequency drift may occur, thus weakening the coherence of the jamming signals. Although the clock used for sampling and reconstruction is the same, the time difference between sampling and reconstruction will be shown as frequency offset of the output signal in the reconstruction process because the time of sampling and reconstruction are generally not the same. For single signal frequency storage, as the maximum jamming delay time or coherent jamming time increases, the time delay between sampling and reconstruction will accumulate with a consequence of a larger frequency drift. When the local oscillator, sampling clock and frequency drift are determined, the coherence of the output jamming signal mainly depends on the coherent jamming time. The quantization error of ADC and DAC will not only produce parasitic signals but also directly affect the coherence of output signals.

In summary, DRFM can relatively accurately and completely retain the signal waveform of threat radar. The generated jamming signal is coherent with the radar signal, and it can also obtain the same processing gain as the real target echo signal. Compared with incoherent jamming, this technology can obtain a power advantage of up to tens of decibels and is often used to generate deceptive jamming or coherent suppressive jamming to radar. Here are the main working modes and technical indicators of several typical DRFM products at home and abroad, as shown in Table 6.3.

6.5 Development Trend of Jamming Sources

6.5.1 Digitalization, Miniaturization, and Multifunctional

The digital system has the characteristics of high stability, strong programmability, and good controllability, which is very suitable for the complex signal waveform generation and flexible jamming modulation of radar jamming source. The digital jamming source is also very beneficial to further integration and miniaturization of

Table 6.3 List of typical DRFM products

Number	Name	Working bandwidth	Instantaneous bandwidth	Quantization method and number of digits	Input dynamic	storage depth	Delay resolution	Doppler resolution	Harmonic suppression
1	Anaren (USA) 45A8000	2.4–5.4 GHz	500 MHz	Phase 8 bit	−35 to + 15 dBm	2000 us	4 ns	3.5 Hz(range30 MHz)	−45 dBc
2	CSIR (South Africa) Wideband single echo	250–2250 MHz	2 GHz	amplitude 10 bit ADC 12 bit DAC	50 dB	3000 us	0.2 ns	0.5 Hz(range ±1 GHz)	>50 dB
3	Mercury (USA) SP030302	3 GHz	600 MHz	Phase 4 bit	50 dB	>1 ms	5.56 ns	< Hz(range > 1 MHz)	>20 dB
4	SPEC (USA) ADEP1300	1–18 GHz	1.3 GHz	Amplitude 10 bit ADC 12 bit DAC	60 dB	500–400 ms	<1 ns	<1 Hz(range ±10 MHz)	−40 dBc
5	MicroSystem(USA) AY09250-1	25–225 MHz	>200 MHz	Amplitude 12 bitADC 16 bitDAC	50 dB	8 ms	2 ns	<1 Hz(range ±10 MHz)	−60 dBc

the system. At present, digital jamming sources have been widely used, and it is an inevitable trend to comprehensively replace analog jamming sources.

From the perspective of function, another development trend of jamming source is to meet the requirements of radar active jamming system for various jamming signals with more powerful hardware performance and richer configuration resources, so as to truly realize the generation and modulation of arbitrary waveforms. In practical applications, this approach cannot only effectively reduce the platform's requirements for the type and quantity of jamming sources and reduce the overall volume and weight of the jamming equipment, but also its versatility and flexibility will be further enhanced.

6.5.2 Large Bandwidth and Refinement

The quality of the jamming signal generated by the jamming source will directly affect the jamming effect on radar. With the continuous progress of radar signal processing technology and anti-jamming technology, the requirements of radar active jamming system for future jamming sources will be as follows: the modulation and control flexibility of the jamming signal must be optimized to respond quickly and realize smart jamming; the frequency range of the jamming source is wide enough to deal with all traditional and non-traditional RF threats in the future; the output jamming signal spectrum must be accurate enough to avoid being identified by radar and interfering with other electronic equipment of our own party, etc.

6.5.3 Synthesis and Intelligentize

From the perspective of signal processing, the future jamming sources can be combined with digital frequency measurement, signal analysis, and waveform synthesis to replace some functions of the receiver and system control unit in the existing radar active jamming system and develop towards integration.

From the perspective of jamming effectiveness, jamming sources can also develop toward intelligence. For example, the instantaneous bandwidth of the jamming source can be automatically selected by high-low matching according to different counter objects; according to the different waveforms of the input radar signals, the internal jamming resources are automatically allocated and the jamming styles are adaptively adjusted.

References

1. Yun H (2007) Design of digital noise source based on direct digital frequency synthesis technology. Master Dissertation of Wuhan University of Technology
2. Xuesong W, Shunping X, Dejun F et al (2010) Modeling and simulation of modern radar electronic warfare systems. Publishing House of Electronics Industry, Beijing

Chapter 7
Jamming Transmitter

In addition to space alignment, time overlap, frequency coverage, and polarization matching, sufficient jamming power is also required for effective jamming of radar. The function of radar jamming transmitter is to amplify low-power jamming signal to the required power level. The signal acts on radar system receiver through antenna, space transmission, and radar antenna and affects radar target detection and data processing.

This chapter introduces the requirements, types, design, performance accounting, and development trend of radar active jamming system transmitters. The first section introduces the requirements analysis of the transmitter, focusing on the analysis of the requirements from the frequency band, power, efficiency, and other aspects. The second section introduces the selection of transmitter type. The third section introduces the design method of transmitter scheme and the design of typical transmitter. The fourth section introduces the performance accounting of the transmitter from the aspects of output power, power consumption, heat consumption, and gain. The fifth section introduces the development trend of transmitter technology.

7.1 Requirements Analysis of Jamming Transmitters

The most important requirements of radar active jamming system transmitters are output power and operating frequency range, as well as output signal quality. Furthermore, taking the power supply capacity and heat dissipation capacity of the system into consideration, the working efficiency of the transmitter needs to be analyzed. Because the transmitter has high-power devices and high-voltage power supply and other vulnerable devices, the requirements of environmental adaptability and equipment reliability are also important to consider. This section will introduce these aspects.

G. Tang et al., *Techniques and System Design of Radar Active Jamming*,
https://doi.org/10.1007/978-981-19-9944-4_7

7.1.1 Working Bands Requirements

The working frequency band requirements of the transmitter matches with the frequency range requirements of the jamming signal, which is mainly determined according to the frequency band of the jamming target. The direct requirement comes from the index requirements such as the working frequency range of the jamming system.

At present, according to the working frequencies of different radar targets [1], the working frequencies of jamming have gradually increased from VHF, UHF, L, S, C, X, Ku bands to K, Ka, V bands, and even W bands. Considering the current capability and level of the actual devices, as well as the requirements of the jamming system, the typical frequency bands of the jamming transmitter are: 0.8–2 GHz, 2–6 GHz, 2–18 GHz, 6–18 GHz, etc.

7.1.2 Output Power Requirements

Output power refers to RF power output from transmitter or transmitter array port. The output power of jamming transmitter is determined by the requirement of the jamming system ERP. The power requirement P_{out} of the transmitter can be obtained by subtracting the antenna gain G_{out} and the insertion loss L_{out} between the antenna and the transmitter from the value of the jamming ERP. The formula of decibel number is as follows:

$$P_{\text{out}} = \text{ERP} - G_{\text{out}} - L_{\text{out}} \tag{7.1}$$

In the previous chapter, the jamming ERP requirements of radar active jamming system have been analyzed, which are mainly related to the protected target RCS, jamming range, JSR, jamming mode and jamming application mode, etc. For self-defense jamming, when the jammer is closest to the radar, the power requirement of the transmitter is the largest. For supporting jamming, the maximum power requirement of the transmitter is when the jammer is farthest from the radar and the protected target is closest to the radar, and the jamming signal usually enters from the sidelobe of the radar antenna.

Compared with the power of the radar transmitter, the output power demand of the jammer in the self-defense radar active jamming system is usually much lower than that of the radar transmitter, ranging from a few watts to several hundred watts. In the case of sidelobe support jamming, the sidelobe loss of jamming signal is generally more than 30 dB, so the jamming transmission power demand is high. Some jamming ERP has even reached the level of megawatt, and the output power of transmitter also needs to reach or even exceed the level of kilowatt. Furthermore, in terms of output duty cycle, most jamming transmitters require continuous wave output capability, while most radar transmitters are pulse output. Therefore, the design of high-power jamming transmitters is more difficult than that of radar transmitters.

7.1.3 Work Efficiency Requirements

The working efficiency of a jamming transmitter refers to the energy conversion efficiency between the system power supply and the RF power output, which is defined as:

$$\eta = \frac{P_o}{P_i} \quad (7.2)$$

η is the working efficiency of the transmitter, P_o is the output RF signal power of the transmitter, and P_i is the power supply to the transmitter.

The energy generated outside the working efficiency of the transmitter is mainly converted into heat energy, which will increase the operating temperature of the power devices in the transmitter, increase the requirements for the heat dissipation capacity of the system, affect the stability and reliability of the transmitter, and cause the energy loss and improve the requirements for the power supply capacity of the system. As a result, transmitters are generally expected to work as efficiently as possible, and there is often a minimum requirement for a particular application. However, the actual situation is that the transmitter efficiency of the same system is restricted by the level of devices, so it is impossible to improve greatly. It will be explained later that the efficiency of transmitters varies from system to system, so the requirement of transmitter efficiency is an important factor in choosing a transmitter system.

7.1.4 Output Signal Quality Requirements

In addition to transmitting output power and working frequency range, the output signal quality needs to be considered when the transmitter amplifies the jamming signal. Nowadays, the output signal quality of transmitter is mainly evaluated from the frequency domain, including harmonic, cross-modulation, stray, out-of-band suppression ability, and so on.

Harmonic refers to a signal whose frequency contained in the output signal of the transmitter is integer times of the frequency of the output fundamental signal. Generally, it can be quantitatively described as the ratio of the power of the harmonic and the fundamental wave in the output signal when the output signal reaches saturation or rated power. The unit is dBc. For jamming transmitters operating in broadband, the maximum harmonics may be between −5dBc and −10dBc when the output is saturated. The adverse effects of harmonics are mainly to disperse the useful output power of the transmitter, lower the efficiency of the transmitter, and affect the electromagnetic compatibility of the system.

Intermodulation refers to the extra signal produced by the nonlinear characteristics of the amplifier after two adjacent signals in the working frequency band pass through the amplifier at the same time. Assume that the two input frequency signals are f_1 and f_2, then their harmonic and differential signals of each order number are intermodulation signals, i.e.,

$$f = m \cdot f_1 \pm n \cdot f_2 \; m, n \in Z \tag{7.3}$$

Thereinto, $2f_1 - f_2$ and $2f_2 - f_1$ have the greatest influence. Because the sum of its order number is 3, it is called third-order intermodulation. In general, when the two signals of equal amplitude input from a wideband transmitter are saturated, the third-order intermodulation will be between −5 and −10dBc at the worst. The adverse effects of the cross-modulated signal can also disperse the useful output power of the transmitter, reduce the efficiency of the transmitter, and affect the electromagnetic compatibility of the system.

Spurious refers to the ratio of useless frequency components unrelated to the expected output frequency signal to the carrier level. The unit is dBc. The reason for the spurious is the signal component that is parasitic on the useful signal generated by the amplification of the jamming source signal impure, power supply ripple, etc. Ordinarily, the spurious caused by the transmitter power ripple is around −30 to −50dBc, while the spurious caused by the impure signal of the jamming source will vary in a larger range. The adverse effects of spurious signals are largely to destroy the coherence of the signal and affect the electromagnetic compatibility of the system.

Out-of-band suppression refers to the ratio of the signal outside the working frequency range produced by harmonics, cross-modulation, and spurious to the normal working signal, which mainly disperses the useful output power of the transmitter, reduces the efficiency of the transmitter, and affects the electromagnetic compatibility of the system.

7.1.5 Self-inspection Requirements

As the function and composition of electronic countermeasure system become more and more complex, and the requirement of system function integrity rate becomes higher and higher. In the design of transmitter products, more and more attention is paid to the internal self-check function (BIT). For example, the design and manufactural of modularization, the detection of module level, the debugging and test of transmitter, etc., the testability design requirements are improving day by day. A reasonable BIT design can realize the detection of key characteristics and main functions, isolation and location of common faults in the extension.

7.1.6 Reliability and Environmental Adaptability Requirements

Due to the large output power, high working voltage, large current, large heat generation, and concentrated temperature of the jamming transmitter, they cause great stress to the power devices in the transmitter, which makes the reliability design of the transmitter face enormous challenges. The reliability of the transmitter is embodied in many aspects such as equipment safety and mission reliability.

With the diversification of combat platforms of radar active jamming system, various jamming transmitters are faced with great differences in application environments, including volume, weight, power supply capacity, ambient temperature, ambient pressure, mechanical vibration, electromagnetic compatibility, etc., which also put forward high requirements for the design of the transmitter.

7.2 Transmitter Type Selection

7.2.1 Types of Transmitter

According to different classification standards, there are different classification methods for jamming transmitters.

On the grounds of the basic time domain characteristics of the amplified signal, it can be divided into continuous wave transmitter and pulse transmitter. Pulse transmitters with a duty cycle of more than 30% are generally classified as CW transmitters. Others are pulse transmitters. The choice of the transmitter is continuous wave or pulse mode mainly depends on the jamming mode and jamming strategy of the radar active jamming system. In most cases, power device selection, platform power supply, transmitter cooling design, and other aspects should be considered in accordance with the needs of continuous wave transmitters.

According to the device type of transmitter to realize the final stage power amplification, it can be divided into electric vacuum transmitter and solid-state transmitter. The most used electro-vacuum device in radar active jamming transmitter is a traveling wave tube (TWT), whose structure is shown in Fig. 7.1. Except for the magnetic field, each part of the TWT is encapsulated in a vacuum tube, and the vacuum degree inside the tube body can reach more than 10^{-7} Pa [2]. The basic principle of TWT to achieve RF amplification is as follows: by heating the heating wire, the cathode with electron emission capability is at the most suitable temperature for electron emission. The electrons emitted by the cathode are constrained to form an electron beam within a certain range under the combined action of the anode, the focusing electrode, and the external magnetic field. In alternating electric and magnetic fields, electron clustering and appropriate speed control are used to exchange energy between electrons in the deceleration field and the high-frequency field. After completing the energy exchange, the electron carries on the final energy recovery in the collecting pole

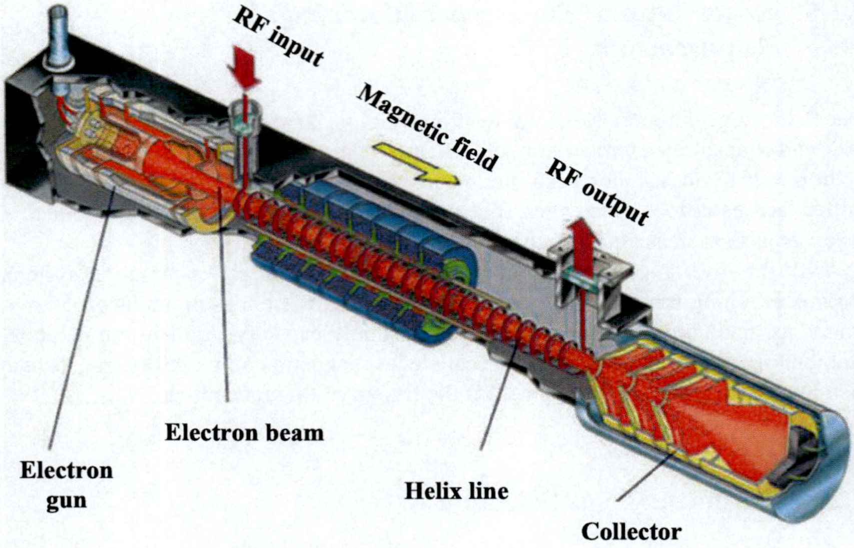

Fig. 7.1 Schematic diagram of TWT structure

part. Traveling wave tube has the characteristics of high output power, high working frequency, large relative bandwidth, high gain, and high efficiency. The working frequency can be from meter wave to millimeter wave, and the relative bandwidth ratio can reach 2:1 or even 3:1. In the following description, the radar active jamming transmitters using electric vacuum devices are illustrated with the traveling wave tube transmitter as an example.

The power devices used in solid-state transmitter are LDMOS, GaAs, GaN and other power chips. The output power of multiple solid-state power chips is synthesized by power synthesis technology to meet the requirements of the system for high-power output. As shown in Fig. 7.2. In the existing jamming transmitters of various power levels, the solid-state power amplifier is generally used as the front propellant amplifier.

Transmitter can be divided into single-tube transmitter and array transmitter according to the number of power components needed to achieve final power amplification. The TWT array transmitter adopts multiple TWTs and match up transmitting antenna arrays to obtain greater equivalent radiation power by means of space synthesis. Solid-state array transmitters generally need to use multiple chips for circuit synthesis at the module level. The transmitter part can either use a single power amplifier module or use multiple power amplifier modules to form a transmitting array in the form of array. The product example is shown in Fig. 7.3.

According to the working platform on which transmitters are installed, they can be divided into ground jammers, airborne jammers, missile-borne jammers, carrier-borne jammers, etc. Due to the different volume and weight requirements and power

Fig. 7.2 Diagram of GaN power tube array

Fig. 7.3 Traveling wave tube array transmitter

supply capacity of each platform, it may be necessary to consider adopting TWT system or solid-state system, single-tube transmitter system or array transmitter system, etc.

It needs to be noted that the above categories of jamming transmitters can be overlapping with each other, such as airborne TWT single-tube CW transmitters and ship-borne solid-state array pulse transmitters.

7.2.2 Selection Method of Transmitter Scheme

The scheme selection of the transmitter is mainly based on the overall technical indicators of the system, such as the operating frequency range, the jamming ERP and the transmitting antenna system, so as to select the working system of the transmitter and the type of power components, etc. The most important and critical thing is the choice of transmitter system and power amplifier components, which is related to whether the system transmitter can finally achieve the required comprehensive performance indicators of the transmitter.

The usual design thinking follows the sequence shown in Fig. 7.4

1. According to the technical requirements of radar active jamming system, the gain of the transmitting antenna is estimated and the minimum power requirement of the transmitter is determined.
2. According to the principle of matching the transmitting power with the power device capability, the device selection conditions are put forward. In addition to the power amplifier, the device also includes the matching high-power circulator, rotating joint, output filter, etc.
3. According to the above device selection conditions, the working system of the transmitter is determined to be single tube, channel synthesis, or space synthesis, and the type of the final power amplifier is a traveling wave tube or a solid-state power amplifier tube.
4. Based on input and output characteristics of power devices and the requirements of the system for the transmitter, carry out the design of the internal radio frequency channel of the transmitter, allocate the internal gain index, determine the specific index of each module, determine the external control interface, internal state detection and protection methods, and estimate the transmitter power consumption and specify the cooling method, etc.
5. According to the installation structure, volume, weight and heat dissipation mode of the transmitter, the specific design of the transmitter is carried out.

In the design process of the jamming transmitter, most of the work focuses on the key issues such as whether the power amplifier device and technical system are consistent with the antenna system and whether they are matched with the jamming ERP. The core of the design is to optimize the system performance and achieve the project requirements at the lowest overall cost. On the premise of realizing the

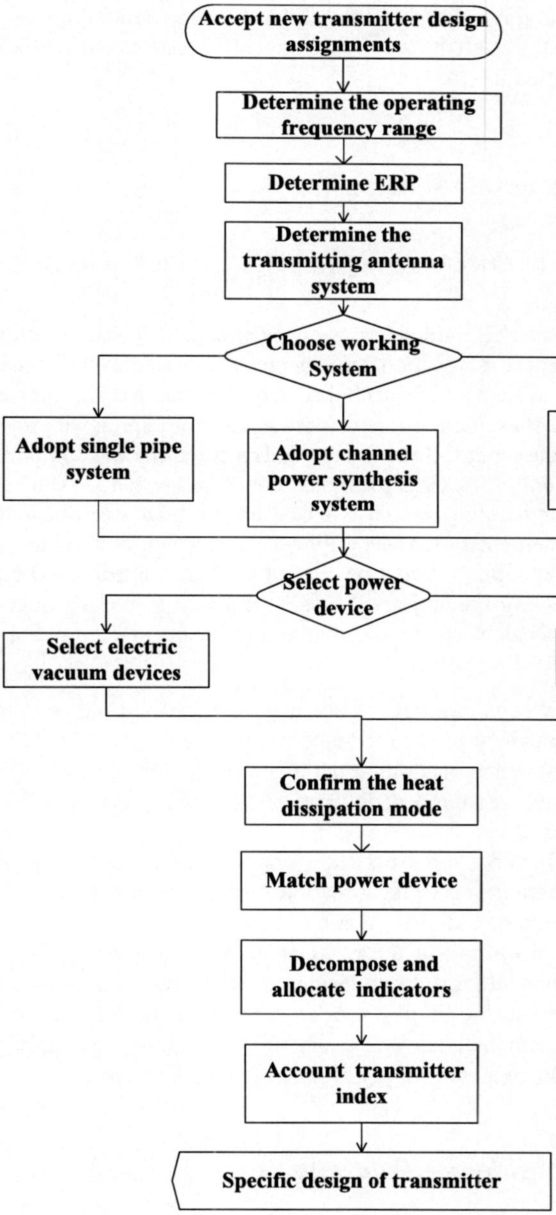

Fig. 7.4 Flowchart of transmitter design

target requirements of the transmitter, the principle of safety, reliability and easy maintenance should be followed all the time, and engineering issues such as efficiency and cost should be paid attention to.

7.3 Design of Transmitter Scheme

7.3.1 General Idea of the Design of the Transmitter Scheme

The design of transmitter can be divided into two stages: scheme demonstration and engineering design. In the previous section, the scheme demonstration has been introduced. It is based on the system's main technical requirements for the transmitter, starting from the basic theories of circuits and microwaves, combining with previous experience, and forming the scheme through repeated calculations and comparisons. This program should not only fully meet the index requirements, but also be reasonable and feasible. The engineering design is based on the final stage power tube selected by the scheme demonstration, according to the device requirements, the requirements for the power supply, excitation control module, control and detection protection circuit and cooling method at all levels are listed in normal operation. After index allocation and calculation, the design requirements of each part are put forward. Focus on the following points:

1. Signal input includes excitation-level matching, gain allocation, amplitude–frequency response equalization, and RF modulation control;
2. Power amplifier components: in addition to the main functions of the amplification and synthesis, it also includes heat dissipation, electromagnetic compatibility and other characteristics;
3. Transmitter control includes power switch control, radio frequency switch control, output orientation and beam direction control, each component state detection, failure detection, and protective functions;
4. The output includes protection in the form of isolators or circulators, power detection functions, and the matching design of power transmission channels which may be involved such as high-power switches, waveguides, cables;
5. The power supply system includes the special high-voltage power supply of the TWT and the low-voltage power supply required by other circuits.

7.3.2 Typical Composition of Transmitters

In this section, the reader has a basic concept and sufficient understanding of transmitter design by describing the functions of typical components of the transmitter and introducing the design of single tube and single machine in detail, and extending it to the synthetic transmitter and array transmitter, etc.

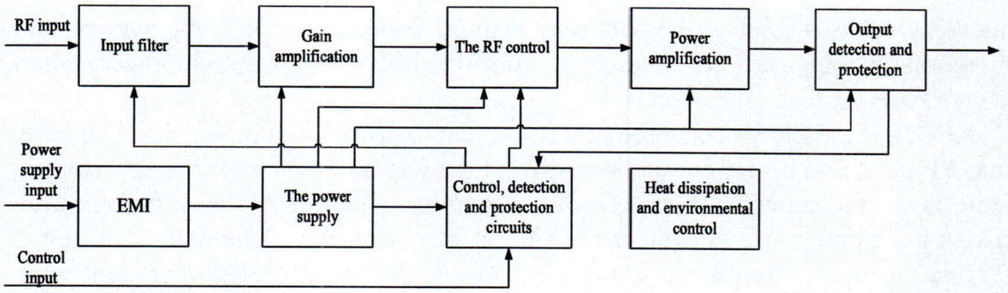

Fig. 7.5 Typical block diagram of a transmitter

The main function of the jamming transmitter is to convert the AC or DC power supply energy of the radar active jamming system into the RF jamming signal energy required by the system and radiate the RF power into the specified space through the power feeder and antenna of the system. At present, most of the transmitters of various radars and radar active jamming systems are of the main vibration amplification type; that is, the transmitter is solely responsible for the power amplification of the signal, but not the generation of the signal. In radar active jamming systems, the generation of the jamming signal is realized by the jamming source in Chap. 6.

Generally speaking, jamming transmitters and radar transmitters are basically the same in composition, usually including radio frequency channel, power supply channel, power amplification, control detection and protection, heat dissipation and environmental control and other parts of the circuit, as shown in Fig. 7.5. It needs to be noted that each part of the block diagram is not necessary in the actual equipment and can be added or reduced according to the actual system functions and equipment requirements.

The RF channel starts from the input filter, goes through gain amplification, RF control, power amplification, and finally the output of detection and protection module, which can be roughly divided into pre-excitation control, last-stage power amplification and back-end power detection and protection. For the selection of each device, the suppression of out-of-band signals and the safety of the final power amplifier should also be taken into account besides the necessary excitation equalization and modulation control.

The power supply path starts from the electromagnetic jamming (EMI) filter, reaches all kinds of power at all levels, and realizes the power conversion required from the system or the extension power supply to each component module. Among them, vacuum tube transmitters generally involve high-voltage power supply. Therefore, it is necessary to focus specifically on job safety and establish circuit protection and personal protection measures, respectively.

Control measurement and protection circuit ensure safe and reliable work of the transmitter through the timing control of the power supply and the RF channel, the state detection of each device, failure determination and protection, etc. Strict timing control of the power supply and control terminals is an important guarantee for the safety and reliability of the transmitter, and it is also the place that needs special attention in the design of the whole transmitter control part. In some systems, the direction of output power radiation is controlled by means of output power switch, array phase control, and so on.

RF power amplifiers consume a large amount of power, part of which is converted into RF power amplifier output, and most of the rest is converted into heat. Temperature is a very important factor for the reliability of power devices; therefore, the rise in the temperature of devices must be strictly controlled. The main task of heat dissipation and environmental control is to ensure the heat dissipation of high-power devices through conduction, convection, radiation, and other means, so as to reduce the thermal stress of devices and ensure the reliability of equipment.

The design of transmitter is to organically combine these circuits and devices with different functions to achieve safe and efficient microwave power amplification.

7.3.3 Single-Pipe Single-Machine Design

Single-pipe single-machine refers to the transmitter unit with only one power amplifier in the circuit. It is the most basic product form of all kinds of transmitters. As the saying goes, "Sparrow, small as it is, is well equipped." So, the basic components and functions of a transmitter are available. In addition, through appropriate cascade and expansion, multiple single tube units can constitute a synthetic transmitter or an array transmitter. Therefore, the design of single-pipe standalone can be easily extended to the design of other types of transmitters and emission arrays.

The composition block diagram of single pipe and single machine is the same as Fig. 7.5. This block diagram is applicable to two kinds of transmitters, traveling wave tube and solid-state power amplifier. Only when the final power amplifier is a traveling wave tube, a special high-voltage power supply needs to be configured, and when the final power amplifier is a solid-state power amplifier, a low-voltage power supply is needed.

The following will take the single-pipe transmitter of TWT system as an example to elaborate the design method and main concern of each part of the transmitter with one pipe and one machine.

7.3.3.1 External Common Interface Design

The external interfaces of the jamming transmitter mainly include power supply input, control input signal, state indication output signal, communication interface and debugging data programming interface.

In terms of power supply, for transmitters with high power consumption such as more than 500 V A, most of them adopt three-phase 400 Hz 115 V, three-phase 50 Hz 380 V or DC270V. For transmitters with relatively low power, AC 50 Hz 220 V, DC28V or DC280V power supply can be used. In practical application, considering the electromagnetic compatibility requirements of radar active jamming equipment on the installed platform, other power supply types are generally not recommended.

For the control signal, the input signal of low-speed control, such as adding low-voltage, adding high-voltage, program control, or manual selection, is recommended to use 28 V voltage, in order to improve the anti-jamming ability. Among them, the signal that needs to control the relay should have the necessary load driving ability according to the impedance requirements of the relay line package, usually designed in line with the load driving ability of no less than 300 mA. High-speed signals need to adopt the way of group difference or differential bus, which can not only ensure the response speed, but also improve the ability of anti-jamming.

The state indication is the signal that transmits the key points and important characteristic quantities inside the transmitter to the outside of the system. The output generally adopts TTL level, and the demand should be put forward separately when there is special driving capability requirement.

The communication interface generally adopts serial communication interface such as RS-422 and Ethernet interface, which mainly undertakes the issuing of general instructions of the system, including the transmission of some low-speed signals with large amount of information and the feedback of state information.

The debugging interface is generally aimed at FPGA, DSP, PowerPC, and other devices, usually using JTAG interface, which can realize the online debugging and programming of the control circuit.

7.3.3.2 Microwave Excitation Channel Design

In the jamming transmitter, the main purpose of the design of microwave excitation channel is to get the best matching excitation power with the TWT input. The design process mainly includes channel equalization design and channel index allocation.

(1) Channel equalization technique

The working states of TWT can be divided into linear amplification, critical saturation amplification, saturation amplification, and over-excitation. When it works in the overexcited state, the harmonic and stray in the output jamming signal will increase more than that in the critical saturation state. When it works in the linear region, compared with the critical saturation state, the harmonic and stray in the output jamming signal decrease somewhat, but the output power and working efficiency decrease more. Therefore, in general, when the TWT of the transmitter of the jamming system works in the critical saturation state or the shallow saturation state, the output power is large and the working efficiency is high, and the output harmonics and stray do not change much, which is the most ideal state.

The equalization of microwave excitation channel is the process of adjusting the power of excitation signal to match the best excitation of TWT. It mainly includes the equalization of amplitude and frequency characteristics, which is usually realized by equalizer.

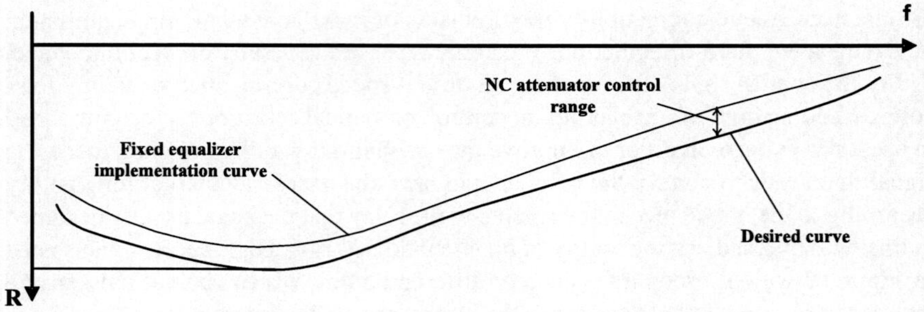

Fig. 7.6 Requirements for the equalization

The working principle of the equalizer is that the attenuation of microwave signals of different frequencies is not same, so that the signals of different frequencies can meet the characteristic requirements of the transmitter according to the most suitable excitation power. Common equalizer can be divided into passive equalization and active equalization. Passive equalization means that the equalizer in microwave channel does not need additional control and can be used directly after debugging. Active equalization normally uses the information related to frequency to assist the selection and control of different attenuation amounts.

In the use of jamming transmitter, the two equalization methods have their own advantages and disadvantages. The advantages of passive equalization are as follows: no additional frequency-related information is needed, the control is simple, and there will not be some local over-excitation due to control errors. However, the disadvantage is that the control accuracy is low, so that all the frequency points cannot achieve the excitation control in the best state, which often requires sufficient gain and power margin. For example, against the equalization characteristic curve in Fig. 7.6, the control accuracy error may cause the transmitter to lower the useful fundamental signal and raise its harmonic power level.

The advantage of active equalization is that the control is very fine and all the frequency points can be controlled optimally. The disadvantage is that additional frequency information is needed. Once the microwave signal and frequency information do not correspond, it is easy to cause serious over-excitation, which threatens the safety of the core power amplifier of the transmitter.

Equalizer dealing with the best solution is to combine the advantages of both active and passive, namely to use passive in the channel equalizer, which can balance RF signal within the relatively rough range, enable the impact minimize the impact caused by inconsistencies in the frequency response of devices, and then use a small range of adjustable active balancing such as digital control attenuator, to achieve more sophisticated balance control. In this way, since the passive equalizer has been used to control the power within a relatively safe range, even if the active equalization control fails, it is still safe for the transmitter. Based on the current state of passive equalizer design, it can match the final power amplifier very well, so only passive equalizer is sufficient unless there is a special need.

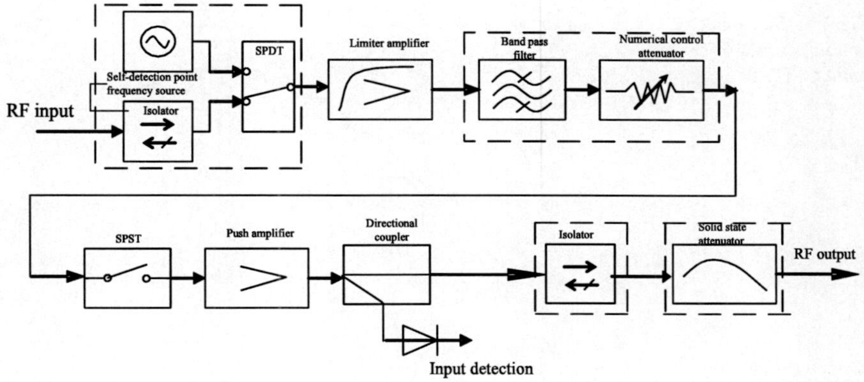

Fig. 7.7 Composition of microwave channels

Therefore, a relatively complete composition block diagram of microwave excitation channel is shown in Fig. 7.7.

According to the above block diagram, the RF jamming signals from the jamming source, after isolator, are selected by the single-pole double-throw RF switch and enter the limiting amplifier for primary amplification, in which the isolator can be omitted when the matching switch is used. In order to keep the system in a certain range change, excitation signal power has a relatively stable and flat amplitude–frequency output to avoid over-driving the device behind it. After amplifying the circuit, a small range of numerical control attenuator is used to achieve fine equalization control or output power control, in which a band-pass filter is often added before equalization to filter out stray and harmonic signals outside the band. Then, through the single-pole single-throw (SPST) switch for transceiver control and the last-stage push amplifier, and then through the coupling detection, the detection of the input signal is finally realized. After the coupler is output by the isolator to the passive fixed equalizer, the RF signal can be equalized in a relatively rough range, so that the impact of the inconsistent frequency response of the device is as small as possible.

During the design, it can be selected or simplified according to the specific requirements of the system and installation structure conditions, but the most basic functions should still include amplification, equalization, switch control and detection. For example, the point frequency signal source used for self-inspection can be provided by the previous stage jamming source, so that the system can be extended to detect multiple extensions, and two devices can be omitted for the transmitter. At present, the new TWT basically has its own equalizer, so the channel cannot set additional passive equalizer. The use of array will generally add a phase adjustment circuit, such as put the CNC phase shifter before the input limiting amplifier or in series.

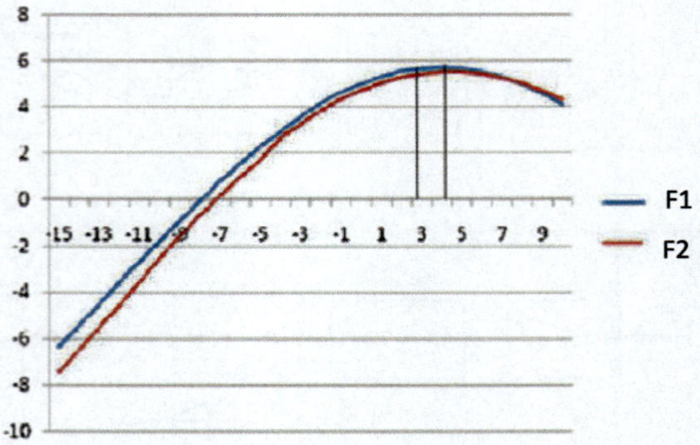

Fig. 7.8 Schematic of optimal excitation power point of TWT

(2) Channel gain and indicator allocation

Through the analysis of the above-mentioned microwave excitation channel, typical technical indicators such as working frequency range, self-check signal frequency, input signal power, output signal power, output signal power during self-check, digital attenuation dynamic range, input and output standing waves can be decomposed. But the design of the channel gain indicator is the most important. The main path of indicator allocation is based on the power and gain of the RF channel.

The index requirements of channel gain should be designed and adjusted according to the excitation requirements of TWT or final power amplifier. When the micro-strip passive equalizer is used alone, the output amplitude–frequency characteristics of the channel can exceed within a few dB of the required amplitude–frequency characteristic curve of the TWT. The optimal excitation power of the TWT should backpedal 1 dB when it is near the saturation excitation point. This not only ensures the realization of the output power index, but also avoids the harm caused by too large over-excitation to the TWT, as shown in Fig. 7.8.

According to the above block diagram of the microwave excitation channel, it is necessary to reasonably allocate the gain index of the channel in each module. A typical example of channel gain allocation is shown in Fig. 7.9.

When the power range of the input RF signal is −5 to +dBm, the TWT requires the input power to be no less than +13dBm, and the specific gain allocation is as follows:

The insertion loss in the front channel of the limiting amplifier is 5 dB. Choose such a limiting amplifier: it can start limiting compression from −15dBm, and the limiting output power is greater than +12dBm, which can ensure that its output can enter the limiting compression state at the minimum input and the output power is greater than +15dBm. The cumulative channel insertion loss of the final propelling

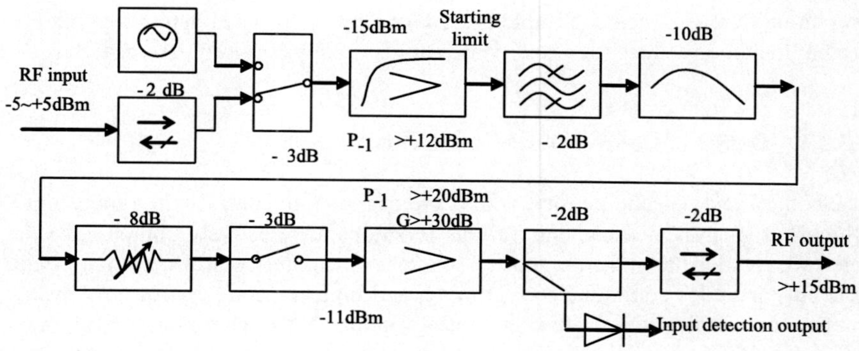

Fig. 7.9 Gain allocation in microwave channels

solid-state amplifier is about 25 dB; that is, its input power is -13dBm. The solid-state amplifier with gain greater than 30 and 1 dB compressed output power greater than $+20$dBm can meet the demand of the TWT input greater than $+15$dBm.

In addition, the microwave excitation channel of TWT should be designed with sufficient state detection function to meet the design requirements of online self-test (BIT), and all these functions should be integrated into one or several microwave modules according to the actual system requirements and limitations of volume, weight, power consumption, and existing microwave integration capabilities. In the design, the designer should consider the versatility of the module structure, so as to facilitate installation and disassembly. The flow direction of the RF signal should be taken into full consideration to avoid the occurrence of annular flow direction resulting in channel self-excitation. For the amplifier, attention should also be paid to the design of heat dissipation. Standing wave adjustment devices, such as attenuators, isolators, should be used between several stages of amplification and when directly connected to the traveling wave tube.

7.3.3.3 Design of RF Output Channel

The RF output channel mainly refers to the device combination between the TWT output and the output panel of the transmitter, generally including high-power directional coupler, detector, isolator, output cable, output adapter block, etc. For transmitters with waveguide output, flexible waveguides should be used at the output end of the traveling wave tube to eliminate fabrication and structural installation errors.

When the transmitter is applied to the airborne platform, it is usually necessary to consider the power derating problem of the power cables and switches at the back end under low pressure conditions to avoid voltage and power breakdown. When applied to high altitude, if the waveguide output is adopted, the waveguide sealing window is usually connected near the output panel of the transmitter to isolate the pressure difference. If coaxial cable is used for output, the panel output seat shall be sealed. In addition, the case of the transmitter is usually required to be airtight.

7.3.3.4 Design of Control Detection Circuit

The control and detection circuit is the core of the logic control in the transmitter, which has many functions, and various functions are relatively complex, so the interface relations are also the most. Common transmitter control detection circuit generally includes communication, timing, and high pressure control, grid switch control (optional), active balancing control (optional), self-checking control, input and output power detection (optional), reflected power detection, temperature detection, fault protection, status display and online debugging and data records, and other functions, its principle as shown in Fig. 7.10.

The timing control and fault protection of transmitter are the most important functions in the control and detection circuit. Involving the necessary operation sequence of power devices, as well as overheating and excessive reflected power, it is necessary to follow the prescribed process and take the fastest and nearest end

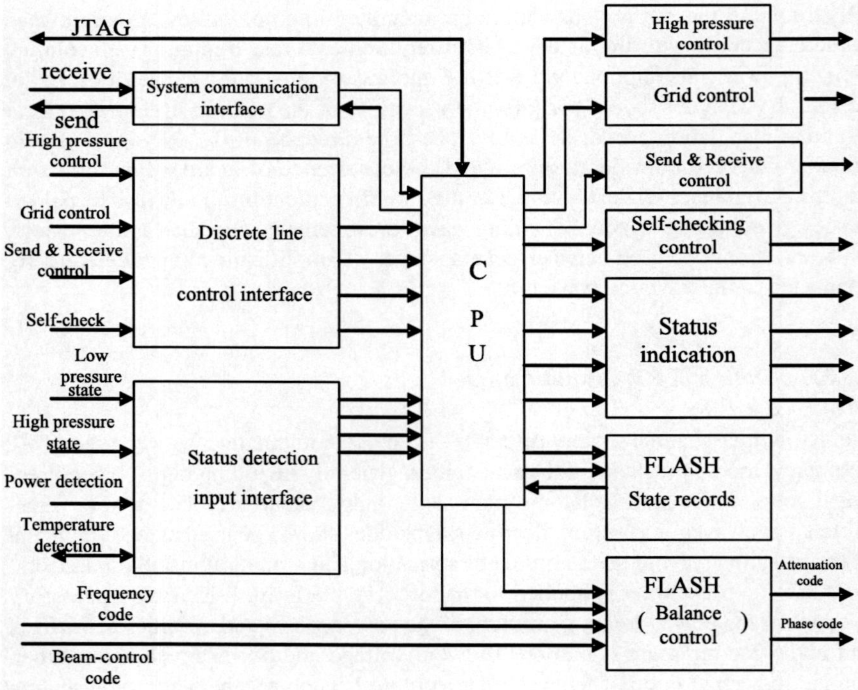

Fig. 7.10 Principle of control and detection circuit

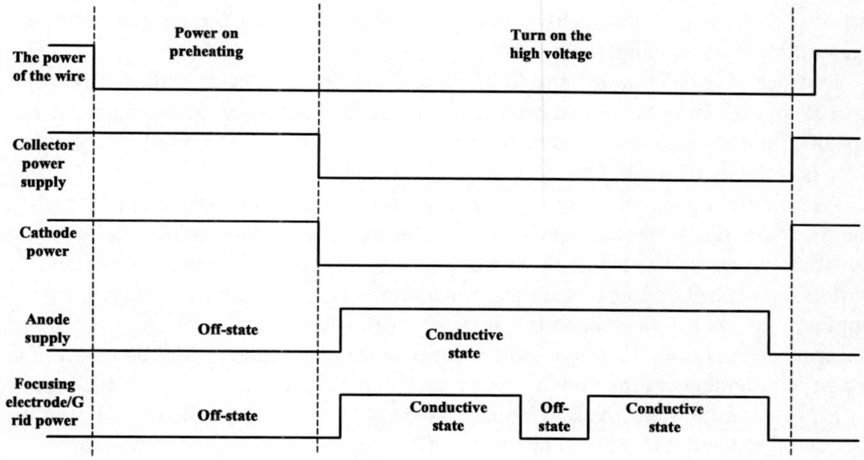

Fig. 7.11 Typical timing diagram of TWT power-on control

control as the criterion. The timing control of TWT is actually the operation of HV power supply, which can become a reality by controlling the output state of each pole of HV power supply.

The control of the transmitter, especially the power-on and power-off control of the power supply, must be in accordance with the requirements of the power device itself, and meet the time sequence relationship strictly; otherwise, it is easy to cause the damage of the power device. Taking the TWT as an example, the timing requirements (timing sequence diagram) for the TWT are shown in Fig. 7.11.

TWT contains high-voltage electrodes such as hot wire, cathode, anode/focusing electrode/gate electrode, and collector electrode. When there is no focusing electrode/gate electrode in TWT, anode is the necessary output control electrode. When the TWT has a focusing pole/gate pole, whether the anode is needed can be selected according to the index requirements and design scheme of the TWT itself. In order to ensure that the TWT can work safely, reliably, and normally for a long time, the voltage range and power-on sequence of each electrode must meet time requirements, which is the control timing sequence of the TWT.

In the fault detection and protective function, the helical overcurrent is performed in the high-voltage power supply. When the helical overcurrent phenomenon occurs, the gate and the high voltage should be closed as the procedure of turning off the high voltage, as well as the provision of helical overcurrent indications. The control detection circuit only needs to maintain the control signal with high voltage. For the protection of overheating, the high pressure should be closed according to the normal steps of closing high pressure when overheating is detected and confirmed. When the reflected power is too large, the RF input should be turned off to confirm the fault elimination, but if there is still such a phenomenon after the shutdown, which ought to deal with the transmitter self-excitation, and turnoff the high voltage according to the operation steps of normal high voltage. All three states need to be reported to the system, prompting manual intervention.

In order to avoid the damage to TWT and the final power amplifier during the process of switching the output orientation, high-voltage power-on and high-voltage turnoff, the RF excitation should be turned off, and it could be turned on when the state is stable at the end of the process.

As a single-pipe transmitter, the above control information is sufficient. But when the array applies it, it is necessary to add the beam control circuit to achieve the equalization control of the phase and amplitude. It is recommended to separate this part of the control circuit so as to maximize the reuse of the control circuit. In array applications, where fault handling may affect the functionality of other channels, the control and protection logic requires special treatment that is consistent with the usage requirements of the system and protective function of the high-voltage power.

PIN switch control in the RF channel, on the one hand, is controlled by the system. On the other hand, PIN should be turned off during high-voltage operation (switch), azimuth switching and other operations that may endanger TWT and final power devices. PIN cannot be opened until the state is stable at the end process. At the same time, in order to ensure the response speed of PIN control, the PIN control signal from the system should pass through the circuit as little as possible to reduce the delay. Other states that need to turnoff the RF should be served as enabling control of the PIN control signal of the system. The circuit principle is shown in Fig. 7.12.

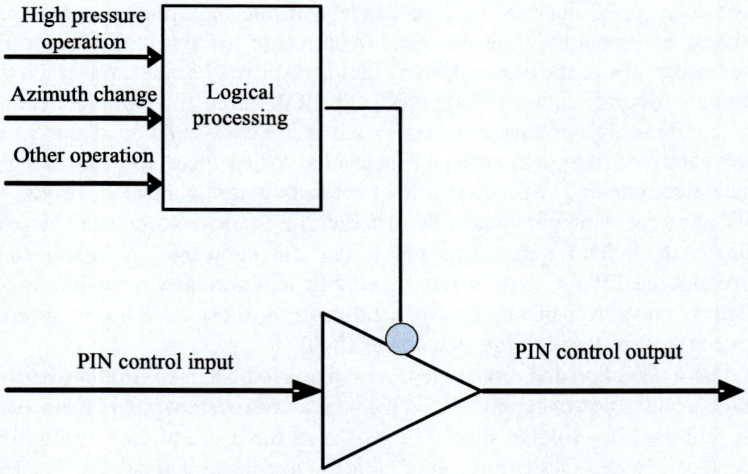

Fig. 7.12 Schematic diagram of PIN control circuit

7.3.3.5 The Design of Low-Voltage Power Supply

The low-voltage power supply is responsible for all the low-voltage DC power conversion from the power supply input to the control detection circuit, microwave channel, etc. Using a solid power device as the final stage amplifier, the low-voltage power supply also includes the power supply part for the final power amplifier tube. This part usually features low-voltage and large current, and there will be special requirements for load stability, impulse response, and so on. Therefore, when designing low-voltage power supplies, sufficient power margin should be given into consideration.

The design of low-voltage power supply usually adopts off-the-shelf DC/DC module and necessary EMI filter. If there are special requirements for output voltage ripple, which should be put forward in detail. The internal low-voltage power supply wiring of the transmitter shall adopt floating power supply and be grounded at the load end.

For the power supply of the solid-state power amplifier, the time sequence of power-on and power-off required by the solid-state power amplifier must be met to ensure the safety of the solid-state power amplifier.

7.3.3.6 The Design of High-Voltage Power Supply

In the TWT transmitter, the high-voltage power supply is the key equipment to provide energy for the traveling wave tube. The reliability, safety, and maintainability of the transmitter depend on the design of the high-voltage power supply to a great extent. In addition to the internal state detection and protection control, the external control and state indicating interface of the HV power supply is an important guarantee for the reliability, testability and maintainability of the transmitter and the system.

In the design of high-voltage power supply, the most important task is protection. Among them, the input phase sequence, temperature, and other protection of the high-voltage power supply are directly detected and output status signals by the high-voltage power supply. When the helix overcurrent is detected, the helix overcurrent fault indication is generated inside the HV power supply, and the fault indication is cleared and reset when the external HV control signal is withdrawn. Therefore, in order to maintain the fault indication signal, the external high-voltage control signal should be maintained, which is the place that needs to be coordinated with the control and detection circuit design during the design.

The output sequence in each pole voltage of a high-voltage power supply plays a decisive role in the safety and long-term service life of the TWT. It is the basic requirement to ensure the safe and reliable operation of the TWT to operate in strict accordance with the power-on and power-off timing requirements proposed by the manufacture of the TWT. Unless there are special requirements, the following steps are generally followed to power up and turnoff the TWT.

For the grid electrode (focusing electrode) control TWT, the control circuit generates the extension turn on high-voltage control signal in time sequence. The high-voltage power supply of each pole of the TWT produced by high-voltage switch power supply, output the first gate cut-off voltage, and then fulfill the cathode, anode, collector electrode. After the transceiver control signal is converted into a gate modulation signal, the gate electrode conducting voltage can be added to amplify and output the input RF signal. During the operation of the high-voltage power supply, if the control signal to turnoff the high-voltage is received or a fault occurs, the high-voltage power supply should first remove the grid conducting voltage, add the grid cut-off voltage, and then close the pole power supply in addition to the filament, finally, close the filament voltage according to the requirements of the system.

For the non-grid control tube, in conditions permitting, the anode power supply should be designed to the cathode mode as far as possible to reduce the generation of helical impulse current in the process of power-on and off, which will affect the life of the TWT.

For different TWTs, the mode of power-on and power-off of HV power supply and the time sequence and the protection threshold of helical overcurrent should be agreed with the TWT supplier to ensure the safety and reliability of the equipment.

7.3.3.7 Structural Design and Six-Performance Design

Structural design and six-performance design as a common part, detailed content refers to Chap. 3 of this book. In view of the particularity of transmitter products in power, voltage, current, temperature, and other aspects, in addition to the common design requirements, for the transmitter, the structure should also value the heat dissipation, vibration resistance, low pressure, three prevention, and other environmental adaptive design. The reliability design, such as reduction and redundancy design, should also be emphasized in the six-performance design.

7.3.4 Synthetic Transmitter Design

Synthetic transmitter refers to a transmitter that uses multiple power tubes and microwave synthesis devices to obtain synthetic power in microwave channels. Synthetic transmitters may adopt multiple TWTs or solid-state power amplifier modules. The basic composition of a synthetic transmitter is shown in Fig. 7.13.

Solid-state power amplifiers generally utilize multistage synthesis. Firstly, the low-power tube core is integrated into a single solid-state power amplifier module by microwave integration technology, and then several solid-state power amplifier modules are synthesized into a transmitter with higher power through high-power combiner. A product example of a solid-state power amplifier is shown in Fig. 7.14.

Compared with the single-pipe single machine, arguably, the main composition of the synthetic transmitter is the front excitation amplifier units, power divider, power

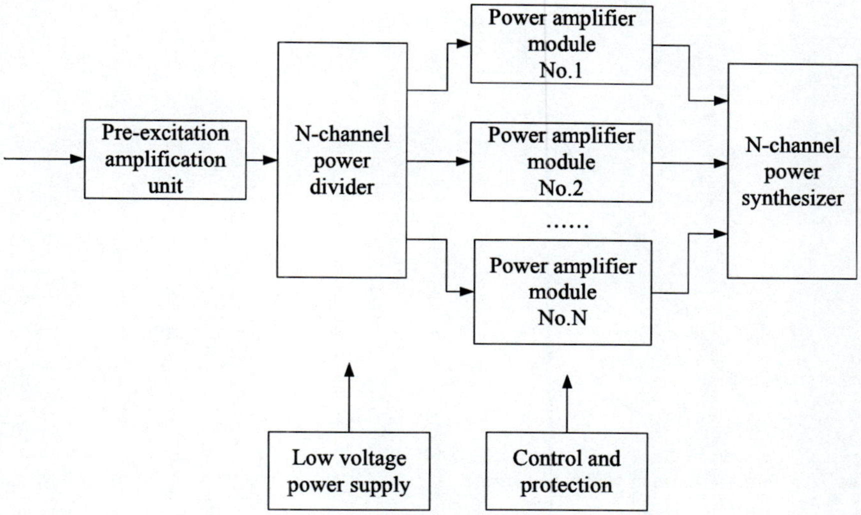

Fig. 7.13 Composition block diagram of the synthetic transmitter

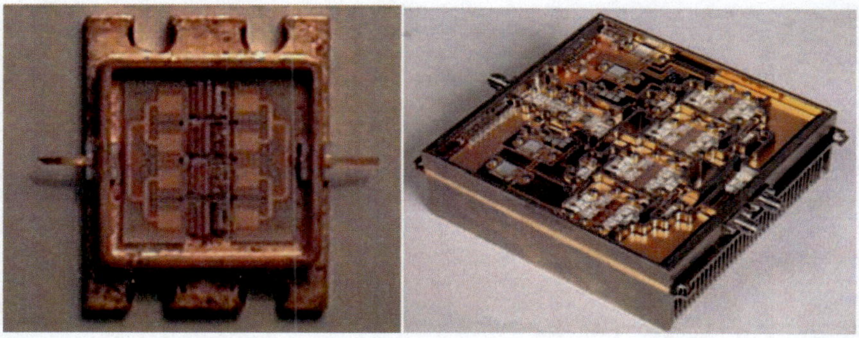

Fig. 7.14 Solid-state power amplifier module

synthesizer, multiple single-tube single machine, as well as the control and protection circuit and low-voltage power supply.

The design of the front excitation amplifier unit is similar to the design of the microwave channel of single-pipe single machine, in which the filtering, excitation amplification, equalization, and control of the RF channel should be considered. Because it is necessary to adapt to the excitation-level demand of the power amplifier unit after the power divider, the output power of the front excitation amplifier unit will be increased.

It is relatively important that according to the requirements of synthesis, the multiple power amplifier units (or single pipe single machine) participating in the

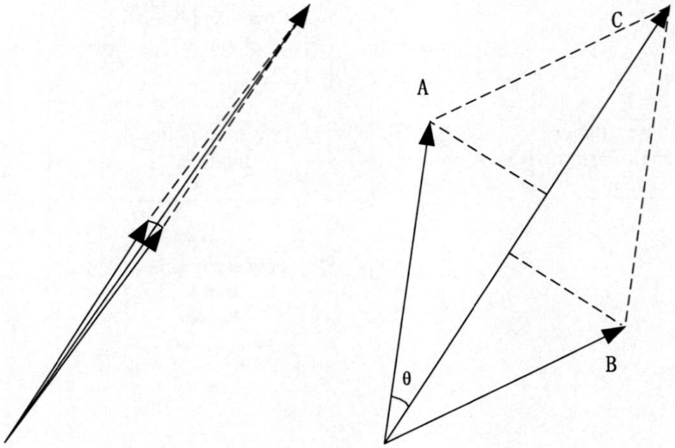

Fig. 7.15 Error analysis of vector synthesis

power synthesis must meet a certain amplitude and phase consistency requirements, so as to carry out effective synthesis in the power synthesis channel, otherwise the efficiency will be very low, or even cannot be synthesized. The diagram of vector synthesis is shown in Fig. 7.15.

Based on the diagram of vector synthesis (Fig. 7.15), it can be seen that the amplitude and phase angle of the resultant vector will be affected when the original vector has phase offset. However, since the signal power of the synthetic transmitter is synthesized inside the channel and has nothing to do with the phase of the final output signal, only the influence on the synthetic amplitude is considered for the synthetic transmitter. The relation between synthetic power and phase deviation can be expressed in the following formula:

$$P = |A\cos\theta|^2 \quad (7.4)$$

When the phase angle deviates by 25°, the synthetic power is reduced to 82% of the original, which is converted to −0.85 dB. When the amplitude changes, it directly shows as the amplitude changes on the synthetic vector. In order to achieve a synthesis efficiency of more than 80% in engineering, the phase consistency of multiple power amplifier units should be controlled within ±25°, and the amplitude consistency should be within ±1 dB.

Due to the limitation of power capacity of output cable or synthesizer, the output power of synthetic transmitter is not too large. It is mainly used in systems where multiple high-gain antennas cannot be installed, and large ERP is achievable with a single high-gain antenna.

The control mode, logic and protection steps in the control and protection circuits of multiple modules in the synthetic transmitter are all the same. Therefore, the repeated functional units in the original single-tube single-machine control can be combined to achieve unified control and protection. Solid-state power devices also have strict timing control requirements expect TWT. In synthetic transmitters, due to the simultaneous action of multiple power amplifiers, strict timing relationship is an important guarantee for the safety and reliability of transmitters.

7.3.5 Array Transmitter Design

Contrast with synthetic transmitters, array transmitter radiates microwave signals into space by using one or more groups of antennas arranged in a certain position relationship and makes them form vector superposition in a certain direction of space by controlling the phase of each microwave signal, so as to achieve the purpose of changing beam direction and realizing power synthesis [3].

According to the principle of space power synthesis, array transmitter can achieve high jamming ERP with low-power amplifier module, provided that the bearing platform of radar active jamming system can meet such scale power amplifier unit and antenna in terms of space volume and weight, which is exactly the advantage of array transmitter.

One-dimensional linear array is shown in Fig. 7.16. In the direction of beam deviation from normal θ, the resultant equivalent power pattern is expressed as [4, 5]:

$$P(\theta) = \frac{P_0 G_0 (1+\cos\theta)^2 \eta}{4} \cdot \frac{\sin^2(\frac{\pi b}{\lambda}\sin\theta)}{(\frac{\pi b}{\lambda}\sin\theta)^2} \cdot \left| \sum_{i=1}^{N} m_i e^{j(i-1)(\frac{2\pi d}{\lambda}\sin\theta - \phi_i)} \right|^2 \tag{7.5}$$

In the formula, P_0 is the output power of power amplifier unit, and G_0 is transmitting antenna unit gain, θ represents the included angle from the normal direction of the beam, η represents the synthesis efficiency of the antenna array, *b* and *d* respectively represent the aperture width of the antenna unit and the spacing of the antenna unit, λ is the signal wavelength, and N is the number of transmitting antenna array elements.

The ERP of each angle is:

$$\mathrm{ERP} = \frac{N^2 P_0 G_0 (1+\cos\theta_B)^2 \eta}{4} \cdot \frac{\sin^2(\frac{\pi b}{\lambda}\sin\theta_B)}{(\frac{\pi b}{\lambda}\sin\theta_B)^2} \tag{7.6}$$

The ERP of antenna array normal direction is:

$$\mathrm{ERP} = P(0^0) = N^2 P_0 G_0 \eta \tag{7.7}$$

According to the above analysis, the amplitude error only affects the jamming ERP and has no effect on other parameters of the power pattern. For example, for a

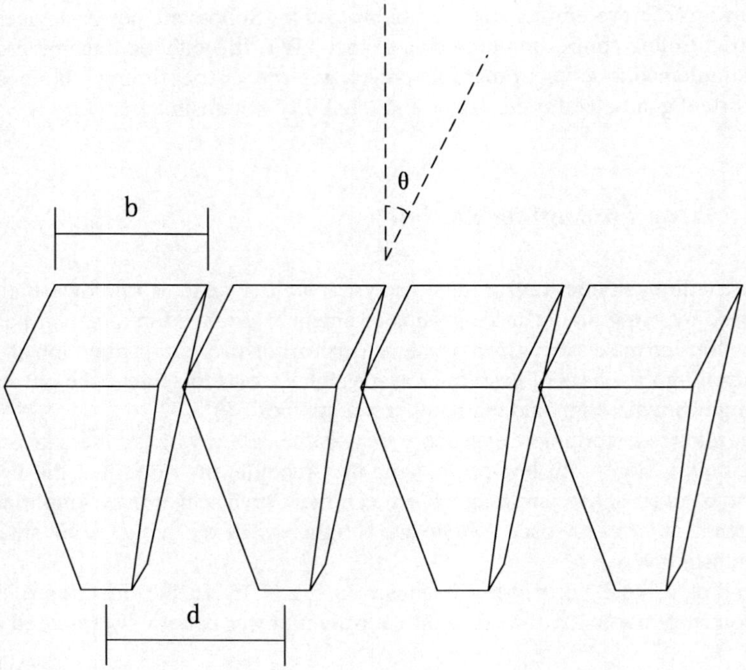

Fig. 7.16 Structural parameters of one-dimensional antenna array

one-dimensional 24-element linear array antenna, when the amplitude difference is 1.5 dB, the jamming ERP of the output is reduced by 3 dB compared with that of the unbiased excitation.

The phase error affects all parameters of the power pattern to varying degrees. Although the effect on ERP is small, it has a great impact on the beam direction, so the power in the actual desired direction will still drop. For example, when the phase difference is less than 15°, the beam pointing deviation is not more than 2 times the width of the half-power beam. However, when the phase difference exceeds 15°, the beam pointing deviation will be more than 2 times the width of half-power beam, which means the ERP drop of 3 dB in the specified direction.

There are several methods to control the phase of microwave signal, such as multi-beam, phased array, and DBF.

(1) Multi-beam system

Multi-beam system utilizes beam selector switch and lens to distribute the phase and amplitude of each power amplifier to realize different beam pointing, enabling the phase of the output signal of each power amplifier to form a specific wavefront in space, complete the power synthesis, and change the beam direction at the same time. The block diagram of the multi-beam transmitting array system is shown in Fig. 7.17.

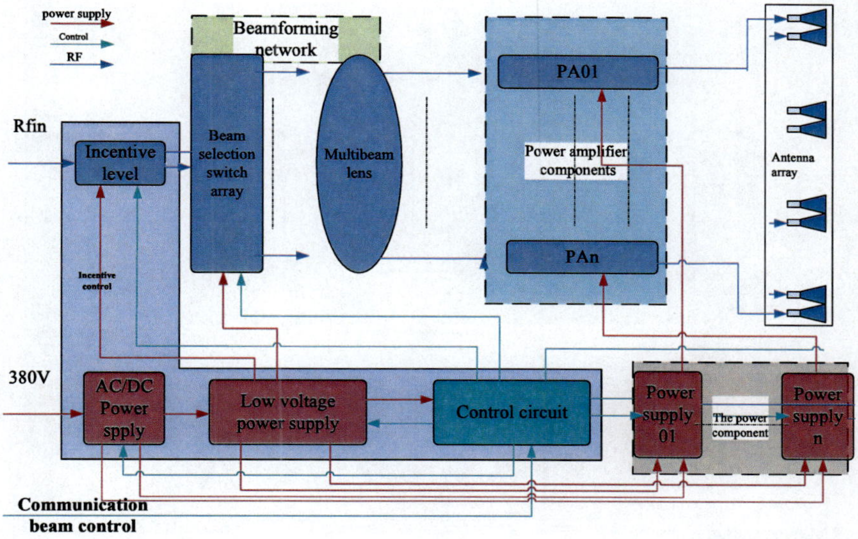

Fig. 7.17 Block diagram of a multi-beam transmitting array system

For multi-beam array, the key to beam pointing is lens. The lens input is gated via the beam selector switch. When output from the lens, the determination in phase relationship of each RF signal depends on the lens, and the power is amplified by these multiple power amplifiers with consistent amplitude and phase. Finally, the RF signal is emitted from the antenna array, which completes power synthesis in space and forms beam pointing according to the phase relationship.

(2) Phased array system

Phased array is to control the phase of output signal of each power amplifier by means of time delay or phase shift device, which can realize the change of beam direction and power synthesis in space. The block diagram of the phased array transmitting array system is shown in Fig. 7.18.

According to the requirements of beam pointing, the wave-controlled computer calculates the corresponding initial phase value of azimuth and pitch angle and then calculates the phase offset of each element according to the relative position of each antenna element, and calculates the phase control code of the phase shifter according to the resolution of the phase shifter. Large arrays can be divided into many subarrays, and the array scale can be extended by hierarchical interpolation control.

For phased array, the key to realize beam pointing is phase shift amplifying unit or time delay amplifying unit, which adjusts the phase or delay of the channel by the phase shifter or real-time delay line, so as to control the output beam pointing of the system. This part is the core of the phased array.

Like a synthetic transmitter, an array transmitter consists of a front-stage excitation amplifier, a beam-forming network, multiple single-pipe units, as well as a

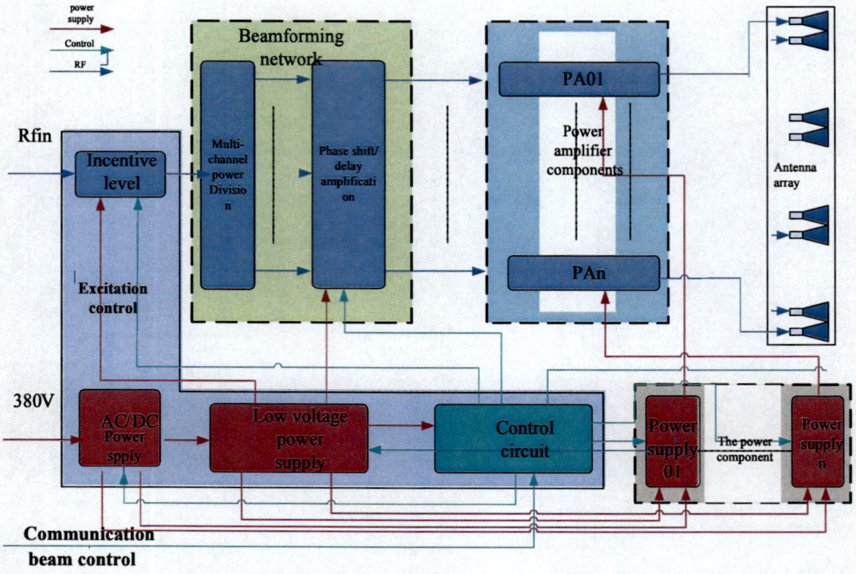

Fig. 7.18 Block diagram of phased array transmitting array system

control protection circuit and a low-voltage power supply. The filtering, excitation amplification, equalization, and control of RF channels are of great importance in the front excitation amplification unit. In order to adapt to the multi-channel power component of the beam-forming network and the excitation-level requirements of the later phase-shifting/delay amplifying unit, the output power of the front excitation amplifying unit will be increased.

There is little difference between multi-channel power amplifier modules in multi-beam and phased array.

In addition, the power supply interface of the array transmitter needs to be grouped to limit the current due to its high power.

7.4 Performance Accounting of Transmitter

Several key indicators in the transmitter design process are accounted to ensure that the system requirements are met.

7.4.1 Output Power

In the process of system design, the influence of each link that the transmitter output signal must pass through should be considered, such as isolator, filter, azimuth control switch, rotating joint and transmitting antenna. According to the requirements of ERP, the output power of the transmitter is derived in reverse. It mainly calculates the difference between the output power of the final stage power amplifier device or module and the input loss of the device and feeder between the power amplifier and the RF output channel of the emitting array port, so as to avoid the omission resulting in insufficient final output ERP.

For single pipe and single machine, it mainly refers to the device combination between the TWT output and the output panel of the transmitter, generally including high-power directional coupler, detector, isolator, output cable, output adapter block, etc.

The output power index should be set according to the output characteristics of the final stage power amplifier (such as traveling wave tube). The TWT or solid-state power amplifier index is not equal to the final output power index of the transmitter, which should take into account the loss of the output channel, the control precision of debugging during mass production and the qualification rate of products.

7.4.2 Power Consumption and Heat Consumption

Power supply includes all input power supply, specify power supply type, voltage range, and current consumed.

The sum of the total feed power includes the total power consumption of DC power supply and AC power supply. Apart from the transmitted power, the remaining part of the power consumption will be converted into heat, which needs to be taken away by the radiator of the transmitter; otherwise, it will accumulate and cause damage to the power device.

7.4.3 Gain

As the basis and condition of transmitter design, the input power or gain plays a very important role in the gain and power allocation of the whole RF channel of the transmitter. For the transmitter design, the key is to decompose the input and output requirements of the transmitter and distribute them to the transmitter to make the indicators of each module and component reasonable and balanced, so as to avoid affecting the working stability of the transmitter due to excessive concentration of gain.

If the input power is given, the excitation power required by the last-stage TWT amplifier should be considered to design the total amplification and gain distribution of the microwave channel. If the design requirements are put forward according to the gain index, the transmitter design should use the backward inference method to calculate the required input power under the output power, and then devise the total amplification and gain distribution of the microwave channel.

According to the influence of system excitation fluctuation on the safety of final power devices, necessary equalization and limiting measures are designed in advance to ensure the safety of final power devices. The amplitude–frequency response characteristics of microwave devices are uneven. In general, the loss of low frequency band is relatively small, while that of high frequency band is relatively large. Therefore, the index should be defined according to the frequency band, or balancing in front should be controlled by the system, so that the input signal power of the transmitter fluctuates ±2 dB at a fixed level.

7.4.4 Array Power Allocation

Normal radiated power is the core index of system emission. After the antenna gain is removed from the normal radiated power, it is the output power index allocated to the transmitter. According to the ERP index and formula (7.5), the output power of the power amplifier unit in the transmitting array can be calculated. This is only a basic requirement, but should also appropriately improve the power amplifier unit output power margin based on the device power, antenna gain, structure layout, cable loss, cost and other information.

7.5 Development Trend of Transmitter Technology

The general trend of the development of transmitter technology is miniaturization, array, and intelligence.

(1) Miniaturization

With the increasingly complex functions of electronic countermeasure equipment and the installation platform more inclined to miniaturization, stealth and other features, the miniaturization, lightweight and high efficiency of transmitters are becoming more and more important to users.

(2) Array

The array of the transmitter can realize large system ERP with small power devices and fast beam switching. It can tolerate the index deviation and fault of individual array elements to weaken the fault, so as to guarantee the system's task reliability significantly improved, and can provide the system with multi-signal and multi-beam direction of simultaneous jamming ability, so it is an important direction and inevitable trend of the transmitter development in the future.

(3) Intelligence

The following aspects can be accomplished through the detection circuit and software of various component in the transmitter and combined with software. Firstly, the internal state of the transmitter can be monitored and detected in real time; secondly, when a fault occurs, it can be alerted and indicated; finally, the fault endangering the safety of personnel and equipment can be protected in time, so that the transmitter can work in a normal and healthy state continuously, and the index and function can be degraded automatically when necessary.

Inside the transmitter, the main power amplifiers are electric vacuum devices (TWTs) and solid-state power devices. Traveling wave tube has the advantages of large power and high efficiency, but its disadvantages are large volume, bulky, and long preheating time (usually 3 min), and high-voltage power supply, which puts forward higher requirements for users to use and maintain it. The advantages of solid-state semiconductor devices are low noise, small size and light weight, while the disadvantages are low power and low efficiency. The characteristics of both determine that they have their values in their own spaces. The development goal of electric vacuum transmitter is to have higher frequency, greater power, higher efficiency, high reliability, miniaturization, and array applications. The miniaturization of vacuum electronic devices is based on its high power, high frequency, and high efficiency. Vacuum electronic devices by lowering high voltage, high current requirements, thus eliminating the cause of ignition, extending life, and enhancing reliability. Compared with TWT, solid-state power devices still lack the available technology or materials to effectively reach power levels of several hundred watts to several kilowatts in terms of high frequency, small size, and electrical performance. With the continuous development of manufacturing technology of solid-state devices and the expansion of application range, many applications require the output power of the power amplifier to reach the kilowatt level, and the working frequency range is also from point frequency to narrowband and even wideband, so that the working frequency is getting a high frequency and the working bandwidth becomes wider. There are still many cases to make the power amplifier work in relatively harsh environment, which promotes the emergence of a new generation of wide bandgap semiconductor materials. The representative ones are SiC, GaN, etc. The third-generation semiconductor materials have the advantages of stable chemical properties, wide bandgap, and radiation resistance than previously, which are especially suitable for the manufactural of high temperature, high frequency, high power, and radiation resistance power devices.

At present, high-power transmitters above 6 GHz generally adopt TWT system, while low-frequency and low-power transmitters adopt solid-state power amplifier system. Increasingly, the frequency at which solid-state power amplifier can be implemented is becoming larger and the power is getting bigger. In the S, C band and even higher frequency band tends to penetrate each other.

The emergence of MPM is an effective organic combination of them, highlighting the importance of the parallel development of vacuum devices and solid-state devices. In the future for a long time, solid-state devices and vacuum electronic devices will continue to develop, which will, respectively, adapt to and meet the different requirements of the whole machine and application environment. The technological development of the two aspects always needs comprehensive balance to make the system optimal.

References

1. Kolnik MI (2008) Radar handbook, 3rd edn. McGraw-Hill Professional
2. Fujiang L, Zhenpeng S, Tiechang Y et al (2008) Vacuum electronics. National Defense Industry Press, Beijing
3. Dechun W (2010) Wideband phased array radar. National Defense Industry Press, Beijing
4. Yang Z, Zhaohui G (1998) Design and research of rideband phased array transmitter for electronic warfare. Elect Warfare Tech 13(1):25–35
5. Yang Z, Qiqin L (2001) Error analysis and performance test of broadband multi-beam array. Elect Warfare Tech 78(3):29–35

Chapter 8
System Controller of Jamming System

System controller is the brain and nervous system of a jamming system. Its duty is to make the compositions of the jamming system properly work according to the predefined method, procedure, and parameter, and realize the overall function of the radar jamming system. Previous electronic countermeasure-related books usually focused on the basic principle and methods of radar jamming [1–4], but seldom discussed the system control. In this chapter, the authors will introduce the method of radar jamming system's control design, which includes the requirement of radar jamming system for system control, the design method of jamming system control, and the developing trend of future jamming system control will be discussed, according to the inspiration during jamming system developing.

8.1 Requirements Analysis of Jamming System Controller

The functions of system controller include time sequence control, working parameter control, system-state monitoring, information interaction, and human–computer interaction. Requirements analysis is the first step of related design works, and therefore, we will demonstrate the requirements analysis for the above functions in this subsection.

8.1.1 Time Series Control Requirements

Time sequence control manipulates the working states of all the system parts from the perspective of time to make sure the system's reception and transmission are working well. The requirements of time sequence control mainly contain the following contents.

G. Tang et al., *Techniques and System Design of Radar Active Jamming*,
https://doi.org/10.1007/978-981-19-9944-4_8

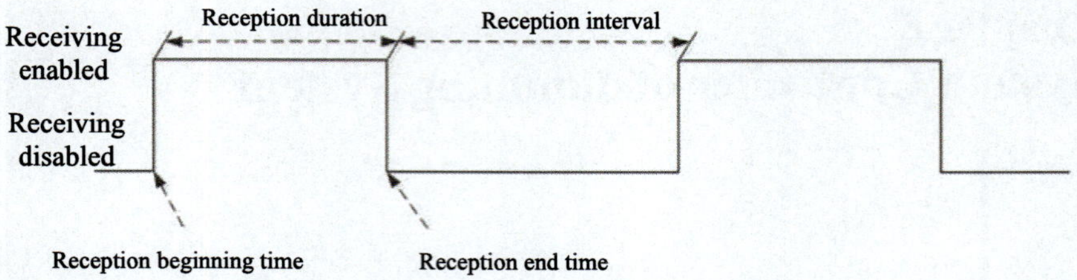

Fig. 8.1 Illustration of reception control parameters

(1) Receiving time control

The mission of active radar jamming system is to perform jamming at the proper time and monitor the environment change to adjust its jamming. Hence, a jamming system needs to receive signals. For most active radar jamming systems, they cannot realize full-time reconnaissance, and time sequence control should be used on demand. The control parameters for reception time include reception beginning time, reception time duration, reception interval, etc., as shown in Fig. 8.1.

Active radar jamming system exploits different reception times for radars with different systems and radiation parameters. For example, pulses received in a short-time window are enough to satisfy the conditions for target sorting and identification, for high PRI and small PW radar. However, for large PW and low PRI radar, a longer time window is required.

The requirements for reception time are different between reconnaissance and jamming modes. When the system is working under reconnaissance mode, it keeps receiving for a long time until a threat target is found and then turns to the jamming mode. When the system is working under the jamming mode, reception time should be assigned so as to be able to intermittently receive the environment signal in order to know the environment and threat changes in order to adjust the jamming.

The requirements of reception time are different when the system is using different jamming modes, such as noise jamming and DRFM jamming. When noise jamming is used, it only needs to intermittently receive the signal to know the radar signal change. If DRFM jamming was applied, it requires regular radar signal reception to update the signal samples digitally stored, besides intermittent environmental signal reception.

(2) Emission time control

Jamming system performs jamming according to the radar and the environment, and therefore, its emission may not be full time. Emission time control is necessary in this case. The objects of emission time control contain emission beginning time, emission time length, emission end time, emission interval, etc., as shown in Fig. 8.2.

The requirements for emission time are different while the jamming system exploits different jamming modes. For example, the emission time length was short

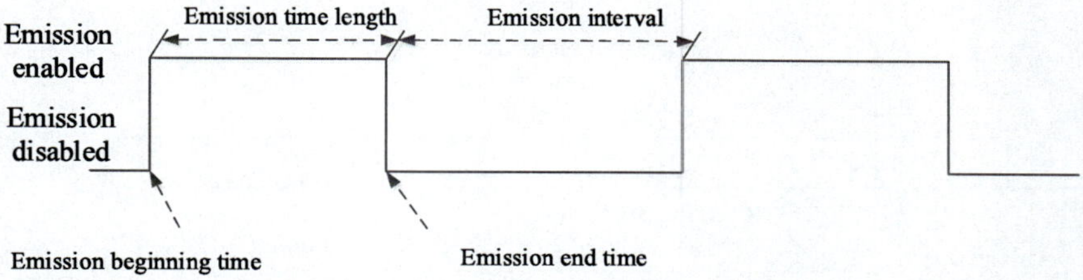

Fig. 8.2 Illustration of emission time control parameters

but the jamming was frequently launched if distance/speed gate pull-off jamming was used. The emission time length would be very long if noise suppressing jamming was used.

(3) Coordination between reception and emission time

Most jamming systems need to use the time gap to solve the reception and emission isolation problem, i.e., the jamming system cannot receive while jamming, and cannot emit while receiving [5]. There should be a reasonable time gap between reception and emission, including the transformation time from emission to reception and that from reception to emission, as shown in Fig. 8.3. The time gap between emission and reception is designed to eliminate the affection caused by the emission to reception and is an extra time consumption of the system. Hence, the time gap should be designed as small as possible. Besides, the transformation time from reception to emission is usually different from that from emission to reception due to device problems.

Most subsystems of active radar jamming systems require time sequence control. The subsystems include antenna, receiving processor, jamming source, transmitter, etc., as shown in Fig. 8.4.

(a) Time sequence control of antenna

For non-phased array antennas, time sequence control is unnecessary. For phased array-based emission antennas, emission control should be performed within the arrays. For phased array-based emission and reception antennas, time sequence

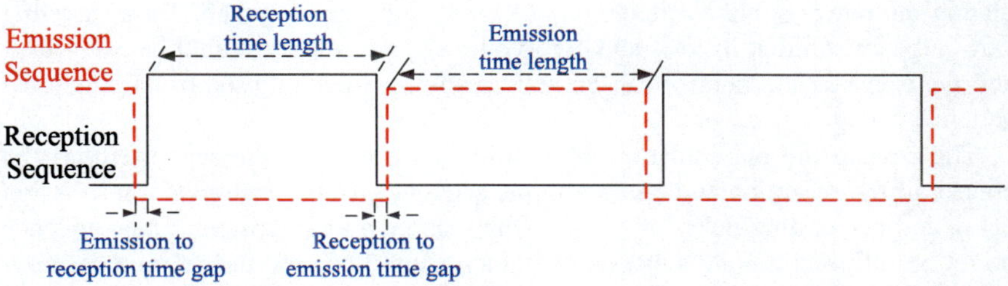

Fig. 8.3 Illustration of emission and reception control parameters

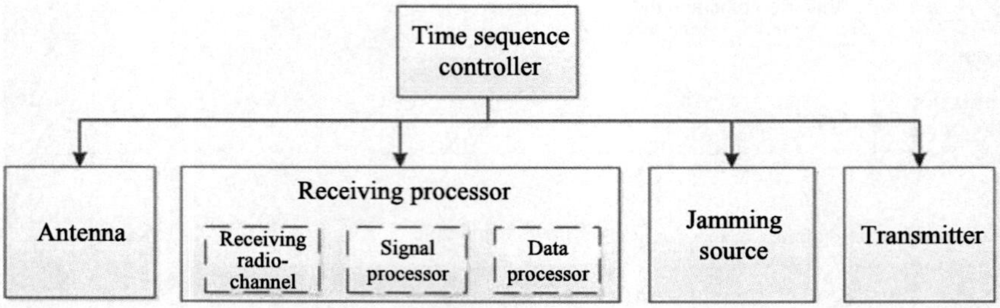

Fig. 8.4 Schematic diagram of main devices of time sequence control

control should be performed for the reception, emission, and switches between them. For antennas with multiple receptions or emission working sets, it probably needs to control the switch matrix to select different working antennas.

(b) Time sequence control of receiving processor

The time sequence control for receiving processor is mainly for the signal receiving time and signal processing time, including radio-channel, signal processor, and data processor. Within the reception time duration, it allows the receiving radio-channel to connect, signal processor and data processor to detect signal, measure parameters, sort, and identify the target. With the jamming time duration, it cuts off the radio-channel by turning off the reception switch and stops the signal processor and the data processor from processing the signals of this period.

(c) Time sequence control of jamming source

For DRFM jamming sources, the radar signal sampling and jamming signal emission are controlled according to the time sequence. For jamming sources exploiting direct frequency synthesis, the jamming signal is transmitted according to the time sequence.

(d) Time sequence control of transmitter

Time sequence controller gives the transmission time sequence to control the output time of the transmitter. For a solid transmitter, the control can be realized by manipulating the power supply voltage, input switch, and output switch. For a traveling wave tube transmitter, its low- and high-voltage power supply should be controlled, and the transmission control can be realized by manipulating its excitation signal and grid voltage.

The core of the manipulation of the time controller for the subsystems is the control of the reception and emission time sequence. Considering the transmission delay and responding delay of devices, the reception and emission time sequence might be different among subsystems, which should be determined on a case-by-case basis while designing the system. Time sequence controller's work should be performed under strict time synchronization conditions, and therefore, the time sequence is usually generated by the same clock to keep the time sequence signal strictly synchronized.

The key indices of system time sequence control include synchronization precision and continuous reception and emission time length. Since the generation of time sequence control at present is usually triggered by a digital timer, the synchronization precision is mainly determined by the clock period. Under certain situations where the synchronization precision demand is extremely high, the influence of time delay of the time sequence from the generator to the controlled device and the responding speed of the controlled device should be considered as well. Assume the timer's clock period is T, and then, the time control is performed with T as the minimal time unit. Continuous reception and the emission time length are determined by the timer's length and the clock period. Assume the timer's length is n bits, and the continuous reception and emission time length will be $(2^n - 1)T$.

8.1.2 Work Parameter Control Requirements

Working parameters control is an important function of the system control subsystem. Through sending control parameters to the subsystems, the subsystems' working states are coordinated, and therefore particular system function is accomplished. The specific control parameters are given in Fig. 8.5.

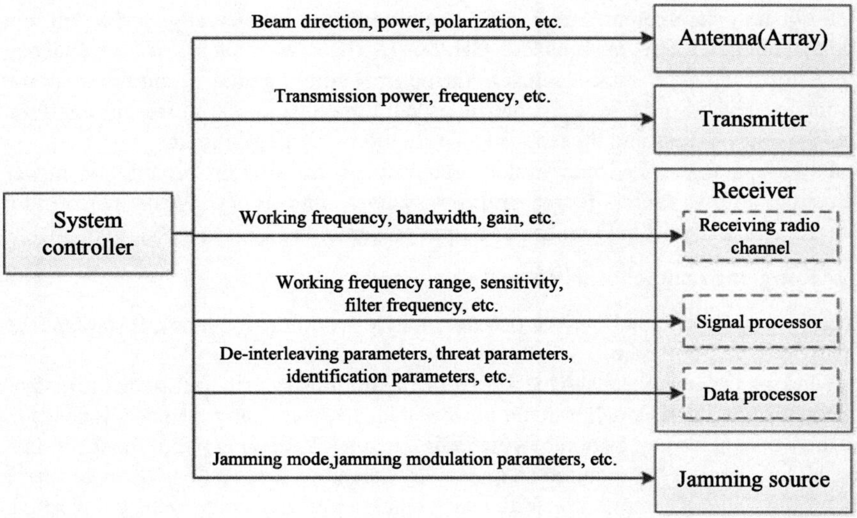

Fig. 8.5 Schematic diagram of working parameters control

(1) Antenna control parameters

Antenna control objects include the system's beam, polarization, and power.

Beam control: For undirectional or wide beam antenna, it is unnecessary to perform beam control if the special coverage requirement is satisfied. For a mechanical scanning antenna, the beam adjustment can be finished by controlling the rotation of the servo. For phased array antenna, parameters such as working frequency and beam direction should be given to realize the specific beam direction of a specific frequency range by the antenna beam position controller [6]. For multiple beams antenna, it needs to give the beam directions to be formed to the antenna array for beam adjustment. The adjustment of a specific signal to a specific direction also can be formed by the modulation of the input signal. While performing requirement analysis, the beam direction control range and the beam control stepped size should be made clear.

Polarization control: Some antenna has the function of selecting its polarization, and the receiving and transmitting antennas' polarization can be controlled by setting polarization parameters.

Power control: For active phased array antenna, the transmission power control is fulfilled at the antenna array whose radiation power can be controlled within certain range. While performing requirement analysis, the number of power levels should be clearly defined.

(2) Transmitter control parameters

Transmitter control objects include frequency and power.

Frequency parameters: Since radar jamming transmitter usually works within a wide frequency range, such as 2–6 GHz, 6–18 GHz. With such a wide frequency range, the transmitter should adjust to the optimal amplification circuit based on the frequency given by the system controller. While performing requirement analysis, the frequency values and the stepped size should be clearly defined.

Power parameters: Some system requires its transmitter capable of power adjusting ability, which is realized by system control [7]. While performing requirement analysis, the number of output power adjusting levels should be given.

(3) Receiving radio-channel control parameters

Receiving radio-channel control objects include working frequency, instantaneous bandwidth, and link gain.

Working frequency: Due to the limitation of jamming system back-end, the instantaneous bandwidth is smaller than the workable frequency range. Hence, it needs to set the working frequency of receiving radio-channel. Receiving radio-channel transforms the signal to the required frequency by adjusting the local oscillator of mixer according to the set frequency. While performing requirement analysis, the working frequency and the stepped size should be clearly defined, and the control parameter is given by frequency parameter or the local oscillator value of mixer also should be made clear.

Instantaneous bandwidth: Some jamming systems have the requirement of using different instantaneous bandwidths. It requires the system controller to give the instantaneous bandwidth parameter to the receiving radio-channel for corresponding bandwidth adjustment. Since instantaneous bandwidth control is usually realized by a switch filter bank, system control is able to directly output the switch signals for different bandwidths.

Link gain: Some systems require having the control function of different receiving power ranges. It needs the system controller to send the gain parameter or amplification link loss value to the receiving radio-channel.

(4) Signal processor control parameters

Signal processor control objects include frequency range, sensitivity, and filter frequency.

Frequency range: Signal processor processes the received signal in the mediate frequency or the baseband to detect the signal of radar and performs parameters measurement for radar signals. Signal processor needs to compute the radio frequency value of radar based on the current working frequency range of the system, which requires the system controller to send the current working frequency range to the signal processor. The current frequency range can be expressed by beginning frequency and end frequency or the combination of center frequency and bandwidth.

Sensitivity: In certain cases, the sensitivity of signal processor needs to be adjusted to fit different signal environments and targets. System controller outputs the sensitivity (or the detection threshold) to the signal processor for executing the adjustment.

Filter frequency: In certain cases, signal processor needs to filter the known signal to reduce the influence of the known signal and so as to reduce the processing load of signal detection and the afterward processing. Signal processor will not process the signal within the corresponding frequency band or output the corresponding results.

(5) Data processor control parameters

Data processor fulfills the sorting and recognition of radar signal, in which sorting parameters, threat parameters (threat target library), and recognition parameters are provided by the system.

Sorting parameters: Sorting parameters include the sorting processing time, frequency tolerance, PW tolerance, PA tolerance, etc. In general, the parameters should be settled before the system goes out. Only people with professional knowledge and skills are allowed to edit the parameters while the system is in use. When professionals need to change the tolerance parameters, they can use the system controller to send the parameters to the data processor.

Threat parameters: Current threat target recognition is based on the threat target library. The threat target library is loaded to the task system before the combat. As the development of technology, the threat parameters can be transmitted in real time by combat network, and the transmission procedure relies on the system controller.

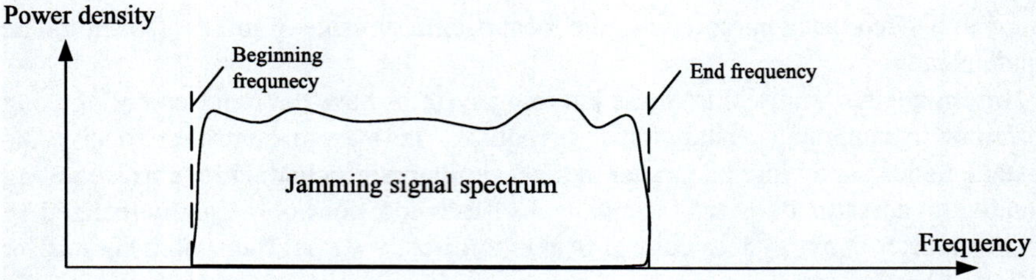

Fig. 8.6 Illustration of barrage jamming's parameters

Recognition parameters: Data processor performs threat recognition according to the target information, frequency tolerance, PW tolerance, and PA tolerance in the library. In general, the recognition tolerance should be adjusted to the proper values before the system goes out. Only people with professional knowledge and skills are allowed to change the parameters while the system is in use. The professionals alter the recognition parameters through the human–computer interface of the system controller.

(6) Jamming source control parameters

Jamming source control objects include jamming modes and the parameters of each jamming mode. Jamming modes include barrage jamming, sweep jamming, comb spectrum jamming, frequency spot jamming, repeater jamming, etc. Parameters of different jamming modes are different.

Barrage jamming emits broadband jamming signals to weaken the detection capability of radars working in this frequency range [8]. The parameters of barrage jamming include beginning frequency and end frequency, as shown in Fig. 8.6.

Sweep jamming's signal continuously varies within a given frequency range to solve the problem of using narrowband jamming signal-to-jam broadband radar, especially frequency agility radar [9]. The parameters of sweep jamming include beginning frequency, end frequency, frequency stepped size, and frequency dwell time [10], as shown in Fig. 8.7.

Comb spectrum jamming's jamming signals distribute at several frequencies, and the jamming signal of each frequency could be either a sinusoidal wave or a narrowband modulated noise signal. This kind of jamming is named comb spectrum jamming because its spectrum shape is similar to a comb. Comb spectrum jamming needs the system controller to output the number of jamming frequencies, the value of each frequency, whether noise modulation should be performed at each frequency, and the modulation bandwidth of each frequency. In certain cases, the frequency values of comb spectrum jamming can be automatically generated according to the input signal, and the frequency parameters sent by the system controller become unnecessary in this case.

Frequency spot jamming's control parameter is the jamming frequency value. System controller sends the information to the jamming source on demand, and the jamming source forms the jamming signal at the corresponding frequency based

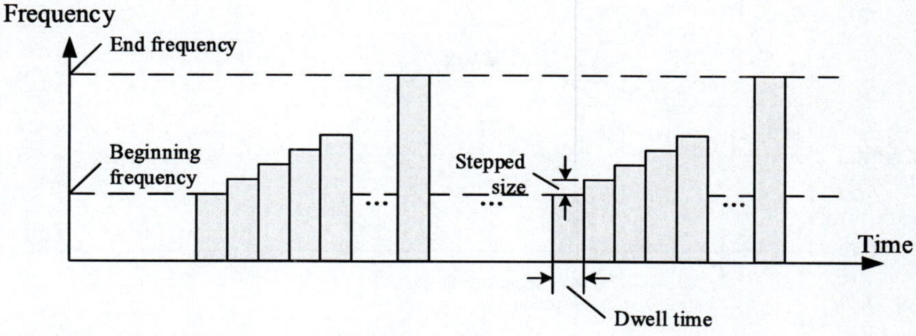

Fig. 8.7 Illustration of sweep jamming's parameters

on the received parameter. Frequency spot jamming mainly exploits narrow noise signals. Some frequency spot jamming has the ability to change the modulation bandwidth of the jamming signal, which requires the system controller to give the jamming signal bandwidth's selection parameter.

Repeater jamming's parameters mainly include repeater signal parameters, replication interval, and modulation parameters [11], as shown in Fig. 8.8. Repeater signal parameters are used to control the jamming source to forward the specific radar signal. There are plenty of different microwave signals including the enemy's signal and our signal, radar signal, communication signal, navigation signal, IFF signal, etc. Some radar signals are of high threat level, and some radar signals are of the low threat level. Only if selecting radar signals according to its threat level and repeating the signal, the jamming can be effective. Repeater signal parameters include radar frequency and pulse width. Most jamming systems do not forward radar signals simply pulse by pulse, but forward the received radar signal multiple times with certain replication intervals, and perform modulation for the radar signal. Hence, the replication interval and modulation parameters are both important parameters for controlling repeater jamming. The range or velocity gate pull-off jamming for pulse Doppler radar is fulfilled mainly by repeater jamming. Range or velocity gate pull-off jamming requires the jamming system to control the delay of Doppler modulation parameters of the forward signals to realize effective range or velocity gate pull-off [8].

8.1.3 System-State Monitoring Requirements

System-state monitor performs state monitoring and state controlling. State monitoring indicates the collection of all system components' data by the system controller. State control is the adjustment of all system components' states by the system controller.

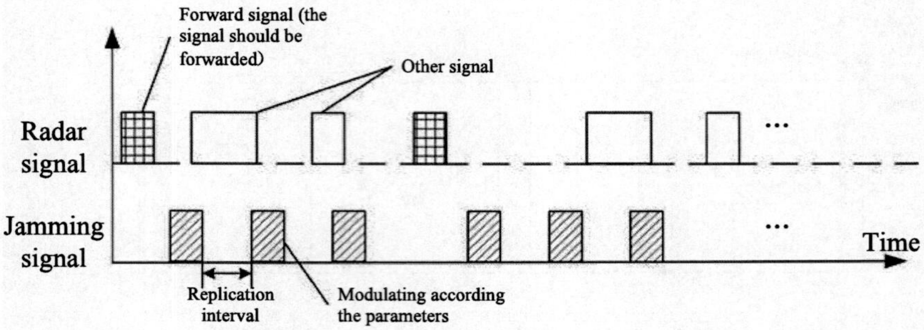

Fig. 8.8 Illustration of repeater jamming's parameters

(1) System-state monitor

The contents of system-state monitor include environment temperature, device temperature, power on state, working state, state of properly working, and working parameters. Every system has its environment temperature boundary, and environment detection is helpful for avoiding system failure due to the use exceeding the system's limit. Device temperature is the temperature state of the device itself, and all the components of the device have their working temperature ranges. Especially for the devices of large heat release, real-time temperature monitoring is necessary to prevent abnormal working or damage due to overheating. Working parameters are the variables that can be changed or adjusted during system's working time and can be referred to when it is needed. Whether the components are working properly affects the function of the whole system. By setting monitor functions at the key points of the system, system controller is able to determine whether some parts of the overall system are working properly.

Most systems should consider the recording of system state. It will facilitate the troubleshooting and close loop when a system failure occurs and play an important role in the analysis of the use and improvement of active radar jamming systems if the system state is recorded.

(2) System-state control

System-state control is to control the system's states such as power on/off, standby, self-testing, working, and rebooting. Some system demands the components to power on in order. It is better to power the components on in order under the control of the system controller. If some subsystem of the system was overheating, it could be powered off by the control. If the system did not need some subsystem to work, it could set the component to standby mode. For example, the transmitter could be set to standby mode if jamming was unnecessary. The system controller could send self-testing commands to the related devices if the overall function was abnormal or it was needed for other reasons.

8.1.4 System Information Interaction Requirements

System controller can be seen as the nerve center of the radar jamming system. In order to realize system-state monitor, system time sequence control, and system working parameters control, it requires the system controller to obtain information from the system and its external places and send control information, time sequence information, and target information to the corresponding devices [6]. In the following part of this subsection, commonly used system information interaction methods will be introduced. In practice, specific systems should make additions or deletions according to their overall requirements.

The interaction between the radar jamming system and its external systems is shown in Fig. 8.9. A radar jamming system obtains information about position, velocity, and attitude, to adjust the direction of antenna in real time. For airborne and shipborne radar jamming systems, the information is obtained from the carrier. Vehicleborne radar jamming system usually works when the car is standing still, and only position and attitude are needed from the localization device of the platform.

In order to assist jamming, an active radar jamming system needs situation information decision such as radar warning, electronic reconnaissance, and radar intelligence as well. For airborne and shipborne self-protection jamming systems, they receive radar warning information to select jamming targets and perform jamming. Support jamming system receives the electronic reconnaissance information from other platforms to make jamming decisions and realize the cooperation jamming between platforms. The radar intelligence is used to compare and merge with electronic reconnaissance by the jamming system, to eliminate the possibility of jamming our radar, and to select the radar with high threat levels to perform jamming.

In order to avoid jamming our electronic devices, such as radar, communication devices, and navigation devices, active radar jamming system needs to obtain the frequency utilization information from the platform.

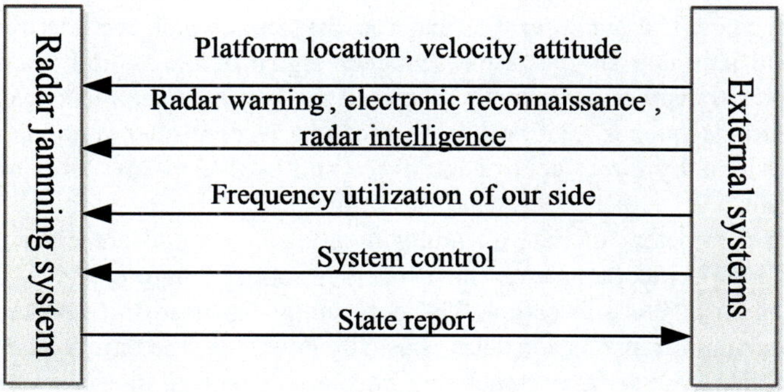

Fig. 8.9 Illustration of external information interaction of radar jamming system

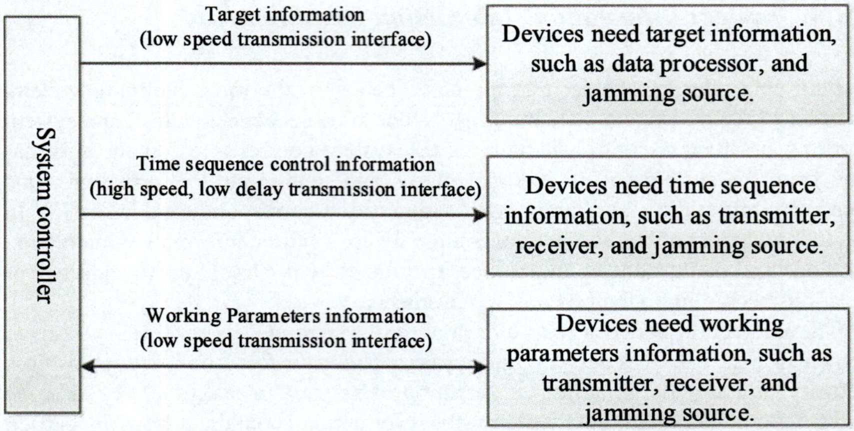

Fig. 8.10 Illustration of the information interaction between system controller and other subsystems

Active radar jamming system receives the control command from its superior system, reports its system state, and cooperates with other operational systems via system control information.

The information interaction between system controller and other subsystems of the active radar jamming system contains target information, time sequence control information, and working parameters, as shown in Fig. 8.10.

Target information is the description information about the radar to be jammed by the system, and its main receivers are data processor and jamming source. Data processor uses the target information to identify the threat target, and the jamming source receives this information to filter out the signal to be jammed.

As described in Sect. 8.1.1, the time sequence control information is used to make sure the system working in specific time order. System controller sends the time sequence control signal to the related subsystems which include transmitter, receiver, and jamming source. Time sequence signal demands high transmission speed and low transmission delay to ensure all the system components work based on a strict time sequence relationship. Hence, system controller is connected with other subsystems by direct control interfaces and uses high–low level to express working state.

Working parameters include a variety of information and are diverse among systems. The specific parameters have been introduced before and will not be discussed again in this subsection. The only thing that need to be made clear is that the demand for transmission delay is usually lower than the time sequence information, and therefore serial port, network port, or customized interface can be used to satisfy the requirement.

8.1.5 Human–Computer Interaction Requirements

Some active radar jamming system needs human–computer interaction functions. Since the content of human–computer interaction is determined based on monitor and control, it is discussed as part of system control other than an independent section in this book. Some kinds of human–computer interaction of active jamming systems are simple while others are complicated [12, 13]. For example, some jamming systems are full-automatic and can fulfill their tasks without manual operation. In this case, there are only a few state indicators for troubleshooting. For airborne self-protection jamming systems, their human–computer interaction contents might only include an emission button and a jamming emission indicator light. For dedicated EW aircraft for support jamming, its human–computer interaction uses several consoles, and several operators perform manual operations according to the task requirements. Since different systems have different human–computer interaction demands, only common human–computer interaction contents will be discussed in this subsection.

(1) System display contents

The human–computer interaction contents displayed to operators by the system contain system-state information, threat targets information, and current working parameters.

System-state information mainly shows whether the system is working properly, and the system's current state, such as standby, receiving, and emitting. For self-protection jamming systems, the state indicator function only displays whether the system is emitting and whether the system is out of order. For support jamming systems which usually have operators, the state information additionally contains the subsystems' states, such as whether the transmitter is in high/low voltage, and the corresponding voltage values.

Threat targets information can be displayed in brief mode or detailed mode. In brief mode, there might be only one indicator light. In detailed mode, the displayed information contains the direction or position of the threat target, frequency, PW, PRI, and PA of the threat target signal. The information can be shown by lists or the combination of figures and lists.

Current working parameters are the key parameters used by the system during its current job and help the operators adjust working parameters and make jamming decisions. Since working parameters include detailed contents, they are displayed only for the systems with manual operation consoles. For example, the working parameters of the antenna include beam direction, polarization, and the transmission power of phased array antenna. The working parameters of receiving state include working frequency range, spatial range, sensitivity setting, as well as the information about threat targets, such as position, threat level, frequency, PW, PRI, PA, etc. The working parameters of jamming state include jamming target information, frequency information, and jamming modulation parameters.

(2) Manual operation contents

The manual operation contents of radar jamming systems may be different due to the different task demands. There may be only a button for the systems requiring minimum operations, and for systems requiring plenty of operations, they need more than one operator to set or adjust the parameters and the state of the system. The operation contents include turning on/off the system, selecting working frequency, choosing the jamming target, selecting jamming mode, and setting jamming parameters. The detailed operation contents have been introduced in the previous part of this section and will not be described again.

8.2 Design of Jamming System Controller

Jamming system controller should be designed according to the system control demands. Function analysis should be performed first and then determine the system controller's software and hardware along with functional decomposition. Finally, the system design, integration, and testing are performed. In this subsection, the design procedure and design methods will be described in general, including design flow, hardware design, and software design.

8.2.1 Design Process

The main design method of jamming system controller is similar to that of other systems, including building the requirements, overall design, detailed design, and integration testing, as shown in Fig. 8.11.

(1) Building up control requirements

In the stage of building up control requirements, the designer should sum up the control system's requirements according to the overall functions, working mode, working flow, and the system indices. The steps for building up control requirements include:

1. Define the overall working flow of the system;
2. Define the controller composition;
3. Define the controller's control parameters;
4. Define the working time sequence requirements of the controller;
5. Define the interfaces between the controller and other subsystems or external systems;
6. Define the performance indices of the controllers.

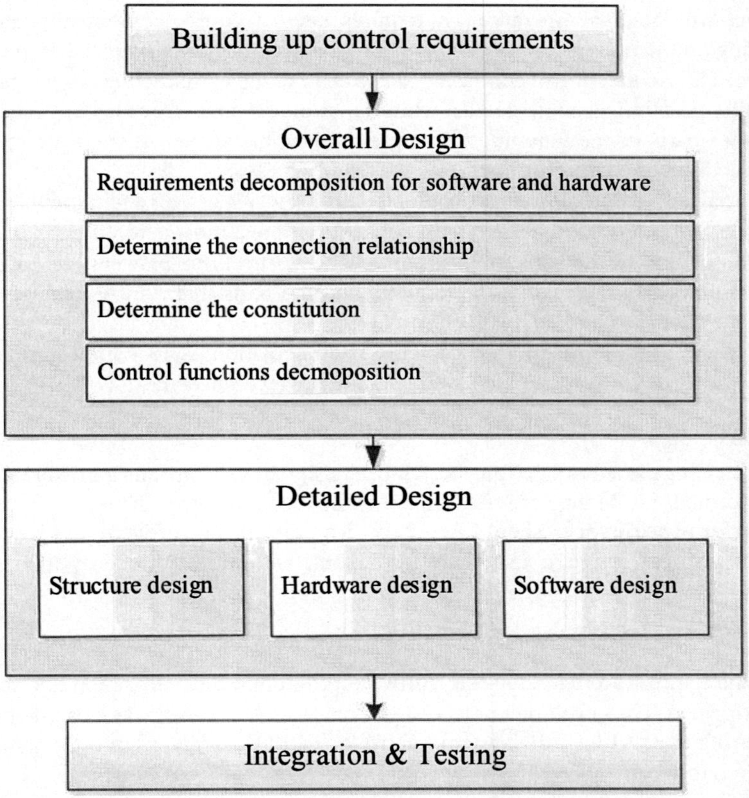

Fig. 8.11 Schematic diagram of system controller design

(2) Controller overall design

After the requirements are determined, the controller's overall design can be started. During the overall design procedure, the control functions are subdivided, the internal and external connections of the controller are determined, and the software and hardware requirements are decomposed.

Control functions division: The overall functions can be divided into human–computer interface functions, software functions, and hardware functions from the perspective of hierarchy; the overall functions can be divided into system-state monitor functions, receiving procedure control functions, and jamming procedure control functions from the perspective of tasks.

Composition of controller: A controller usually consists of console, control processing computer, time sequence controller, and interface devices. For a specific subsystem, its composition should be selected according to the need. For example, a self-protection jammer does not need a console, and a simple indicator light or a simple integral display inside other display devices. For dedicated radar jamming systems with manual operation, it requires several consoles to manipulate the receiving and jamming resources manually so that human intelligence can be fully utilized. The architecture of controller can be centralized or distributed. A centralized controller is an independent device, while a distributed controller may be embedded into other parts of the jamming system; e.g., the time sequence controller can be embedded into the jamming source or the data processor.

Internal and external connections of controller: When the requirements of the controller are determined, the external connections are almost determined too. After the composition of the subsystem is set up, the connections between the different parts of the subsystem should be defined, and the parts that will be connected to external systems should be made clear as well.

Software and hardware requirements decomposition: The compositions and requirements of software and hardware should be decomposed.

(1) Software and hardware detailed design

After finishing the overall design, the detailed design of software and hardware should be performed. While the software and hardware detailed design stage, the functions and performance indices should be made clear for all the software and hardware according to the software and hardware decomposition determined by the overall design.

(2) Controller integration and testing

After the hardware processing and software programming, integration and testing are performed. The development of system controller is finished when its integration and testing end. At last, the system controller will be integrated into the jamming system to test.

In order to directly illustrate the controller of jamming system's overall composition, an example including remote control, local control, underlying control, and interface circuit is given in Fig. 8.12. The hardware of a remote control terminal consists of control computer and communication devices. The remote computer is usually a desktop or laptop computer. Remote control terminal software includes remote control software, communication software, etc. The hardware of a local control device includes control computer and communication device, and the software includes local control software and communication software. The control computer is usually a desktop computer or a Power PC. The hardware of an underlying control device is a control circuit, and the software is embedded software. The control circuit is realized usually by FPGA or micro-processor. The external interfaces of system controller include serial port, parallel port, network port, and self-defined interfaces.

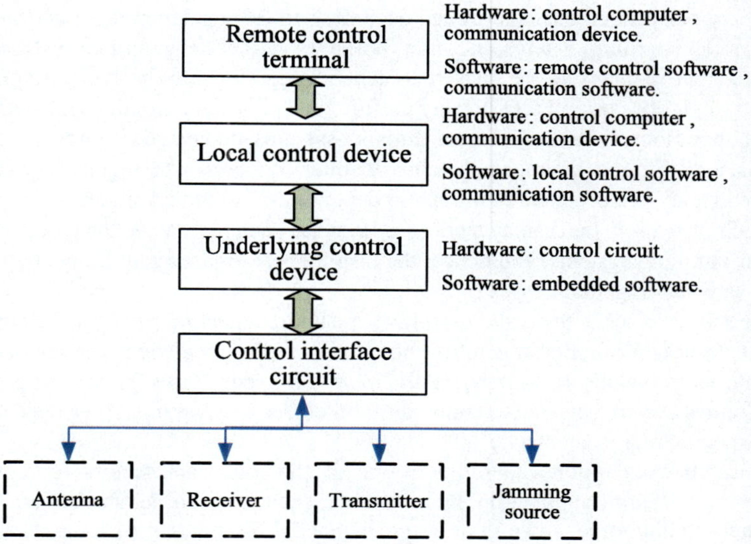

Fig. 8.12 Schematic diagram of system controller overall composition

8.2.2 *Hardware Design*

The overall composition of jamming system's controller has been discussed in the previous subsection, in which the hardware classification is introduced. The specific hardware composition should be determined according to the exact requirements.

(1) Control computer

The factors that should be considered while designing the control computer include computation performance, storage performance, display performance, external interface, and environment adaption capability.

The computation performance should be determined according to the amount of computation controlled by the system. In order to satisfy the demand for real-time computation, computers with high performance should be chosen for massive calculation. Engineers choose the control computers by the means of the computation amount estimation, personal experience, and simulation based verification.

Storage performance includes running memory and external storage. Running memory is the storage for running operating systems and application programs. Running memory works after the system is powered on, but the program and data will be lost after the system is powered off. External storage is used for saving procedure data and loading programs, and data will not lose after the power is off. Since the solid hard disk has good shock-resistant capability, it is widely used in airborne, vehicleborne, and shipborne storage. Engineers choose the suitable storage volume according to the operation system, program size, and storage data size.

Display performance is determined according to the display content and mode. Certain radar jamming systems, such as airborne self-protection jamming systems, do not have video displays, or the display content is integrated into the avionics system. In this case, it does not require display performance. For the jamming systems with manual operations, such as shipborne jamming systems, dedicated airborne jamming systems, and ground jamming systems, display functions are required, and the specific display performance requirement depends on individual situations. Pushed by the 3D animation, the display technique level has been far beyond the requirement of radar jamming systems, and hence, the display performance requirement always can be satisfied in general cases.

External interface is the path controlling the interconnection between the control computer and control circuit, another subsystem, or display device. The commonly used interfaces include serial port, parallel port, audio port, video port, and network port. Control system selects the computer with corresponding interfaces depending on the specific requirements.

Environment adaption capability is one of the important factors for control computers. Different application environments, such as airborne, shipborne, vehicleborne, satelliteborne, and indoor usage, demand different control computers. For airborne cases, it requires the computer to satisfy the shock, vibration, temperature, and pressure conditions for aviation equipment.

Besides, watchdog function is also needed to keep the control computer's functional robustness. With this, the system is able to automatically recover from an exception.

(2) Control circuit

Control circuit receives the control commands from the control computer and generates the control signals for subsystems according to the predefined logical relationship. The generated control signals include time sequence control signal, state control signal, and jamming control signal.

Time control signal is generated by the control circuit according to the specific logical computation and is used to control the working time sequence of the subsystems and modules to ensure the orderly working of the signal receiving, processing, jamming signal generation, and transmission of the jamming system, as given in Sect. 8.1. Time sequence control circuit is usually realized by a programmable gate array or DSP [14, 15].

State control signal is the signal controlling the states of subsystems. Control circuit is generated based on the real-time need. For example, the switch selection of radio-channel, the output frequency of the frequency synthesizer, the beam direction of receiving or transmission array, and the loss code of the receiver are all controlled by the control signals.

Jamming control signal is the signal controlling the jamming source, and different jamming sources have different functions, designs, and control signals. For example, pull-off jamming needs to generate jamming range and speed control signals; noise jamming needs to generate noise frequency, and bandwidth control signals; false targets jamming needs to generate false target distance spacing, and targets number control signals.

(2) Interface circuit

Interface circuit directly connects the system controller and other systems. It is the interface between the system controller and other subsystems. Some connections between the system controller and other subsystems can be built directly by control computers or control circuits, while others should be designed by exploiting special interface circuits.

8.2.3 *Software Design*

The developing procedure of system control software is given in Fig. 8.13 and contains five stages which are software requirements analysis, software preliminary design, software detailed design, software programming, and software test and acceptance.

During the software requirements analysis phase, the software requirement specification and the development plan are outputted based on the analysis of software requirements. The main contents are software functional requirements, external interface requirements, internal interface requirements, computer resource requirements, security and secrecy requirements, and so on.

Software preliminary design describes how the system executes to satisfy the requirements from the point of view of the customer. The selections of software units and design decisions are given. The functions are assigned to software units during the overall architecture design.

Software detailed design further refines the software design based on the preliminary design, and documents such as detailed design description, unit test description, and unit test plan are outputted during this phase.

Software programming is performed based on the software detailed design description and outputs software code modules that will be integrated into the software.

During the software test and acceptance phase, software functions and reliability are tested, and the acceptance test report is outputted.

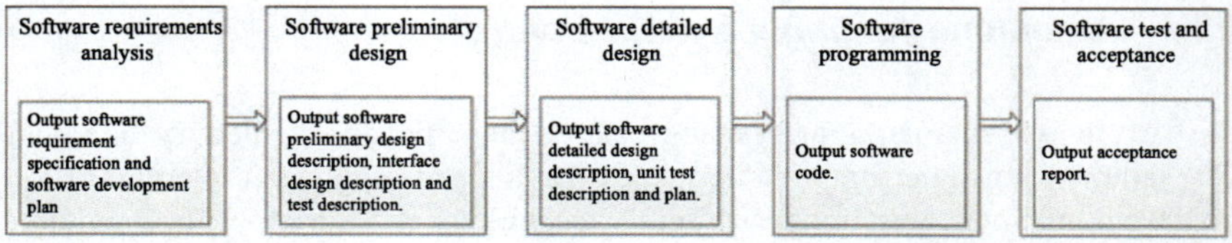

Fig. 8.13 Schematic diagram of system control software developing procedure

The software of the system controller includes upper machine control software and control circuit software. Upper machine control software mainly finishes the complicated human–computer interaction and computations which are flexible and have low timeliness. Embedded software mainly fulfills the high-speed computation and generates the control signal with low delay.

(1) Upper machine control software

The human–computer interaction functions of upper machine software include the graphical display for the system state, reconnaissance results and jamming state, human input acceptance, input information processing, and control command generation.

Upper machine software also records the process data including system state, reconnaissance results, jamming state, and manual operation command. The exact items to be recorded differ based on the system's requirements.

(2) Embedded control software

Embedded control software is developed according to the requirement for specific high-speed computation and is usually coupled with the control circuit. Embedded control software contains the control program running in microprocessors such as single chip microcomputers, and the code loaded to the programmable gate array, which will not be further introduced.

8.3 Development Trend of Jamming System Control

Automation, intelligence, and cooperative control are the important development trend in radar jamming system control. Automation means the reconnaissance and jamming can be fulfilled by a machine based on the predefined program, in which the manual operations can be significantly reduced. Intelligence is the result of utilizing artificial intelligence (AI) technology, and the performance of jamming system can be improved if AI technique is adopted. Cooperative control helps multiple active radar jamming systems form enhanced system jamming capability through cooperation among them.

8.3.1 Automatic Jamming System Control

The early active radar jamming systems required their operators to identify the signal to be jammed which the jamming frequency would be manually adjusted to [16]. As the development of radar frequency agility technology and the application requirements of countering missiles guided by radar, the previously mentioned manual way does not fit modern war, especially for self-protection jamming systems. Modern weapon systems guided by radar can fulfill their attack within dozens of seconds or

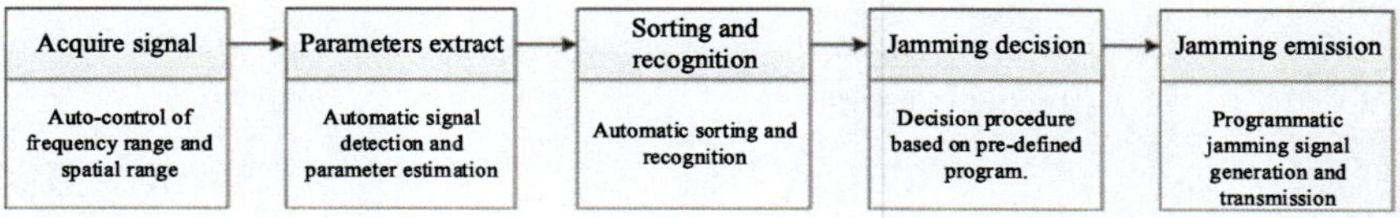

Fig. 8.14 Procedure and features of automatic active jamming

even less than twenty seconds, and therefore self-protection jamming systems have the highest demand for automation and have almost realized. The main feature of automatic active radar jamming is programmed; that is, it finishes its work according to the predefined program whose procedure is given in Fig. 8.14.

1. Signal acquisition: jamming system acquires environmental signals according to the programmed space and frequency scan strategy. For systems that can cover all concerned space and frequencies, the scan is unnecessary.
2. Parameter extraction: CFAR-based algorithms are exploited to detect the signal in the time or frequency domain and then estimate parameters by automatic algorithms.
3. Sorting and identification: use a specific algorithm to classify the detected signals and compare them with the threat database of the jamming system to identify known threats and mark unknown targets.
4. Jamming decision: the automatic jamming decision is made according to the predefined strategy embedded in the jamming system, to select jamming targets and jamming parameters.
5. Jamming: programmed generate jamming signal according to the selected jamming targets and parameters.

8.3.2 Intelligent Jamming System Control

Artificial intelligence is a new technique science that researches and develops the theory, method, technique, and application systems to simulate extend, and enlarge human intelligence. AI is a branch of computer science. It tries to understand the nature of intelligence and produces a new intelligent machine that is able to react in the way similar to human intelligence. AI techniques include machine learning, pattern recognition, intelligent support system, expert system, and artificial neural network.

The theory and technique of AI have become matured since it was born, and its application domain also keeps increasing. AI techniques can be used in active radar jamming systems [17]. The US has started advanced technique research in this area in recent years. In 2012, Defense Advanced Research Projects Agency (DARPA) launched a five years project named Adaptive Electronic Countermeasure. At the end of 2012, the US navy research office published the road agency announcement named EW invention and discovery, whose theme was cognition and adaptive processing.

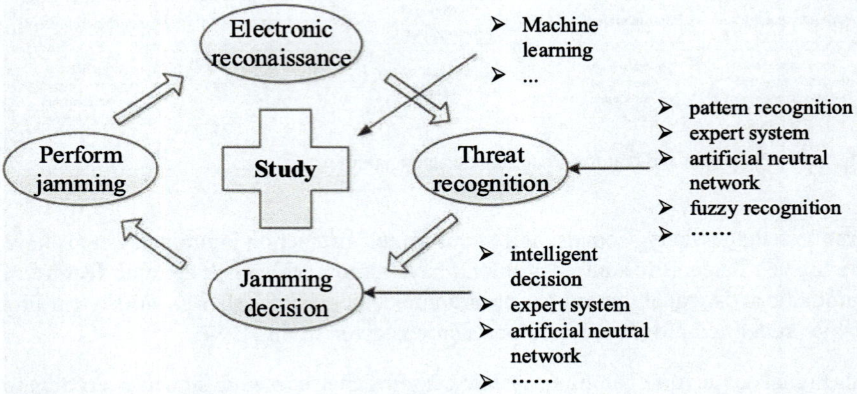

Fig. 8.15 Schematic diagram of radar jamming system applying AI

Cognitive and adaptive EW enabling technique was the top topic, and its target was to apply the adaptive algorithm and machine learning algorithm to EW to deal with flexible threats [18]. Besides, in 2010, the US air force laboratory started research on cognitive EW technology.

In active radar jamming systems, pattern recognition, expert system, artificial neural network, and fuzzy recognition can be used in the radar recognition phase to improve the environmental sensing capability. In the jamming decision phase, intelligent decision, expert system, and artificial neural network can be used to make intelligent decisions to improve timeliness and accuracy. Embedding machine learning function into a jamming system can improve the observation, orientation, decision, and action (OODA) loop's level, as shown in Fig. 8.15.

8.3.3 *Coordinated Jamming System Control*

As the development of equipment technology, the systematic application level of military equipment in all the countries becomes higher. During the procedure of systematic operation, cooperation is the inevitable trend of radar jamming equipment's systematic development. In 2012, a public document of the US navy research office described the future development of EW, where cooperation is one of the most important aspects [18]. The cooperation of future EW will be used among the devices inside and outside the platform, between manned and unmanned platforms, and between mobile and non-moving platforms, and its purpose is to maximize the EW effectiveness [19].

The application of cooperative jamming system control mainly solves the problems of threat situation awareness, global jamming situation awareness, jamming targets assignment, and jamming resource coordination.

There are two main types of cooperative control, which are centralized control, as shown in Fig. 8.16, and temporary cooperation without a center, as shown in Fig. 8.17. Centralized control needs a control center that acquires the reconnaissance information and state information from different platforms and gets the global situation after the fusion processing. Next, the control center performs jamming targets assignment and strategy adjustment according to the global situation and sends the adjustment command to all the jamming systems which report their reconnaissance results and systems states on receiving the jamming control command. The cooperative control without a center is usually exploited among several jamming systems of equivalent relationship or inside a jamming system. One jamming system requests cooperation, and another jamming system responds to the request and executes the requested functions.

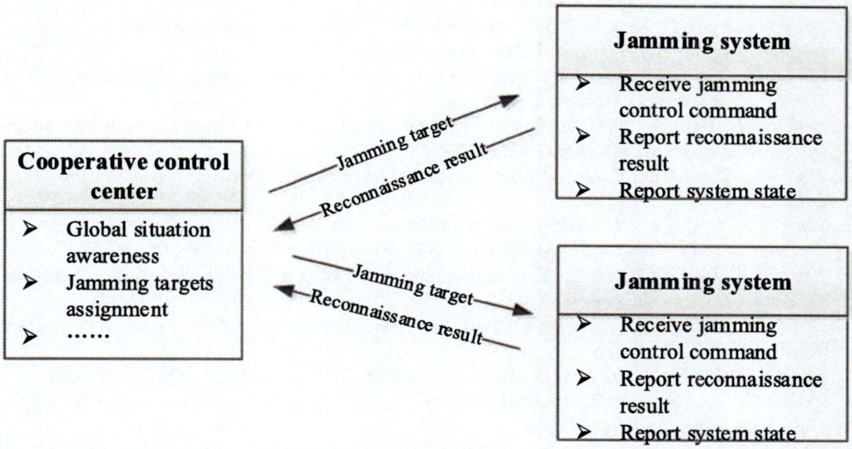

Fig. 8.16 Centralized cooperative jamming control

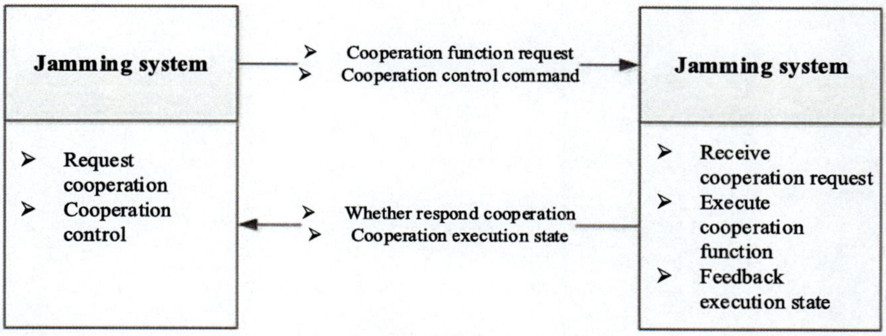

Fig. 8.17 Cooperative jamming control without a center

References

1. Xiangping L (1982) Electronic countermeasure principle. National Defense Industry Press, Beijing
2. Guoqing Z (2003) Radar countermeasure principle. Xi'an University of Electronic Science and Technology Press, Xi'an
3. Adamy DL (2004) EW102: A second course in electronic warfare. Artech House Inc., Norwood
4. Neri F (2001) Introduction to electronic defense system. Artech House Inc., Norwood
5. Maosheng Z (2008) Research into the methods of improving receiver-transmitter isolation of EW system. Shipboard Elect Counter 32(1):54–55
6. Wei L, Guodong L (1995) Beam control system of ECM phased array receiving antenna. YingYong KeJi 2:41–46
7. Ruijia W, Wangxing, Siyi C (2015) Power control of defensive electronic countermeasures based on radio frequency stealth. Telecomm Eng 55(1):19–26
8. Kai X, Yongguang C, Liandong W (2006) Analysis, modeling & evaluation of range gate pull off designs. Syst Eng Elect 8:1158–1163
9. Yanjie C, Jinbao X, Xing W (2015) Jamming effectiveness analysis of sweep jamming against frequency agile radar. Command Cont Simul 37(2):98–101
10. Shiyan Z, Guoping X, Juhua D (2012) The technique of sweep jamming signal on the base of DDS. Aerospace Elect Warfare 28(4):59–61
11. Donghai L, Jixin X (2009) A kind of practical control circuit in electronic countermeasure system. Shipboard Elect Counter 32(2):32–34
12. Zheng L, Changyong C (2009) User interface design in EW software. Elect Warfare Tech 24(5):51–56
13. Rong S, Chang L, Jun L (2015) Analysis of man-machine interaction mode evolvement and driving factors in electronic countermeasure. Shipboard Elect Counter 38(1):1–6
14. Xiaoxia X, Qixiang F (2012) A radar jamming controlling and processing system based on DSP chip. Microprocessors 4:89–91
15. Bin Z, Juhong Z (2010) A design method of embedded system control in avionic device. Elect Warfare Tech 25(3):70–76
16. Price A (1978) The history of US electronic warfare. Charles Scribner's Sons, New York
17. Zixing C, Guangyou X (2010) Artificial intelligence and its application, 4th edn. Tsinghua University Press, Beijing
18. ONR. Electronic Warfare Technology (2012) BAA-13-005[R/OL], 11 September. http://www.fbo.gov
19. Zhenyu F, Lei W, Jianchun S (2010) Technology of track deception for cooperative control of multiple electronic combat air vehicles. Info Elect Eng 8(3):265–268

Chapter 9
Evaluation of Radar Jamming Effect

Evaluation of radar jamming effect is to evaluate the effect of jamming system on specific radar system under specific conditions, and its purpose is to get the influence degree of jamming on radar target detection ability. It consists of a set of evaluation criteria, indicators, and methods. It can evaluate the technical and tactical capability of radar jamming system and find the system defects, which has important guiding significance for the improvement and perfection of the system [1].

Radar jamming effect evaluation is a very complex problem, because the differences of radar object, jamming technology, electromagnetic environment, operational mode, and evaluator's perspective will lead to different evaluation methods. But in general, the evaluation process of radar jamming effect is shown in Fig. 9.1. First, according to the relationship between radar and jamming, put forward the jamming effect evaluation criterion, establish the correct jamming effect evaluation model, and then based on different jamming mode to build evaluation index system, choose the appropriate methods of evaluation for jamming effect evaluation.

This chapter introduces the basic principle of radar active jamming effect evaluation, puts forward the latest jamming effect evaluation index system, and summarizes the methods of outfield evaluation, hardware-in-the-loop simulation evaluation and digital simulation evaluation. The first section introduces the criteria of jamming effect evaluation, including information criteria, power criteria, and tactical application criteria. The second section introduces the evaluation indexes of jamming effect, including suppressing jamming, dense false target jamming, and dragging deception jamming. The third section introduces the evaluation methods of jamming effect, including outfield evaluation, hardware-in-the-loop simulation evaluation, and digital simulation evaluation.

G. Tang et al., *Techniques and System Design of Radar Active Jamming*,
https://doi.org/10.1007/978-981-19-9944-4_9

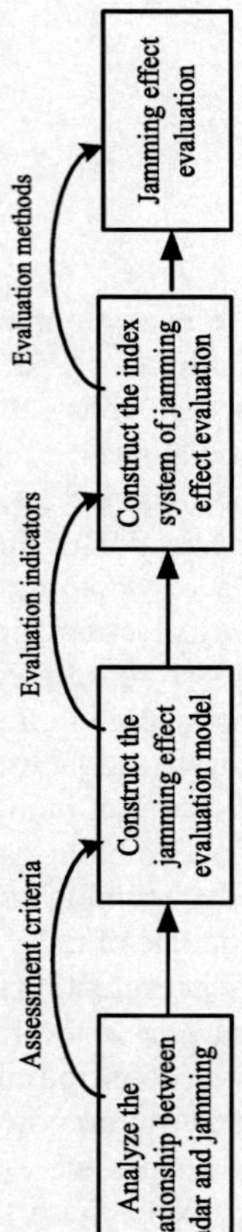

Fig. 9.1 General method of radar jamming effect evaluation

9.1 Evaluation Criterion of Jamming Effect

Evaluation criterion of jamming effect is the necessary basis for jamming effect evaluation. After determining the evaluation criteria of jamming effect, by checking the value of the evaluation index after the implementation of electronic jamming, and then comparing with the classification standard of jamming effect, we can determine whether the jamming of electronic countermeasure equipment is effective to the corresponding jamming object and the jamming effect level achieved. From different dimensions, this criterion can be divided into information criterion, power criterion and tactical application criterion [2].

9.1.1 Information Criterion

The information criterion measures the jamming effect from the perspective of information loss. According to this criterion, the working process of radar is an information transmission process. The target information is contained in the radar echo signal, and the radar receiver can obtain various parameters of the target by extracting the target information from the echo signal. Therefore, the target information contained in radar signal determines the target detection capability of radar. From this point of view, the information criterion measures the jamming effect by the change of target information contained in the radar signal before and after jamming. In general, information is expressed in terms of entropy. Entropy is a measure of the uncertainty of a random variable or random process.

The radar echo signal is set as $S(t)$, which contains information such as target position, velocity, intensity, and polarization, and its entropy is set as $H(S)$. The jamming signal is, $J(t)$ and its entropy is $H(J)$. Then, the mixed signal received by the radar is

$$\xi(t) = S(t) + J(t) \tag{9.1}$$

According to the information theory, the amount of information that radar can extract from mixed signal is expressed in the following formula

$$I(\xi, S) = H(\xi) - H(J) \tag{9.2}$$

where $I(\cdot)$ represents the amount of information. It can be seen from the formula that in order to reduce the amount of information obtained by radar it is necessary to increase the entropy of jamming signal and reduce the entropy of mixed signal. In extreme cases, when the entropy of the jamming signal is equal to the mixed signal, the amount of information that radar can obtain is 0.

The entropy of the random variable J (where it referred to jamming signal) is $H(J)$

$$H(J) = -\sum_{i=1}^{n} P_i \log P_i \tag{9.3}$$

P_i is a probability density function, and the finite total probability matrix of the discrete random variable J is

$$J = \begin{pmatrix} J_1, \ldots, J_n \\ P_1, \ldots, P_n \end{pmatrix} \tag{9.4}$$

J_i is the value of the random variable, and P_i is the probability of J_i. The higher the entropy of suppressing jamming signal is, the better the quality of jamming signal is. It is convenient to use entropy to express the uncertainty of random variables. Entropy can be obtained as long as the probability distribution of random variables is known. If the random variable is continuously distributed, entropy can be expressed by probability distribution density.

$$H(J) = -\int_{-\infty}^{\infty} p(J) \log p(J)\, \mathrm{d}J \tag{9.5}$$

The entropy of multi-dimensional random variables can be expressed by multi-dimensional probability density [3].

$$H(J) = -\int_{-\infty}^{\infty} \ldots \int_{-\infty}^{\infty} p(J_1, \ldots, J_n) \log p(J_1, \ldots, J_n)\, \mathrm{d}J_1 \ldots \mathrm{d}J_n \tag{9.6}$$

By calculating the entropy of jamming signal to evaluate the quality of jamming signal, so as to evaluate the possible jamming effect. The operation is simple, and the theory is clear, but it is necessary to know the probability distribution of jamming signal, which is sometimes not easy to achieve. The information criterion is suitable for the design of suppression jamming waveform, but not for the test of radar jamming system capability. It can only evaluate the quality of the jamming signal itself and does not consider the influence of radar anti-jamming measures and other factors. Figure 9.2 shows the schematic diagram of the jamming waveform with different entropy. The jamming signal with higher entropy power has better coverage on target echo.

Due to the jamming signal, the effective information of the jammed object suffers a large amount of loss, which is mainly reflected in the following ways: the signal has deviation, the signal is actively simulated, the signal is covered, etc. [4, 5]. When the characteristics of the jammed object are consistent with the jamming signal, the amount of information loss of the jammed system is large, if not, the amount of information loss is small, or even the jamming signal has no effect. Therefore,

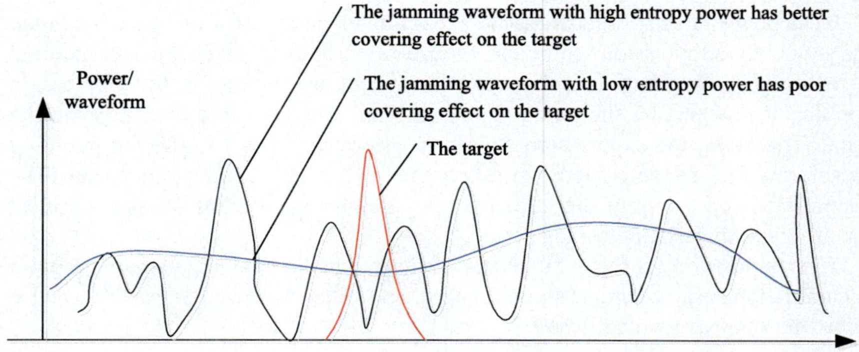

Fig. 9.2 Schematic diagram of jamming effects of jamming waveforms with different entropy power

the characteristics of information loss depend on the characteristics of the jamming signal and the jammed object.

If the jamming effect is considered directly from the perspective of radar information acquisition and taken as the criterion, then it can be evaluated from the accuracy and resolution of radar information acquisition. When the jamming signal makes the radar unable to detect the target, the amount of target information acquired by the radar will be zero. If the jammed radar can still detect the target, but the detection accuracy and resolution are reduced, the jamming effect can be evaluated by the decline degree of detection accuracy and resolution.

9.1.2 Power Criterion

The power criterion, sometimes known as the energy criterion, is generally represented by the suppression coefficient, that is, the power (or energy) ratio of the minimum jamming signal required at the input of the radar receiver to the radar signal when effectively suppressing jamming the radar or causing the suppressed radar to produce a specified information loss [6].

The meanings of effective jamming vary from types of radar. For target search and indication radar, effective jamming means to reduce the detection probability of radar to a certain value. For tracking radar, it refers to the increase of its tracking error to a certain extent, making the radar lose tracking capability.

Power criterion, also known as jam-to-signal ratio criterion, is expressed by the suppression coefficient K_s. The ratio between the minimum jamming power P_j required by the receiver input and the target echo power P_s at the radar receiver input when characterizing the given jamming effect, that is:

$$K_s = \min(P_j/P_s) \tag{9.7}$$

It can be seen from the above formula that when it comes from the same jamming, the greater the suppression coefficient, the greater the jamming signal power required to realize effective jamming to the radar; on the contrary, the smaller the suppression coefficient, the smaller the jamming signal power required to effectively jam the radar. Therefore, the suppression coefficients corresponding to different jamming waveforms for the same radar can reflect the quality of jamming waveforms. The suppression coefficient of the same kind of jamming to different radars can reflect the anti-jamming ability of radar.

The suppression coefficient is related to radar signal processing gain, radar pulse accumulation loss, jamming signal quality, and other factors. For coherent pulse radar, the suppression coefficient is:

$$K = \frac{2N}{q} \cdot \frac{K_n}{K_k} \tag{9.8}$$

N is the accumulated pulse number of radar, q is the threshold of radar detection SNR, K_n is the loss coefficient of radar signal processing, and K_k is the quality coefficient of jamming signal.

For incoherent pulse radar, the suppression coefficient is:

$$K = \frac{2N^{\gamma}}{q} \tag{9.9}$$

γ is the radar incoherent accumulation loss. When N is greater than 30, γ is 0.7.

Power criterion is a widely used jamming effect evaluation criterion, which is usually only applicable to the evaluation of suppressive jamming effect of radar system.

9.1.3 Tactical Application Criterion

The tactical application criterion is based on the completion of tactical tasks by electronic jamming equipment. For support jamming, take the ability of covering fighter penetration as the criterion, and for self-defense jamming, take the improvement of the survivability of its own platform as the criterion [7, 8]. The ratio of an index before and after jamming is usually used as the quantitative evaluation standard:

$$\eta_i = \frac{w_{ij}}{w_{i0}} (i = 1, 2, \ldots, n) \tag{9.10}$$

w_{ij} is the ith performance index under jamming, w_{i0} is the ith performance index without jamming, n is the number of performance index of the weapon system.

Different evaluation criteria are used for different tactical application scenarios. For example, for support jamming, the evaluation indexes concerned are the suppression distance of jamming to radar or the decline coefficient of radar found target distance, the decline coefficient of guidance probability or guidance probability of air defense system to fighter under jamming conditions, the reduction coefficient of the number of ground-to-air missiles under jamming conditions, etc. For self-defense jamming, the most direct indicator is the probability of its own platform (aircraft or ship, etc.) being hit or the decline degree of hit probability under the jamming conditions.

For example, in the Vietnam battlefield in 1966, when the US military did not carry out electromagnetic jamming at the beginning, the killing probability of the ground-to-air missile weapon system of the Vietnamese army was up to 90%. Later, after the US military implemented electromagnetic interference, the killing probability of missile weapon system decreased to 0.7%. The decreasing degree of killing probability directly reflects the contribution of jamming to combat. After that, when the Vietnamese ground-to-air missile weapon system took corresponding anti-jamming measures, the killing probability of the missile weapon system increased to 30%.

On the premise of large sample statistical data, the tactical application criterion can directly reflect the anti-jamming ability of radar, and is an ideal criterion for jamming effect evaluation. However, to obtain large sample statistical data, multiple repeated tests need to be carried out under the same conditions. In many cases, due to the uncontrollable test environment, or various factors such as test conditions, test costs, and time constraints, it is impossible to carry out multiple repeated tests, so the tactical application criterion are also limited.

9.2 Evaluation Indicators of Jamming Effect

The effects and manifestations of different types of jamming are different, and the effect evaluation indicators are also different. The evaluation indicators are introduced from three aspects: noise suppressing jamming, intensive false targets jamming, and dragging deception jamming.

9.2.1 Evaluation Indicators of Noise Suppressing Jamming

By radiating jamming signals to the radar, the target echo signal is submerged by the jamming signal, resulting in the decrease of radar detection probability, thus destroying radar's detection and tracking of the target [9, 10].

9.2.1.1 Suppression Coefficient

The suppression jamming is mainly aimed at searching radar; its jamming effect is manifested as the reduction of target detection probability of radar. Usually, the detection probability $P_d = 0.1$ is taken as the measurement standard of effective jamming, so the suppression coefficient is defined as

$$K_d = \left(\frac{J}{S}\right)_{\min, P_d=0.1} \tag{9.11}$$

J represents jamming signal power, S is target signal power, and K_d is the minimum jam-to-signal ratio when the detection probability of radar drops to 0.1. The suppression coefficient can be used to compare the quality of various jamming signals. The smaller the suppression coefficient is, the less the power of jamming signal is required for effective jamming to radar.

9.2.1.2 The Loss Degree of Detection Probability

Radar detection probability is an important tactical and technical index of radar. Detection probability is related to false alarm probability, jam-to-signal ratio, and target fluctuation type. Jamming equipment emits suppressing jamming signal to reduce the detection probability of radar, so the loss degree of detection probability E_{jd} can be used to quantitatively evaluate jamming effect, and its expression is

$$E_{\text{jd}} = \frac{P_d - P_{\text{dj}}}{P_d - P_{\text{dl}}} \tag{9.12}$$

In the formula, P_d is the detection probability standard required by the radar to complete tactical tasks, P_{dj} is the detection probability achieved by the radar under jamming conditions, and P_{dl} is the detection probability in case of radar functions failure.

9.2.1.3 Radar Detection Range

Radar detection range is the main power index of radar, so the jamming effect can be characterized by the influence of jamming on radar detection range. The radar detection range is related to the signal-to-jam ratio received by the radar. The suppression jamming to the radar can reduce the signal-to-jam ratio received by the radar, thus reducing the detection ability of the radar to the target.

(1) The loss degree of maximum range E_{jr}

The maximum detection range of radar is the basic index of radar operational capability. According to the basic definition of weapon effectiveness, the jamming effectiveness of jamming equipment against radar system can be measured by the maximum detection range loss of radar under the condition of electronic jamming. It is expressed by the maximum detection range loss degree E_{jr}, and its calculation formula is

$$E_{\text{jr}} = \frac{R_{\max} - R_{\max j}}{R_{\max}} \tag{9.13}$$

$R_{\max}$ is the maximum detection range of the radar in the free space environment, and $R_{\max j}$ is the maximum detection range of the radar in case of suppressing jamming.

(2) Minimum jamming range

In tactical application or field test, the minimum jamming range is usually used to indicate the effect of self-defense suppressing jamming. The minimum jamming range refers to the distance from the protected target to the radar when the jam-to-signal ratio just meets the requirements of the suppression coefficient or detection probability. The minimum jamming range is generally not equal to the maximum detection range of radar under jamming conditions, because the requirements of radar countermeasure on radar detection probability are different from those of radar, but they have a certain relationship. In jamming conditions, the smaller the minimum jamming range is, the smaller the maximum detection range of radar has.

(3) Effective jamming sector

If the size or activity range of the target or target group protected by jamming is smaller than the angular resolution unit of the jammed radar, and the minimum jamming range can reflect the jamming effect; otherwise, it is necessary to use the effective jamming sector to represent the jamming effect. When the antenna of the jammed radar is scanning, the jamming area formed by the point jamming source on the radar PPI display presents a sector, which is called the jamming sector. In the jamming sector, the area where the probability of radar finding targets meets the operational requirements also presents a sector, which is called the effective jamming sector (or suppressing area). The effective jamming sector should meet the requirements of minimum jamming range and radar detection probability. The minimum jamming range is a special case where the effective jamming sector is equal to the azimuth beamwidth of the radar antenna.

9.2.1.4 Parameter Measurement Accuracy of Radar

Suppressing jamming can not only interfere with radar target detection and reduce the probability of target detection, but also jam with radar target parameter measurement, increase parameter measurement error and reduce radar guidance probability or acquisition probability. Parameter measurement accuracy is an important tactical and technical index of all radars, which can evaluate the suppressing jamming effect.

Radar utilizes multiple measurement parameters to describe the target attribute, and the measurement error of any one parameter will affect its judgment of the target attribute, so the suppressing jamming effect can be expressed by the measurement error of specific dimensional parameters or the comprehensive measurement error of multi-dimensional parameters.

The loss degree of parameter measurement accuracy after jammed is

$$\overline{\sigma} = \frac{\sigma_{\mathrm{ij}} - \sigma_i}{\sigma_i} \tag{9.14}$$

σ_{ij} represents the measurement error of the *i*th specific dimensional parameter by the radar in the case of jamming, and σ_i is the measurement error without jamming. Specific dimensional parameters can be the target range, azimuth, pitch, speed, etc. The larger the $\overline{\sigma}_i$ is, the greater the measurement accuracy loss is, indicating that the better the jamming effect. On the contrary, the smaller $\overline{\sigma}_i$ is, the smaller the measurement accuracy loss is, indicating that the jamming effect is worse.

The comprehensive measurement error of multi-dimensional parameters to represent the suppressing jamming effect can be expressed as:

$$\overline{\sigma} = \sum_{i=1}^{N} a_i \overline{\sigma}_i \tag{9.15}$$

a_i is the weighting coefficient of measurement accuracy of each dimensional parameter, and $\sum_{i=1}^{N} a_i = 1$. In formula (9.15), the measurement accuracy of parameters in each dimension is integrated into a total measurement accuracy by a certain weighting to evaluate the jamming effect.

9.2.2 Evaluation Indicators of Intensive False Targets Jamming

As an important jamming mode, intensive false targets jamming has both suppression and deception effects. The evaluation indicators of intensive false targets jamming mainly include false target quantity, false target density, and true target recognition guidance rate.

9.2.2.1 False Targets Quantity

Different from the noise suppressing jamming, intensive false targets jamming does not win by power, but by quantity. The radar can only select true and false targets with equal probability if the radar does not have the ability to distinguish true and false targets, then it is better that more false targets have. If the radar boasts this capacity, the more the number of false targets, the more likely the radar data processing system will be overloaded and saturated, resulting in failing to work normally. Therefore, the number of false targets N_j is an important evaluation index of intensive false targets jamming.

9.2.2.2 False Target Density

False target density is the number of false targets in unit space:

$$\rho_j = \frac{N_j}{V} \tag{9.16}$$

V is the size of space, which can be one-dimensional space such as range and velocity, two-dimensional space such as area, or three-dimensional space such as volume. N_j is the number of false targets in space V. Jamming effects is varied from the density of false target, as shown in Fig. 9.3. If the false target density is large, multiple false targets will fall into the reference units of radar CFAR detection, which will raise the threshold of radar target detection and reduce the detection probability of radar to targets. If the false target density is small, the probability of falling into the reference units of radar CFAR detection is less, and then the radar will detect multiple false targets, which affects radar target discrimination and tracking.

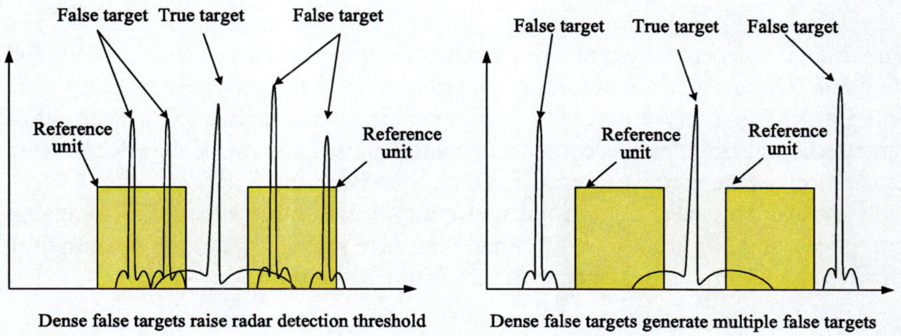

Fig. 9.3 Schematic diagram of different effects of false target jamming

9.2.2.3 True Target Recognition Guidance Rate

Search radar or target indication radar is used to guide the tracking radar of weapon system. Therefore, it is not necessary to judge whether the jamming is effective only from the jamming effect of one link, but from the impact of jamming on the whole process of target detection, recognition, and guidance.

N experiments were carried out under a typical combat situation of intensive false targets jamming, in which the radar system correctly identified and guided the real target M times, then the guidance rate of real target recognition is

$$P_{\text{yin}} = \frac{M}{N} \tag{9.17}$$

According to Eq. (9.17), it can also be converted into success rate of jamming

$$P_{\text{dec}} = \frac{N - M}{N} \tag{9.18}$$

9.2.3 *Evaluation Indicators of Dragging Deception Jamming*

The purpose of dragging deception jamming is to make the range gate (or velocity gate) of the radar tracking system unable to lock the target effectively, resulting in the weapon system missing the target [11, 12]. Its evaluation indicators mainly include capture success rate, tracking error, and kill probability.

9.2.3.1 Capture Success Rate

The dragging deception jamming is generally divided into two steps: capture and dragging. Capturing is the process of jamming to seize the control of tracking gate, and dragging is to move the centroid of jamming signal relative to the centroid of target echo, making that it is completely controlled by jamming. Hence, the success of capturing is the premise of dragging deception jamming.

N experiments were carried out under a typical combat situation of dragging deception jamming, of which M times the radar tracking gate was successfully captured by the dragging signal, then the capture success rate is:

$$P_{\text{bu}} = \frac{M}{N} \tag{9.19}$$

9.2.3.2 Tracking Error

The dragging deception jamming is mainly aimed at the tracking stage of radar to increase the target tracking error of radar. When the tracking error is large enough, the working state of radar can be changed to make it enter the searching state again, which can make the radar unable to track the target stably and make the weapon miss the target. The effect of dragging deception jamming is related to the target tracking error of radar, so tracking error is an important indicator to evaluate it.

Based on the jamming effect evaluation method of tracking error, Sect. 9.2.1.4 can be referred to.

Kill Probability

The most direct index for evaluating the effect of the radar seeker's implementation of the dragging deception jamming is the decreasing value of the target's kill probability, that is, the kill probability of weapon system without jamming minus the kill probability of weapon system after jamming. The greater the decrease value of kill probability is, the better the jamming effect will be; otherwise, the worse the jamming effect will be.

N experiments are conducted under a typical battle situation of dragging deception jamming, in which the target is destroyed by radar-guided weapon system for *M* times; then, the kill probability decrease value is:

$$P_{\text{cui}} = P_0 - \frac{M}{N} \tag{9.20}$$

where P_0 is the kill probability of the weapon system without jamming.

9.3 Evaluation Methods of Jamming Effect

The evaluation methods of radar jamming effect can be divided into external field evaluation, hardware-in-the-loop simulation evaluation, and digital simulation evaluation.

9.3.1 External Field Test Evaluation

External field test evaluation is to carry out electronic countermeasure test in the outdoor test site by using real equipment, real countermeasure scene, or equivalent equipment to simulate the real countermeasure scene, using the way of electromagnetic wave radiation, and process the collected experimental data to obtain the evaluation results [13]. The electromagnetic environment under the external field evaluation method is relatively realistic, and the test results feature high reliability. But there are some shortcomings, such as small sample size, difficult coordination, high test cost, hidden dangers of confidentiality, etc.

9.3.1.1 The Composition of External Field Evaluation System

The basic composition of the external field test system is shown in Fig. 9.4, which consists of the radar and platform cooperating with test, the jamming equipment and platform to be tested, the electromagnetic environment simulator, the control and data processing center, the data communication network, etc. The tested radar jamming equipment can be reconnaissance equipment, jamming equipment, or reconnaissance jamming integrated system. The cooperative test radar can be one or more, the radar can be early warning radar, fire control radar, guidance radar, and so on. According to the application environment, combat object, and evaluation purpose of the tested equipment, the platform can be ground platform, flight platform, and offshore platform, etc. In order to evaluate the effectiveness of radar countermeasure equipment, the external field test must have accurate timing system, positioning system and data acquisition, transmission and processing system.

During the test, the command and control center uniformly conducts the working procedure, configuration relationship, and platform motion state of the test equipment through the data transmission communication network. The data acquisition system collects the data of the key parts in the test system. If it is real-time processing, the collected data will be transmitted to the data processing center in time through the data transmission communication network, and the center will process it in real time. In case of post-processing, the test data shall be recorded on the storage medium in

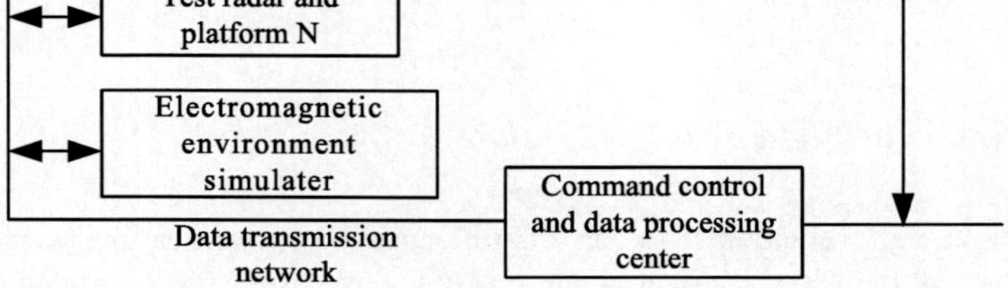

Fig. 9.4 Block diagram of the external field evaluation system

real time according to the unified time standard and format, and it shall be sent to the data processing center for processing when necessary. The time standard for recording data is usually the timing system of the test base.

9.3.1.2 Test Method for External Field Evaluation

(1) Test procedure

The test procedure of external field evaluation is shown in Fig. 9.5, including six steps. The first step is to propose the test task. The test demand is determined by the test task requirements or the requirements specified by the jamming system. The test task determines the test scheme. The second step is to develop the test scheme, including determining the test situation and scenario, proposing the test environment and condition requirements, test equipment performance and index requirements, evaluation criteria, indexes and algorithms, etc. The third step is to organize test resources, including personnel division, test scene construction, test equipment debugging and calibration, performance recovery, etc. The fourth step is to carry out the field test according to the test scheme, and record and collect relevant performance index data. The fifth step is to process the original data collected in the experiment. The final step is identification and evaluation, summarizing the processed test data, analyzing the performance of the jamming system, and giving the evaluation results.

(2) Alternative equivalent calculation

In the modern battlefield, the electromagnetic environment faced by radar EW system is extremely complex, and the types, systems, and functions of radar are very various. In order to evaluate the performance indexes of a jamming system against various radars, it is impossible to use the exhaustive method to carry out real equipment tests, whether in terms of efficiency or cost. On the other hand, in many cases, it is very difficult to select the desired matching radar for testing. Therefore, the alternative equivalent calculation test method is often used in the external field evaluation test.

Alternative equivalent calculation is to use one or several sets of alternative real equipment similar to the performance and working state of the expected test object

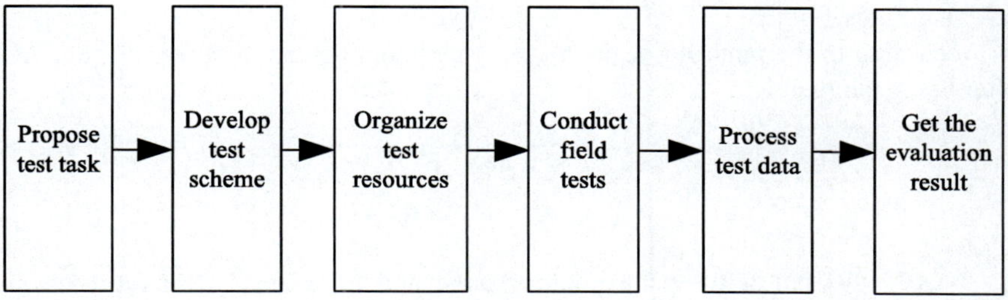

Fig. 9.5 Procedure of external field evaluation test

to equivalent the expected test system for external field evaluation test. Using the external field test data, the measured technical data of the substitute test system and the technical data of the expected test system, the mathematical simulation test is carried out on the basis of the established mathematical model, and the evaluation results of the tested system are finally given [14]. The purpose of alternative equivalent calculation is to solve the problem of lack of test system and difficult to test with the expected test system.

Three prerequisites must be met in the application of alternative equivalent calculation test: first, the mathematical model must be correct and feasible. Second, the system data obtained is accurate. Third, the test scenario conforms to the actual combat situation.

The basic idea and implementation method of alternative equivalent calculation test are as follows. Design a typical radar countermeasure situation or scenario, select a representative test radar system, and obtain test data through real equipment countermeasure test. Theoretically analyze the performance parameters of the substitute test radar and the expected test radar, as well as the jamming system parameters and various other influencing factor. Using the established alternative equivalent calculation model and the verification evaluation data under the conditions of typical test scenarios, the jamming performance of the tested jamming system against the expected test radar or any other radar of the combat object can be calculated equivalently.

Figure 9.6 shows the conceptual block diagram of the alternative equivalent calculation test, where $R_t^{'}$ is the external field test indicator of the tested jamming system relative to the substitute test matching system, R_z^l is the theoretical index of the tested jamming system relative to the expected combat object test matching system, $f_{z1}, f_{z2}, \ldots, f_{zN}$ are various factors affecting R_t^l. $R_z^{'}$ is the equivalent calculated value of the external field assessment index of the tested jamming system against the real combat object systems.

The following examples illustrate the application of alternative equivalent calculation in the evaluation of jamming effects.

X radar is far away from us, with less startup time, so it is difficult to intercept its signal. In addition, X radar is a non-cooperative target, so it is difficult to carry out confrontation tests with it, and it is impossible to obtain its confrontation data. In order to figure out the jamming effect of the jamming system against X radar, Y radar with similar functions to X radar is selected to carry out alternative equivalent calculation test.

According to the jamming equation, the maximum detection range of the radar after being jammed is:

$$R_T = \sqrt[4]{\frac{K\gamma_J \sigma R_J^2}{4\pi} \bullet \frac{P_T G_T G_T}{P_J G_J G_{rj}}} \tag{9.21}$$

R_T is the distance between the protected target and the radar; K is the suppression coefficient required for jamming; γ_J is the polarization loss of jamming, and the

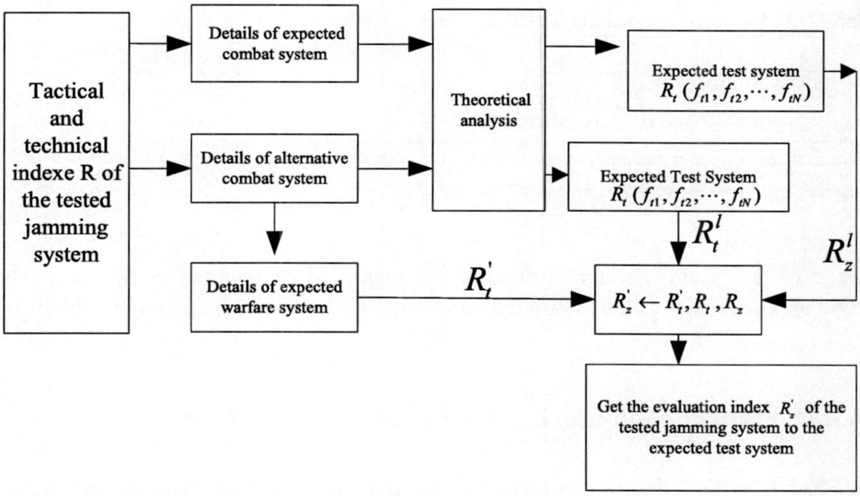

Fig. 9.6 Principle of alternative equivalence calculation test

sidelobe is 0 dB; σ is radar cross section; R_j is the distance between jamming platform and radar; P_T is radar transmitting power; G_T is radar main lobe gain; P_J is jamming transmitter power; G_J is jamming antenna gain; G_{rj} is radar antenna gain in jamming direction. Assuming that RCS of protected target, suppression coefficient K and polarization loss γ_J remain unchanged, and the distance between jammer and radar is reduced to half, the relationship between jamming distance against x radar and Y radar can be obtained as follows (subscript 1 means for radar X, and subscript 2 means for radar Y):

$$\frac{R_{T1}}{R_{T2}} = \sqrt[4]{\left(\frac{R_{J1}}{R_{J2}}\right)^2 \bullet \frac{P_{J2}G_{J2}}{P_{J1}G_{J1}} \bullet \frac{G_{rj2}}{G_{rj1}} \bullet \frac{P_{T1}G_{T1}G_{T1}}{P_{T2}G_{T2}G_{T2}} \bullet \frac{\sigma_1}{\sigma_2}} \tag{9.22}$$

The actual range between the jamming system and X radar is 300 km, in alternative equivalent calculation test, the range between the jammer and Y radar is 150 km, so the ratio R_{J1}/R_{J2} of them is 3 dB. The maximum detection range of X radar is 400 km, and that of Y radar is 100 km. Under the same conditions of radar loss and detection SNR, $\frac{P_{T1}G_{T1}G_{T1}}{P_{T2}G_{T2}G_{T2}}$ is equal to 24 dB. The equivalent radiation power ratio $\frac{P_{J2}G_{J2}}{P_{J1}G_{J1}}$ between the jammer used in the test and the actual jamming system is about -16 dB. $G_{\text{rj1}}/G_{\text{rj2}}$ between Y radar and X radar is -20 dB. The target RCS in the actual test is about 0.1 m^2, while the target RCS covered by the jamming system is about 10 m^2; that is, σ_1/σ_2 is about -20 dB.

Thus, it can be calculated that after being jammed, the relationship between the maximum detection range of Y radar and X radar is $R_{T1} = 0.708R_{T2}$. In order to reduce the maximum detection range of X radar to 100 km, the maximum detection

Table 9.1 Radar countermeasure test effect calculation table

	Jamming X radar	Jamming Y radar
Target RCS (m^2)	0.1	10
Distance between jammer and radar(KM)	300	150
Radar detection range before jamming(KM)	400	317
Radar detection range after jamming(KM)	100	186

range of the tested jamming system to Y radar shall be reduced to 186 km. The comparative results of the alternative equivalent calculation are shown in Table 9.1.

9.3.2 *Hardware-in-the-Loop Simulation Evaluation*

Hardware-in-the-loop simulation evaluation is to add some physical objects in the electronic countermeasure system, carry out the electronic countermeasure test in real time by means of electromagnetic wave radiation or injection, and process the collected experimental data to obtain the evaluation result [15].

Compared with the external field test evaluation, the hardware-in-the-loop simulation evaluation has the advantages of low experimental cost, repeatability, and good confidentiality. In addition, although the fidelity of electromagnetic environment and the reliability of experimental results are inferior to the external field test evaluation, they are also relatively high due to the addition of some physical objects in the countermeasures system. Hardware-in-the-loop simulation is widely used in radar countermeasures, missile guidance, rocket control, satellite attitude control, etc. [16].

9.3.2.1 The Composition of a Hardware-in-the-Loop Simulation Evaluation System

To evaluate the effectiveness of radar countermeasure equipment by hardware-in-the-loop simulation test, it is necessary to build a radar countermeasure hardware-in-the-loop simulation evaluation system. The system consists of signal environment simulator, general radar simulator, radar target and background clutter simulator, data acquisition and system control computer and its network, as shown in Fig. 9.7.

The signal environment simulator is used to generate the required electromagnetic environment, test the signal interception and processing capability, power management capability, and response capability to the specified density signal environment and its changes of the tested system. The general radar simulator simulates the working mechanism of radar, detects, and evaluates the jamming effect of radar jamming equipment and various jamming technologies on specified radar and specified anti-jamming technology. Radar target and background clutter simulator simulates the working environment of radar, generating radar echoes and background

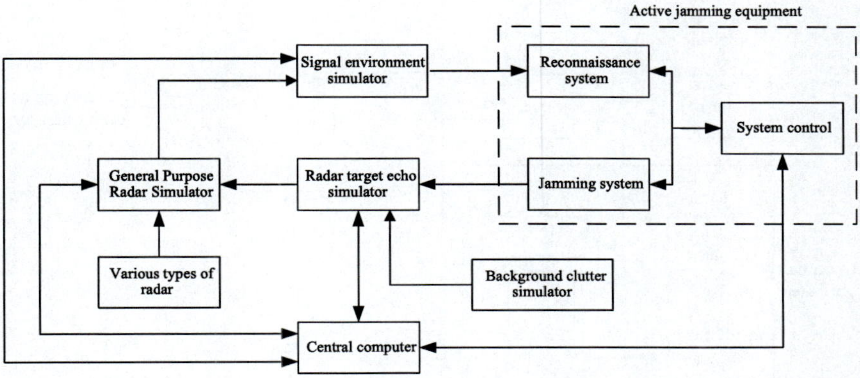

Fig. 9.7 Composition diagram of radar countermeasure hardware-in-the-loop simulation system

clutter of various targets. Computer networks are used for data acquisition, system control, and performance analysis and evaluation of the tested system. The system under test receives the output signals from the signal simulator and radar simulator, and sends the data to the jamming subsystem and the system control subsystem after being intercepted and processed by the reconnaissance equipment. The jamming subsystem sends jamming signals to the general radar simulator according to the power management instructions. The data acquisition system collects the response of the general radar simulator after being jammed. The central computer analyzes and processes the collected data and provides the quantitative evaluation results of jamming ability.

(1) Signal environment simulator

The basis of hardware-in-the-loop simulation method to evaluate the effectiveness of radar countermeasure equipment is to establish the simulated battlefield electromagnetic environment. The characteristics of the battlefield electromagnetic environment, such as signal density, signal complexity, and variation, can be determined according to the performance and the number of the electronic equipment put into use by both sides, which can be seen as the basis for setting the parameters of the signal environment simulator. The signal environment simulator consists of several signal simulators. A signal simulator can simulate multiple radiation sources and multiple signal simulators work in parallel to form the required signal density and signal type. All signal simulators operate effectively under the control of high-speed computer to form an organic whole.

(2) General radar simulator

The general radar simulator adopts the combination of hardware module and software control. It is generally composed of antenna beam forming unit, RF front-end, receiver, radar signal generator, signal processing unit, computer, and display unit. Its composition block diagram is shown in Fig. 9.8.

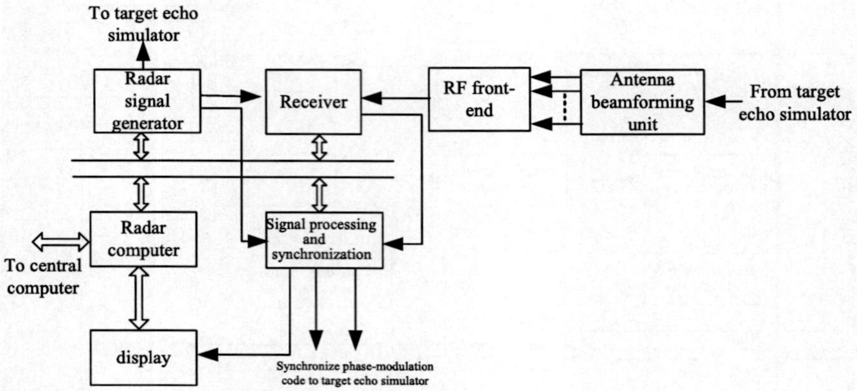

Fig. 9.8 Composition block diagram of general radar simulator

Antenna beam forming units simulate the functions of radar antennas, such as beam shape, main lobe and side lobe characteristics, antenna scanning, and tracking characteristics. The radar signal generator generates stable local oscillator and coherent signals for internal use of the receiver and sends them to the target echo simulator at the same time. The target echo signal from the target echo simulator and the background clutter generated by the background clutter simulator are synthesized in the synthesis network together with the external input jamming signal and then added to the antenna beam forming unit, transformed into a signal with radar antenna characteristics, and finally sent to the radar receiver. The receiver simulates the working mechanism of various radar receivers. The signal processing unit completes the simulation of various radar signal processing modes. The synchronizer simulates the radar transmitting waveform and parameters, and outputs the synchronous signals for the use of general radar simulator and target echo simulator. The computer controls the range, Doppler frequency and angle of the target, manages the working mode of the radar, completes the calculation of jamming effect, and transmits data to the display, which presents the picture of the radar before and after jamming.

(3) Target and background clutter simulator

The target echo simulator consists of radio frequency sources and modulators. The target echo simulator forms the corresponding target echo according to the radar working system and parameters, target characteristics, range, speed, etc. The target echo simulator is started by the synchronous pulse output from the general radar simulator. After appropriate delay, the target echo is transmitted to the general radar simulator.

The background clutter simulator also consists of RF sources and modulators. Background clutter refers to radar echoes reflected by clouds, rain and snow on the ground, sea and air. Its characteristics are clutter intensity, fluctuation and spectrum spread. Different background, frequency and incident angle will affect its characteristic. In use, it should be able to control the characteristic parameters of clutter in real time according to the simulation scene. In order to realistically simulate the characteristics of background clutter, it is necessary to carry out a lot of experimental and theoretical research and establish the mathematical models and simulation algorithms of various background clutter.

The target echo simulator and the Background Clutter Simulator set the signal parameters according to the instructions of the central computer, such as RF frequency, pulse width, echo intensity, fluctuation, and spectrum spread characteristics.

9.3.2.2 Hardware-in-the-Loop Simulation Evaluation Method

The evaluation process of hardware-in-the-loop simulation is similar to that of the external field evaluation, which is divided into six steps: putting forward the test task, making the test scheme, organizing the test resources, carrying out the test, processing the test data, and getting the evaluation results. Most of the steps are similar to the external field evaluation. In the third step, the mathematical model of electromagnetic environment, including the mathematical model of target signal, jamming signal, and clutter signal, should be established depending on the test scenario in the test scheme. Then, according to the scheme, the model parameters are loaded to generate electromagnetic environment. Secondly, the generated electromagnetic signal is tested to verify the correctness and effectiveness of the model. After the test, the test equipment is connected for adaptive debugging, laying a good foundation for the formal test.

The hardware-in-the-loop simulation evaluation test verifies and evaluates the performance of the radar jamming system with high reliability, real time, and repeatability and can comprehensively evaluate the reconnaissance capability, jamming capability, system response capability, and power management capability of the jamming system. The evaluation method is as follows:

(1) Evaluation method of reconnaissance capability

The signal parameters set by the signal environment simulator are sent to the central computer as standards. The signal parameters measured by the reconnaissance equipment and the sorting identification results are also sent there. The central computer processes the data of multiple tests, calculates parameter measurement errors, signal sorting and recognition, processing time, etc., compares these parameters with the specified requirements, counts the probability that various parameters meet the specified requirements, and finally integrates them into the reconnaissance capability or reconnaissance effect of radar countermeasure equipment.

(2) Evaluation method of jamming capability

To evaluate the jamming capability of radar countermeasure equipment, it is necessary to determine the required jamming effect, and then send the output signal of the jammer under the set working environment to the general radar simulator. The central computer analyzes the output of the general radar simulator, obtains various parameter indexes of the jammed radar, compares these parameter indexes with the required jamming effect, counts the multiple measured values, and integrates them into the jamming capability or jamming effect of the radar countermeasure equipment.

(3) Evaluation methods of other abilities

System response capability, electromagnetic environment adaptability, and power management capability are also important operational capability of radar countermeasures system. The evaluation method is to take the signal type and parameters set by the signal environment simulator as a reference, predetermine the correct response of the system within the specified time, that is, the judgment criteria, and then send the set signal to the tested system. The central computer analyzes and processes the final response of the signal processing unit, power management unit of the system under test, and compares the analysis and processing results with the predetermined judgment criteria. Through multiple tests, the proportion of various test data meeting the criteria is calculated, which is the signal processing capability, power management capability, and system response capability of the tested system. Setting different signal environment and repeating the previous test content can obtain the electromagnetic environment adaptability of the system.

9.3.3 Digital Simulation Evaluation

Digital simulation is a comprehensive technology based on similarity principle, model theory, system technology, and information technology, using computer and relevant physical effect equipment as tools, using models to study, analyze, evaluate, make decisions and participate in system operation. Digital simulation evaluation is a way to evaluate the radar jamming effect by using digital simulation technology. It has the advantages of easy realization, controllability, non-destruction, safety, repeatability, low cost and high efficiency. It has a wide and important application in the development demonstration, product development, qualification evaluation, training, and use of electronic warfare equipment.

There are two ways of digital simulation evaluation: one is digital simulation evaluation based on energy equation. The other is digital simulation evaluation based on signal/data stream.

9.3.3.1 Evaluation of Digital Simulation Based on Energy Equation

The simulation method based on energy equation is to test the operational capability of radar countermeasure equipment under various conditions by deploying and simulating the engagement of weapons, forces, radars and countermeasure equipment of both offensive and defensive sides in the set test environment. The set test conditions generally include different working parameters, different use modes, different engagement procedures, different configuration relations, and different cooperation relations with other systems of the equipment.

The energy equation in radar countermeasures includes radar equation, reconnaissance equation, jamming equation, etc. Different energy equations are selected for different detection and evaluation objects and purposes. If the equipment reconnaissance effect is evaluated, the radar reconnaissance equipment effect simulation model is established based on the reconnaissance equation. If the radar jamming effect is evaluated, the radar jamming equipment effect simulation model is established based on the radar equation and radar jamming equation. For the integrated radar countermeasure system including reconnaissance and jamming functions, the effect simulation model and command-and-control system model of the countermeasure system should be established. At this time, both reconnaissance equation and radar equation and radar jamming equation should be used.

To build a simulation system based on the energy equation, first, the combat scenario should be set according to the equipment and configuration relationship of the opposing sides and the engagement process of the confrontation. According to the combat scenario, the required models should be integrated to form the engagement models of the two sides. Input the relevant data of both sides, run the simulation system according to the predetermined algorithm software, and obtain the effect of the radar countermeasure system.

The computer simulation models for evaluating the effectiveness of radar countermeasure equipment may vary greatly. The scale of the simulation model mainly depends on the tested system, the combat object system, and the combat scenario. Figure 9.9 is a computer simulation system based on energy equation, which is used to test and evaluate the effectiveness of the radar countermeasure system. The system adopts distributed simulation structure and consists of four parts: the simulation model of the tested radar countermeasure system, the simulation model of the combat object system, the radar countermeasure master control, and the distributed simulation network.

(1) Simulation model of the tested system

The tested system in Fig. 9.9 is an integrated radar countermeasure system, including radar reconnaissance equipment, jamming equipment and radar countermeasure command and control center. Therefore, its simulation models include radar reconnaissance equipment simulation model, jamming equipment simulation model and radar countermeasure command-and-control center simulation model. In the simulation, in order to obtain the data for effect evaluation, it is necessary to output the running results of key simulation modules in real time.

(2) Simulation model of combat object system

The simulation models of combat object system include combat command and control center simulation model, early warning radar simulation model, tracking

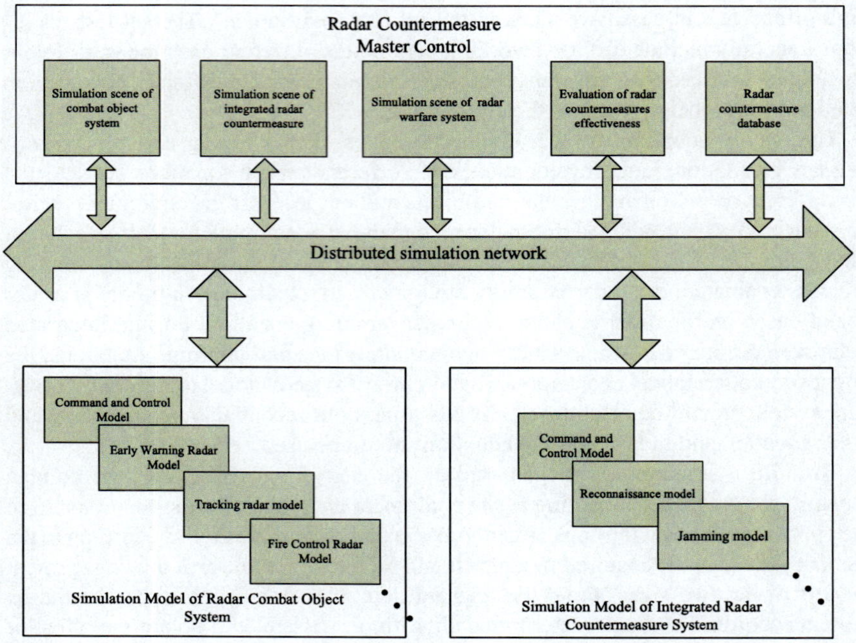

Fig. 9.9 Function block diagram of radar countermeasure effect evaluation simulation

radar simulation model and fire control radar simulation model. The radar simulation model and the command-and-control center simulation model can output the operation results in real time to evaluate the combat capability.

(3) Radar countermeasure master control

Radar countermeasure master control completes the general functions of radar countermeasure simulation, including the simulation scenario of combat object system, radar countermeasure system and integrated radar countermeasure, radar countermeasure effect evaluation, and radar countermeasure database.

The combat object system simulation scenario envisages the number, type, position, functional performance index, predicted route, combat procedure, etc., of the radar and its platform participating in the battle. Under the overall planning of the integrated radar countermeasure simulation scenario, the dynamic simulation scenario of the combat object is formed by advancing according to the simulation clock. In the process of radar countermeasure simulation, the radar countermeasure simulation information of each part of the combat object is collected and merged into the countermeasure effect, which is displayed dynamically and visually.

The simulation scenario of radar countermeasure system assumes the quantity, type, position, function, and performance index of the combat equipment and its platform, presets the route of jamming platform, reconnaissance platform and protected platform, and sets the workflow of each equipment and the operation procedure of the whole countermeasures system. During the simulation, the relevant computer collects the countermeasure information of various parts of the radar countermeasure system and merges them into the countermeasure effect, which is displayed dynamically and visually together with the track of various platforms.

Integrated radar countermeasure simulation scenario not only controls the simulation operation of each part of the entire simulation system and "direct" the whole engagement process according to the combat scenario, but also integrate the working status, working process and platform route of the participating radar and radar countermeasure equipment, that is, integrate the dynamic simulation scenes of both the combat object system and the countermeasure system, so as to visually and dynamically display, so that the radar countermeasure effect evaluator or referee can visually see the whole engagement process.

In the process of radar countermeasure simulation, the information of both sides is collected and processed, and the countermeasures effect is evaluated according to the predetermined criteria, and the comprehensive evaluation results are given. Compare the reconnaissance results with the set radar signal data to obtain the signal interception capability, threat identification capability, and jamming guidance capability of the radar reconnaissance equipment. By comparing and analyzing the maximum range of radar target detection, jamming area, jam-to-signal ratio, and so on, the jamming ability of the integrated radar countermeasure system to radar target detection and parameter measurement is obtained. The power management capability and system response capability of the integrated countermeasures system can be evaluated by analyzing the time of radar signal discovery, the time of jamming, the jamming mode, parameters, and angle.

The radar countermeasure database includes four aspects of data or mathematical models: equipment and platform parameters of both sides of the countermeasure; data of the protected platform; meteorological simulation model or data, battlefield geographical environment simulation model or data; system initialization data.

(4) Distributed simulation network

The distributed simulation network makes the electronic warfare simulation modules interconnect, and connects the computers scattered in different places to complete different simulation functions to form an organic whole.

9.3.3.2 Evaluation of Digital Simulation Based on Signal/data Stream

To evaluate the effectiveness of radar countermeasure equipment with high fidelity, radar countermeasure simulation system based on signal/data stream processing mechanism can be used. This simulation system tests and evaluates the reconnaissance capability and jamming capability of radar countermeasure equipment, or analyzes and evaluates the anti-jamming capability of radar by simulating the signal/data stream processing process of each key link or functional module in the radar countermeasure system, that is, by simulating the generation and transmission

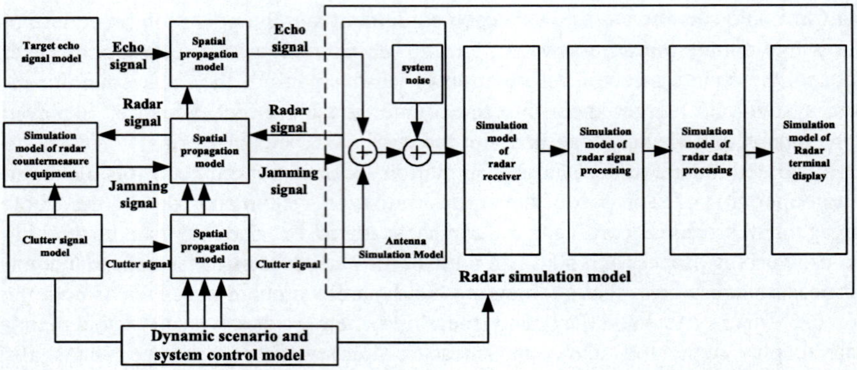

Fig. 9.10 Block diagram of effectiveness evaluation simulation system based on signal/data stream processing mechanism

of jamming signal, radar echo, and radar clutter, as well as the whole process of radar reception, signal processing, data processing and display. In the simulation system, the signal and the transfer function of its correlation processing link are described digitally, which is approximately equivalent to the digitization of the actual system, so the simulation results have high fidelity.

Figure 9.10 is the block diagram of radar jamming simulation system based on signal/data stream processing mechanism. The simulation system includes target echo model, radar jamming equipment model, clutter model, propagation path model of different signals, radar antenna model, radar receiver model, signal processing model, data processing model, display model, dynamic scene, and system control model.

The radar simulation model generates radar transmitting signal, target echo signal, and clutter signal according to the current dynamic scene. The simulation model of radar countermeasure equipment processes the received radar signal, determines whether to jam and what jamming technology and parameters to adopt, and generates corresponding jamming signals through its jamming signal generation module. After signal propagation model processing, radar antenna pattern modulation and linear superposition, the jamming signal, target echo signal, and clutter enter the radar receiver and signal processing module, where automatic gain control, pulse pressure, coherent accumulation, CFAR, and other processing are performed. Finally, after signal detection and data processing, the target and its track are displayed on the radar terminal display simulator. Jamming effect can be evaluated manually on radar terminal display simulator. Similar to the simulation based on energy equation, the output data of relevant modules can also be collected, and the quantitative test and evaluation results of radar countermeasure effect can be given through computer processing. If the platforms and protected targets of both sides are moving, their position, moving speed, and moving direction are functions of time. In the whole simulation process, dynamic simulation can be realized as long as the platform position, speed, movement direction, and radar antenna scanning change with the advance of the simulation clock according to the scenario.

Using the modeling and simulation technology based on signal/data stream processing mechanism, a more complex radar countermeasure effectiveness evaluation simulation system can be established. According to the composition of a typical radar system and the signal/data processing process of radar, the basic structure of the corresponding radar simulation model is extracted. Many simulation modules, especially the signal processing and data processing modules, basically correspond to the actual radar system one by one. According to the radar countermeasure process, the simulation model of radar signal reconnaissance, jamming guidance, and jamming signal generation is established. The jamming signal model is the digital modeling of various jamming signal waveforms that can be generated by the jammer, that is, using mathematical methods and considering the influence of some links in the jamming equipment to generate signals with similar characteristics and forms to the actual jamming signal.

References

1. Guoyu W, Liandong W (2004) Mathematical simulation and evaluation of radar electronic warfare system. National Defense Industry Press, Beijing
2. Li XQ, et al (2008) Integrated electronic warfare—killer mace in information war, 2nd edn. National Defense Industry Press, Beijing
3. Yi W, Yongle L, Qingjun H (1995) Applied mathematical statistics. National University of Defense Technology Press, Changsha
4. Guopei S et al (1998) Analysis of electronic warfare operation effectiveness. PLA Press, Beijing
5. Guoqing Z (2003) Principle of radar countermeasures. Xidian University Press, Xi'an
6. Yongshun Z, Ningning T, Guoqing Z (2005) Principle of radar electronic warfare. National Defense Industry Press, Beijing
7. De Bao C (2002) Modern radar anti-countermeasures technology. Aviation Industry Press, Beijing
8. Peiyao H (1980) Radar anti-jamming technology. National Defense Industry Press, Beijing
9. Wanxing Z (2013) Research on evaluation model and index system of radar anti-jamming effectiveness. Modern Radar 35(11):1–5
10. Xue-song W et al (2010) Modeling and simulation of modern radar electronic warfare system. Publishing House of Electronics Industry, Beijing
11. Xiang-ping L (1985) Principle of radar countermeasures. Northwest Telecommunications Engineering Institute Press, Xi'an
12. Deshu L (1989) Basic theory and technology of radar anti-countermeasures. Beijing Institute of Technology Press, Beijing
13. Ying Z, Xuesong W, Zhenhai X, et al (2004) General thinking on effectiveness and effeciency evaluation of radar electronic warfare. Syst Eng Elect 26(5):617–620
14. Guoyu W et al (2002) Principle and method of alternative equivalent calculation in radar countermeasures test. National Defense Industry Press, Beijing
15. Jiaqi L, Yan Z, Guoyu W et al (2006) Hardware-in-the-loop simulation system of radar electronic warfare. Missiles Space Vehicles 6:29–32
16. Qiang Z (2005) Hardware-in-the-loop simulation system of electronic warfare. Ship Elect Countermeasures 28(1):50–53

Chapter 10
Frontier Technology of Radar and Radar Active Jamming

In recent years, with the widespread application of technologies, such as microelectronics, digitization, and intelligence in radar, radar technology has made great progress, and the performance of radar has also been significantly improved [1]. As a pair of contradictions, radar technology and radar jamming technology spirally push each other forward in the process of dynamic game. The continuous advancement of radar technology also requires continuous development of radar active jamming technology.

This chapter briefly introduces the frontier technology of radar and radar active jamming. The Sect. 10.1 introduces the frontier technology of radar, including cognitive radar, digital array radar, software radar, and ultra-wideband radar, etc. The Sect. 10.2 introduces the frontier technology of radar active jamming, including cognitive electronic warfare, distributed cooperative jamming, micro-miniature, and integrated countermeasures, etc.

10.1 Frontier Technology of Radar

10.1.1 Cognitive Radar

Cognitive radar (CR) perceives the environment through prior knowledge and interactive learning of the environment. On this basis, it can adjust the transmitter and receiver to adapt to changes in the environment in real time, so as to achieve the predetermined goal effectively, reliably, and robustly.

The concept of cognitive radar was first proposed in 2006 by the famous signal processing expert Simon Haykin. Before the concept of cognitive radar was put forward, people had been exploring methods and technologies to improve performance by changing the working mode of radar and had made many achievements in

G. Tang et al., *Techniques and System Design of Radar Active Jamming*,
https://doi.org/10.1007/978-981-19-9944-4_10

matched illumination-receiving technology, waveform diversity technology, intelligent signal processing technology, etc. The development of these technologies has laid a solid foundation for the realization of cognitive radar.

Cognitive radar integrates brain science and artificial intelligence into the radar system, giving the radar system the ability to perceive the environment, understand the environment, learn, reason, and make decisions, so that the radar system can adapt to the increasingly complex and changeable electromagnetic environment, thereby improving the radar system's performance. For example, by adjusting the transmission waveform to effectively avoid the jamming spectrum to improve the anti-jamming capability of the radar; by adaptively adjusting the transmission waveform to achieve the given performance requirements in a shorter time, thereby greatly reducing the possibility of radar being discovered and attacked; the cognitive radar network can effectively counter various stealth aircraft. Cognitive radar adapts to the trend of radar intelligence, gradually weakens the role of operators in the closed-loop system composed of human and radar, gradually increases the intelligence of the radar itself, and can effectively reduce casualties in the battlefield environment.

According to the concept of cognitive radar, the essence of cognitive radar is a closed-loop radar system that understands and adapts to the environment through continuous interaction with the environment. Therefore, cognitive radar must have:

(1) Ability to perceive the environment.
(2) Intelligent signal processing capability.
(3) Memory and environmental database.
(4) Closed-loop feedback from receiver to transmitter.

Figure 10.1 provides a basic framework for the implementation of cognitive radar. Cognitive radar is a closed-loop structure. Firstly, the radar transmitter radiates electromagnetic signals to the surrounding environment. Secondly, after the environmental electromagnetic signals enter the receiving system, they enter the radar environment analyzer and Bayesian target tracker (detection, tracking, identification, etc.) at the same time. Thirdly, the radar signal processor uses the information provided by the scene analyzer and prior knowledge to improve performance. At last, the processing results are analyzed to guide the next transmission waveform.

The proposal of cognitive radar is inspired by the biology of echolocation in bats; but how to simulate the echolocation behavior of bats and make radar have cognitive function? In addition to the research in the field of bionics, the radar community is also thinking about and discussing this issue. According to people's expectations and conception of cognitive radar, cognitive radar needs to realize the following key technologies: scene perception and description, waveform optimization, adaptive beam forming, autonomous operation and management [2–6].

(1) Scene perception and description
The radar environment analyzer is an important part of the cognitive radar, whose function is to provide information about the environment (including the target and radar working environment) for the receiver, so it plays an important role in determining the possible interested targets for the receiver. The information provided by the radar environment analyzer to the receiver mainly comes

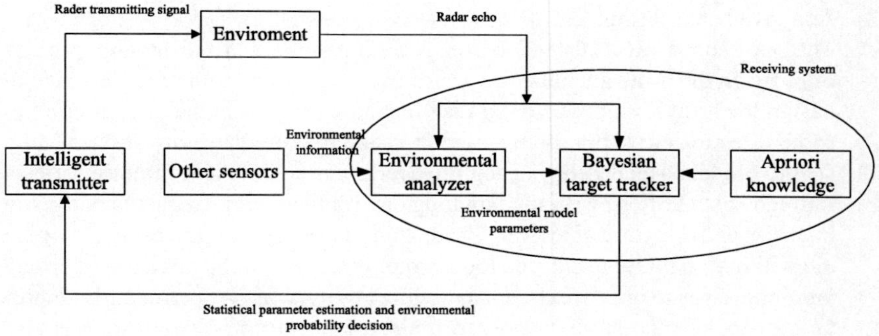

Fig. 10.1 Typical framework of cognitive radar

from two channels: one is radar echo; the other is other information about the radar environment provided by other sensors, such as temperature, humidity, air pressure, and sea conditions. The radar processing the information provided by the echo must be based on a certain mathematical description; that is, the model of the information obtained by the radar environment analyzer must be determined first, including the clutter model, the background perception method, and the target model.

(2) Waveform optimization technology
Waveform optimization technology is to obtain radar waveform that meets specific conditions by adopting specific optimization criteria for the current electromagnetic and target environment faced by radars. The application of waveform optimization technology in practical radar includes two types: optimized waveform selection and optimized waveform design. The optimized waveform selection requires a set of waveforms or waveform parameters to be designed and a waveform library to be established before the radar works. When the radar is working, based on the current working environment, a certain waveform or waveform parameter value is adaptively selected from the waveform library. The optimized waveform design is to design waveform or calculated waveform parameters in real time according to the current environment, and the radar can generate qualified waveform in real time and transmit. The disadvantage of optimized waveform selection is that the library of waveform or waveform parameters must be designed before work. When the radar works, the waveform selection algorithm must search for the optimal waveform in the whole library to obtain the optimal performance. The waveform design is more convenient than the waveform selection and can better use the information of dynamic environment. However, under the existing technical conditions, there is still a contradiction between the waveform adaptive design and the real-time requirements of radar detection. Of course, if the waveform library is large, it will also cause a great computing burden.

(3) Adaptive beam forming
The beam shape affects the resolution, detection range, and anti-jamming ability of radar. In the array antenna, the vector matching filter method can be used to design the array weight vector so that the zeros of the antenna pattern point to some directions to suppress the jamming sources from these directions. Using adaptive antenna technology in cognitive radar can achieve anti-jamming effects through spatial filtering. In the jamming environment with frequent changes in intensity and direction of arrival, by adjusting the array weight vector, changing the width of main lobe and side lobe, a null is formed in the direction of strong jamming, so as to improve its anti-jamming ability, which is particularly important for today's increasingly upgraded electronic warfare. Therefore, it is also one of the key technologies for cognitive radar to achieve intelligent illumination to estimate the statistical characteristics of the environment and targets through prior knowledge and current measurement signals, adaptively adjust the intelligent transmitter according to the environment model, and give the beam illumination suitable for the current environment.

With the introduction of array signal processing technology in the early 1990s, adaptive antenna technology has been fully developed [7]. The adaptive beamforming algorithm is the core content of adaptive antenna technology. According to whether the transmitter needs to transmit the reference signal, the adaptive beamforming algorithms can be divided into two categories, blind and non-blind. The non-blind algorithms are based on the time domain reference signal sent by the transmitter, including LMS algorithm, NLMS algorithm, RLS algorithm, etc. The blind algorithms do not require the transmitter to send the reference signal, such as MUSIC algorithm based on DOA estimation, ESPRIT algorithm, CMA algorithm based on signal characteristic recovery, etc.

At present, adaptive antenna technology has been widely used in the field of new radar and communication industry, and scholars are also gradually exploring its combination with artificial intelligence algorithms. Literature [8] proposed a sidelobe cancelation method based on adaptive neural network. The simulation results show that its optimization effect is better than the traditional weight selection method and can meet the real-time requirements. In addition, Romero [9] introduced the method of generating efficient waveform illumination adaptively and cooperatively in the search and tracking of multiple radar platforms in the cognitive radar system.

(4) Autonomous operation and management
Radar is an important means of long-range sensing. To realize cognitive function and become an independent system, autonomous operation, and management is a key technology. On the one hand, adaptive algorithms and intelligent algorithms are introduced into many modules of cognitive radar, but these algorithms cannot achieve the desired performance independently. Each independent function must cooperate with each other, and effective work can further improve performance, so it is necessary to study how to realize the coordination and cooperation technology of each intelligent processing link. On the other hand,

we hope that in a working system composed of man and radar, the role of man will become smaller and smaller, and even be completely replaced by radar in the future. Radar itself will become a closed-loop system, which requires radar to have the ability of inference, decision making, etc. This requires that cognitive radar can independently complete task deployment and conversion and can reconfigure resources.

10.1.2 Digital Array Radar

Digital array radar (DAR) is a kind of all digital phased array radar whose receiving and transmitting beams are realized digitally. Unlike traditional array radars, which use phase shifters, delay lines, power synthesizers, and distributors to form beams at radio frequencies or intermediate frequencies, digital beamforming uses digital signal processing technology to form synthetic beams, so multiple beams can be formed at the same time. It has the characteristics of low power consumption, high stability, and high precision. It is an important development direction of future radars. It has broad application prospects in the fields of ground-based air defense and anti-missile multi-functional radar, advanced multi-functional airborne early warning radar, shipborne integrated electronic system, digital weather radar, and digital air traffic control radar, microwave remote sensing synthetic aperture radar, etc.

Digital array radar has the following characteristics [10–13]:

(1) Reduce the dynamic range requirements for digital receivers. For the radar system using analog beam forming, the beam forming network will produce great gain to the signal. In order to match it, the digital receiver connected to the output of the analog beam forming network is required to have a large dynamic range. For digital array radar, each digital receiver is only connected to the output of the unit or subarray, so the requirements for its dynamic range can be greatly reduced.

(2) Ultra-low sidelobe can be achieved. Since the array error, the amplitude and phase inconsistency of each element and the mutual coupling effect can be accurately corrected, ultra-low sidelobe can be realized.

(3) Easy to realize simultaneous multi-function. Generally speaking, radar systems have different detection accuracy and data rate requirements for targets at different ranges. Therefore, the signal processor can design corresponding detection methods according to different ranges, so as to meet different detection requirements for short-range, medium-range, and long-range targets at the same time and realize simultaneous multi-function.

(4) Easy to realize softwareization. With the continuous improvement of digitalization and the application of high-performance general-purpose hardware platforms, the main functions of the radar can be realized by software programming.

It can change the radar function or improve the detection capability of the radar system through software loading and upgrading without replacing the hardware platform.

(5) Improve the detection ability of weak small targets in strong clutter background. Multiple beams at the same time increase the beam dwell time, so the Doppler filter can be improved through long-time coherent accumulation, so as to achieve the effective separation of moving targets and clutter.

(6) Low probability of interception. The radar transmitted signal can be expanded in multi-dimensional space such as time domain, spatial domain, and spectral domain, which can increase the difficulty for the electronic intelligence system to intercept the radar signal without affecting the radar detection power.

(7) Strong anti-jamming ability. The adaptive array signal processing technology can be used to adaptively adjust the array parameters according to the space–time characteristics of the radar echo, so as to achieve the effect of suppressing jamming and maintaining target information, thereby improving the anti-jamming capability.

(8) Wide-angle scanning of wideband phased array is easily realized. The conventional wideband phased array radar adds real-time delay lines and phase shifters at the level of each element or subarray of the array to realize wide-angle scanning, which is complex and has limited delay accuracy. As for digital array radar, direct digital frequency synthesis (DDS) technology can be used to accurately control the delay during transmission, and digital signal processing technologies such as fractional delay filter can be used to achieve accurate delay at baseband during reception.

(9) It is conducive to non-cooperative target recognition. It takes longer continuous observation time to realize non-cooperative target recognition than to realize target detection. Conventional phased array radar can only increase the observation time of a certain target for recognition at the expense of search data rate, while digital array radar can observe the target for a long time without sacrificing other functions, which provides favorable conditions for target recognition.

Digital array radar is a fully digital phased array radar that adopts digital beamforming technology for both receiving and transmitting. It realizes amplitude and phase weighting in the digital domain to form a digital beam. The basic structure of digital array radar is shown in Fig. 10.2, and its core technology is the all-digital T/R module. The basic working principle of the digital array radar is: in the transmitting mode, the signal processor gives the amplitude and phase values required for the scanning of the transmission beam and sends them to the digital transmitting/receiving module (DTR). The DTR presets the phase and amplitude when the waveform is generated, and then transmits them through the radiating unit for space synthesis after up-conversion and amplification. In the receiving mode, the signal received by

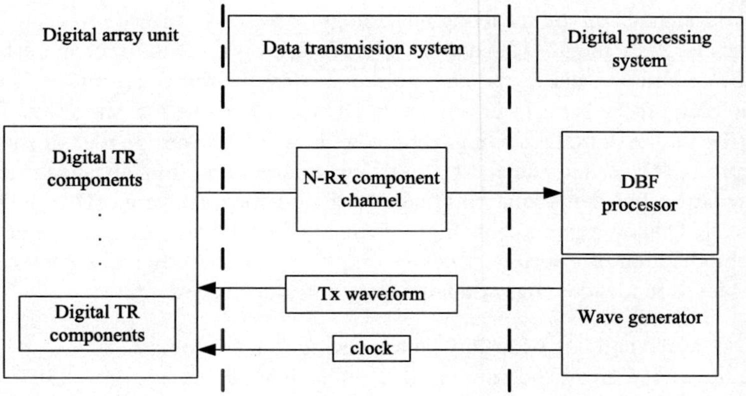

Fig. 10.2 Basic structure of digital array radar

each unit is down-converted, digital down-converter, and then the I/Q signal is sent to the signal processor for DBF processing and conventional signal processing to form the target trace. The digital array radar does not have a complicated feeder network and only has two parts: signal processor and array antenna (composed of DTR). The DTR and the signal processor are connected by optical fiber, so the system has high reconfigurability.

The key technologies for realizing digital array radar mainly include digital T/R module, multi-channel digital receiving based on DDC, high-speed and large-capacity data transmission, integrated transceiver channel design, and high-performance software signal processor.

(1) Digital T/R module
The digital T/R module is the core component of the digital array radar. The digital beam forming module based on direct digital frequency synthesis, which integrates waveform generation and amplitude and phase control, is one of the key technologies of the digital T/R unit. It not only realizes the generation of complex radar signals, but also realizes high-speed, high-precision amplitude, and phase control. On the one hand, the multi-functional working mode requires the radar signal to have a variety of waveforms. In a repetition period, it is often necessary to transmit multiple pulses. For example, pulses of different lengths are emitted in a repetition period to irradiate different long-distance and short-distance targets, respectively. In order to adapt to the energy management of more complex radar signals, it is necessary not only to change the pulse width of the signal, but also to reasonably allocate energy by changing the signal amplitude. This requires that the formation of radar waveform is very flexible, and direct digital frequency synthesis can meet this requirement. On the other hand, the digital array radar is required to transmit digital beams with short beam switching time.

(2) Multi-channel digital receiving technology based on DDC
Because the analog I/Q quadrature phase detector is difficult to achieve satisfactory amplitude balance and phase orthogonality, the mirror signal level is very high, and there is an insurmountable DC drift. For this reason, in the digital T/R unit, the digital receiving technology based on the design idea of software radio can be used to realize I/Q digital quadrature demodulation, and the device used to realize demodulation is called a digital down-converter (DDC). Due to the NCO and programmable high-efficiency digital filter inside the device, the demodulation of a variety of center frequencies and bandwidth signals can be achieved in a wide range when the sampling clock is determined.

(3) High-speed and large-capacity data transmission technology
In the digital array, the number of digital T/R modules is large, so it is a key technology to solve the problem of large-capacity data transmission. There are usually two methods for solving high-speed and large-capacity data transmission: LVDS method and optical fiber transmission method. LVDS is a low-swing differential signal technology that allows signal channel data to be transmitted at the rate of hundreds or even thousands of megabits per second. Its low swing and differential drive mode has the characteristics of low noise and low power consumption, so it is suitable for larger capacity data transmission. However, the transmission range of this method is limited, so LVDS is generally used inside the module, and optical fiber transmission is used outside the module. Optical fiber transmission has the advantages of long transmission range, high transmission data rate, low delay, lightweight, and good security performance. Its transmission data rate can reach more than gigabit.

(4) Integrated transceiver channel design technology
In the digital T/R module, the frequency span is designed from digital to radio frequency. The circuit form includes both small signal and low noise and high-power circuits, both analog circuits and high-speed data acquisition and generation. The traditional circuit has a large volume, which does not meet the requirements, loses the advantages brought by the flexible control of digital T/R, and limits the application range of digital T/R modules. Using integrated radio frequency to digital design technology, the radio frequency circuit and the digital circuit are connected seamlessly, because there is no cable, the impact of size and uncertain factors are reduced. Microwave monolithic integrated circuits, new materials, and new processes should be widely used in circuit realization.

(5) High-performance software signal processor
Digital array radar requires a powerful processing platform for task control, timing generation, correction processing, beam control, target tracking, and display processing. Therefore, it is very necessary to study the high-performance signal processor with bus structure.

10.1.3 Software Radar

The concept of "software radar" originates from the concept of "software radio". The design of the traditional radar signal processing system is based on the specific tasks. The designer determines the algorithm flow and the corresponding algorithm structure, which is aimed at the application background, and then, according to the structure, the designer designs the circuit. This determined circuit structure brings great inconvenience to the reconstruction and function expansion of radar. According to the requirement of radar versatility, a software-based radar signal processor that is easy to reconstruct and expand functions must be established.

Software radar is a new type of multi-band, multi-user and multi-system general and modular radar system that uses broadband smart antennas, RF front-end and general high-speed DSP/CPU system to form a new and standard interface general hardware platform for detection and tracking. It uses software to complete the functions of warning radar, guidance radar, aviation radar, and meteorological radar.

The core idea of the software radar signal processing system is to digitize the received signal at radio frequency or intermediate frequency and realize various m digital filtering, direct digital frequency synthesis, digital down conversion, modulation, and demodulation through software programming. At present, due to the limitation of the device level, most of the implementation of software radio is based on IF digitization and baseband software. The working frequency of radar is generally higher than that of radio station. Therefore, it is difficult to realize RF software-based radar under current technical conditions. In this structure, the more mature IF digital technology is still adopted.

The general structure of software radar is shown in Fig. 10.3. The basic workflow is as follows: the received echo signal is processed by the front-end radio frequency circuit and mixer into the intermediate frequency radar echo preprocessing signal, converted into digital signal by A/D, and formed into the baseband signal to be processed by digital down-converter (DDC). Digital signal processor (DSP) processes various radar signals. In addition, in the signal transmission process, DSP generates a control signal to control the radar modulation signal required by direct digital frequency synthesizer (DDS), and then transmits it after D/A conversion, mixer and RF back-end processing into RF signal. The field programmable gate array (FPGA) in the figure is used to assist the design. The transmission control and analysis processing of radar signals are mainly completed by DSP. If the DSP program is written according to the requirements, the required radar function requirements can be realized.

Radar software is the development trend of radar technology. The key technologies to realize radar software are mainly as follows [14–18].

(1) Broadband RF and antenna technology
An important feature of software radar is reconfigurability; that is, different functions can be realized by configuring different software on the same hardware platform. This requires that the software radar must work within a very wide frequency range; that is, the antenna system and RF components of the

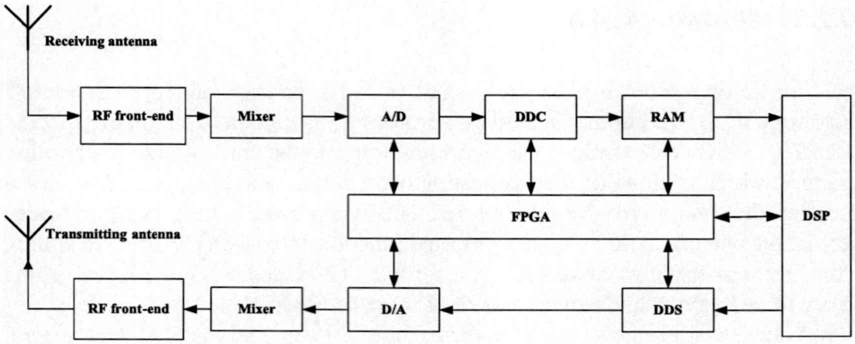

Fig. 10.3 General structure of software radar

software radar must be able to work normally within a wide frequency band, which is technically quite difficult. In addition to researching and developing broadband antennas and radio frequency components, a feasible way to solve this problem is to adopt the combined multi-band antenna and radio frequency system, the reason is that the general radar operating frequency does not cover all frequency bands, but covers several frequency windows in different frequency bands. Therefore, it is feasible to use a combined antenna and radio frequency system.

(2) Real-time data acquisition and storage technology
The software of radar means defining and controlling its work by computer and software, so the real-time acquisition and storage of radar signals becomes a key technique for software radar. Due to the powerful function, versatility and complex form of transmitted signal of software radar, the A/D converter is required to directly digitize the intermediate frequency signal, even the radio frequency signal directly within a wide band range. Besides, in order to ensure the accuracy of the signal processing, A/D converter is required to have higher quantization bits.

(3) General high-performance radar signal processing technology
The reconfigurable characteristics of the software radar determine that its signal processing system must meet the requirements of universality to meet the needs of different functional definitions. In addition, the signal detection of software radar is realized through software. In order to process the digital signal output from the A/D converter in real time, the digital signal processing system must have high-speed processing capabilities. The signal processing of radar system can adopt general computer system or special digital signal processor (DSP).

(4) System software technology
The function of software radar is mainly defined by the corresponding software. Therefore, the system software technology is very important to software radar.

On the one hand, the function of the system hardware needs to be set by the underlying device driver module; on the other hand, the intelligent processing of the radar signal also needs to be completed by the corresponding software. The detection performance of the radar depends on the signal processing capability to a certain extent. Therefore, in order to improve the performance of software radar, it is necessary to research and develop advanced digital signal processing methods and abundant functional software modules.

10.1.4 UWB Radar

Ultra-wideband (UWB) radar is a new system radar. It is defined as: the radar with fractional bandwidth (FBW) greater than or equal to 25% is ultra-wideband radar. Among them

$$\mathrm{FBW} = \frac{2(f_H - f_L)}{f_H + f_L} \tag{10.1}$$

In the formula, f_H and f_L, respectively, represent the upper limit frequency and lower limit frequency of the energy band. The energy bandwidth is a frequency range, which accounts for 90% of the total energy of the signal. Ultra-wideband radar has the advantages of high range resolution, strong penetration, low interception rate, and strong anti-jamming. It is widely used in military, commercial, environmental protection, and other fields.

A typical ultra-wideband radar system is composed of waveform generator, transmitter, receiver, transceiver antenna, and signal processor. Among them, the waveform generator generates ultra-wideband signal waveforms, such as chirp, nonlinear FM pulse, impulse pulse, random noise, and so on.

Compared with conventional narrowband radar system, ultra-wideband radar has the following advantages [19–22]:

(1) Strong anti-jamming performance. Most ultra-wideband radar systems have large processing gain (except impulse radar). When transmitting, the weak signals are scattered in a wide frequency band, and the output power is even lower than the noise generated by ordinary equipment. When receiving, the signal energy is restored, and the spread spectrum gain is generated in the dispreading process. Compared with conventional radar system, it has stronger anti-jamming performance.

(2) Ultra-wideband radar has the characteristics of both low frequency and wide frequency. It also has strong penetrating ability to the ground and leaves and can detect hidden targets in the forest or under the ground.

(3) Radar signal has extremely high range resolution. Due to the large relative bandwidth of UWB radar, it can distinguish many scattering points of the target.

Accumulating the echo signals of these scattering points can improve the signal-to-noise ratio, and its resolution can reach centimeter level.

(4) Good target recognition ability. Due to the short time of radar transmitting pulse, the response of different regions of the target can be separated, and the characteristics of the target can be highlighted, so that it can identify the target.

(5) Ultra-short-range detection capability. The traditional radar has a blind zone when detecting short-range targets. However, ultra-wideband radar using impulse pulse has the ability of ultra-short-range detection because of its short pulse duration.

The key technologies to realize UWB radar mainly include the design and generation of UWB signal, UWB electromagnetic scattering, signal processing algorithm, and receiver design.

(1) UWB signal
The carrier-free signal based on impulse pulse is one of the earliest signals used in UWB radar. Although the range of UWB radar is limited due to the limited average power, this kind of signal is particularly suitable for the study of UWB radar target characteristics. It is necessary to develop high-power nanosecond pulse oscillators to generate impulse signals, such as silicon oscillators with fast recovery photodiodes. The pulse front is about 1.2ns. The power of array units made of these oscillators can reach thousands of watts to hundreds of kilowatts. The disadvantage of this oscillator is that the output pulse is unstable. The oscillator made of gallium arsenide (GaAs) switch tube can overcome this shortcoming. It is an important subject in the research of impulse radar to develop a stable and controllable UWB signal that can produce high power pulse. LFM signal is another UWB signal form, which can achieve hundreds to thousands of gigabytes of bandwidth by frequency modulation within the pulse. LFM signal has been widely used in radar because of its ultra-wide bandwidth and ultra-high processing gain.

(2) Target echo modeling
For UWB radar based on non-carrier frequency, when the target is moving, because the number of scattering centers on the target and the observation angle of the target are uncertain, the multi-time-shifted echo reflected by the target is a series of echoes, rather than a narrowband concentrated echo. The signal form is random, resulting in the unpredictability of the received signal. Therefore, it is not possible to describe the received signal of UWB radar with the concept that the received echo signal is a known signal like narrowband radar. Therefore, it is necessary to model the target echo to study the UWB radar echo.

(3) Signal processing
Signal processing is the key to the development of ultra-wideband radar, and advanced signal processing technology must be used to detect targets in clutter and noise. The echo signal form of narrowband radar is known, and

the maximum output signal-to-noise ratio can be obtained by using matched filtering. In UWB radar, the echo of complex target is the synthesis of time-divided echoes. These echo signals come from each scattering center of the whole target, and each target has unique echo characteristics. Therefore, it is difficult to realize the matching filter processing of UWB radar echo signal. For this reason, it is necessary to study the signal processing methods of UWB radar. In recent years, some new signal processing methods such as wavelet analysis, high-order spectrum analysis, time-frequency analysis, and neural network have been gradually applied to UWB radar and achieved good results.

10.2 Frontier Technology of Radar Active Jamming

Electronic warfare has become an indispensable and important part in modern high-tech warfare, which can often influence the situation of war. The future development of radar active jamming system mainly depends on the development of radar active jamming technology and electronic devices.

Radar jamming technology and anti-jamming technology are always fighting and promoting each other. Modern radars are changing and developing rapidly. Therefore, the development of radar active jamming technology will inevitably be affected by the development of radar and radar anti-jamming technology. By speculating on the combat form of future war and the characteristics of electronic warfare, combing the development trend of radar active jamming systems may include the following aspects.

10.2.1 Cognitive Electronic Warfare

With the continuous progress of science and technology, in the face of the increasingly complex and changeable battlefield electromagnetic environment, the combat effectiveness achieved by conventional electronic warfare means is gradually declining. Focusing on improving the intelligent level of electronic warfare systems, cognitive electronic warfare technology with autonomous perception capabilities, real-time response capabilities, accurate strike capabilities, and evaluation feedback capabilities will inevitably become the development trend of electronic warfare in the future.

Combining cognitive science and electronic warfare technology, cognitive electronic warfare involves a wide range of disciplines. The idea of cognition was first embodied in the field of cognitive radio. Its technical core is to be able to perceive the surrounding environment and adjust and optimize its own working parameters in real time according to the characteristics of the environment. Subsequently, American scientist Simon Haykin integrated the idea of cognition into the field of radar design and proposed the concept of cognitive radar. Since 2009, the US military has gradually introduced the concept of cognition into electronic warfare equipment in order to improve the effectiveness of electronic warfare, marking the formation of the concept of cognitive electronic warfare.

In general, cognitive electronic warfare can be understood in this way: in order to cope with the intelligent development of threat targets, the increasing complexity of battlefield electromagnetic environment, and the continuous emergence of new waveforms, it is a new way and new idea of electronic warfare. Cognitive electronic warfare can meet the needs of independently predicting, discovering, identifying, confronting, and evaluating any threat at any time and any place in a harsh defense environment. Cognitive electronic warfare system should have the ability of real-time battlefield environment perception and learning, intelligent selection of the best jamming measures, and real-time evaluation and feedback of the effectiveness of jamming measures. It is an intelligent, dynamic, large closed-loop adaptive system that people can participate in.

Compared with conventional electronic warfare, cognitive electronic warfare has significant advantages in real-time and accurate perception of battlefield electromagnetic situation, dynamic learning, experience accumulation, intelligent decision making, effectiveness evaluation, effective countermeasures against cognitive systems, and concealment and survivability. Cognitive electronic warfare has extremely broad application prospects.

The realization of cognitive electronic warfare system requires three functional modules: cognitive reconnaissance module, countermeasure synthesis module, and countermeasure effect evaluation module, as shown in Fig. 10.4. The cognitive electronic warfare system first conducts adaptive reconnaissance and perception of the target object and the surrounding environment, quickly and accurately analyzes the available knowledge from the massive data received, and then intelligently selects or synthesizes the best electronic attack measures, after that, evaluates the attack effectiveness through further perception, and finally guides the next electronic attack of the system according to the evaluation results.

By perceiving the electromagnetic environment of the battlefield, the cognitive reconnaissance module intercepts the signals of the target and its surrounding environment, carries out signal processing and analysis processes such as measurement, classification, feature extraction and recognition, and then extracts the core parameter features describing the current environment, forms feature description data, and transmits them to the countermeasure synthesis and effect evaluation module.

The countermeasure synthesis module searches for the best jamming strategy by analyzing the signal characteristics and combining the learning information in the knowledge base; at the same time, it allocates jamming resources and optimizes the jamming waveform, and then implements jamming to the target.

The countermeasure effect evaluation module quantitatively analyzes the countermeasure effect according to the changes of target signal characteristics before and after the implementation of jamming, obtains the effectiveness evaluation result of the current jamming measure, optimizes the countermeasure strategy, and then promotes the synthesis of the next round of countermeasures.

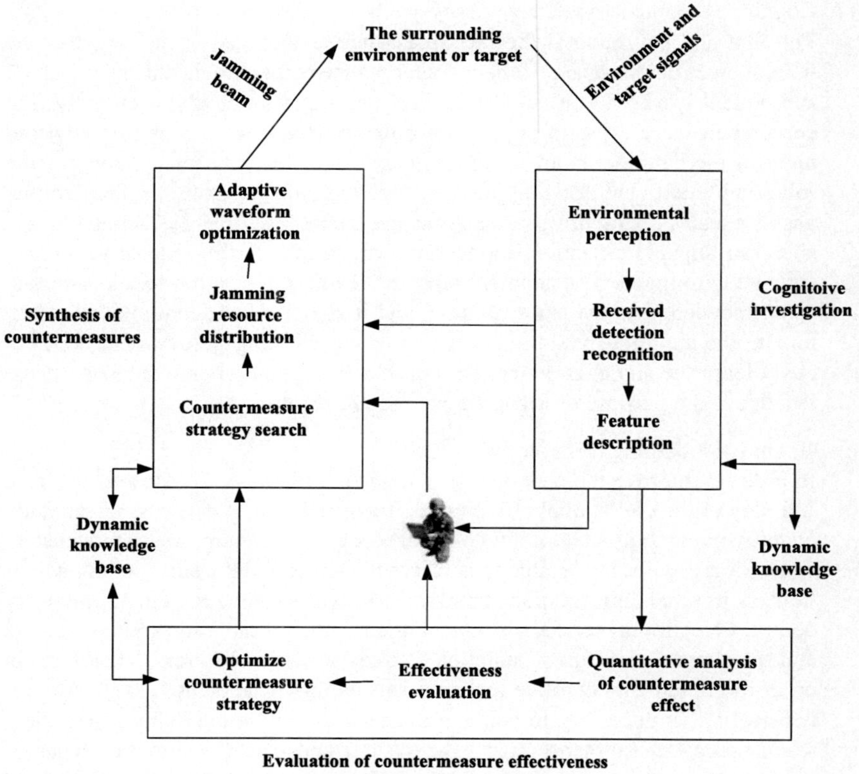

Fig. 10.4 Composition diagram of cognitive electronic warfare

Although the cognitive electronic warfare system has certain intelligence capabilities and can autonomously adjust the best countermeasures according to different target states, the battlefield environment is intricate and complex, and the system needs to be controlled according to the actual situation of the battle or the set tactical strategies, which requires the operator to achieve the overall control of the cognitive electronic warfare system through man–machine interaction.

In addition, the construction of knowledge base is also indispensable in cognitive electronic warfare system, and the three functional modules must have corresponding dynamic databases that can update knowledge in real time, so that the system can use these knowledge bases to quickly obtain information in the working process, and use feedback and newly captured information for cognitive learning to dynamically update the knowledge base.

The following key technologies are required to realize cognitive electronic warfare [23–25].

(1) Cognitive reconnaissance technology
The first and key thing is the reconnaissance technology of the target in the field of electronic warfare. If the reconnaissance perception is not sufficient, all subsequent work will be aimless. Therefore, only by quickly, accurately, and comprehensively capturing useful information from the surrounding environment of the battlefield can we provide necessary information support for the follow-up intelligent decision making, confrontation generation, and effectiveness evaluation. Cognitive reconnaissance technology mainly includes threat signal sorting, identification, and feature extraction in the high-density complex signal environment. The adaptive machine learning algorithm requires certain prior knowledge as the basis training, and it also needs to continuously accumulate the captured new threat signals in the working process and continuously learn the signal knowledge accumulated in the dynamic database, so as to achieve the purpose of improving cognitive ability.

(2) Cognitive modeling technology
In order to improve the perceived efficiency of cognitive EW system and give full play to the operational effectiveness of the system, it is necessary to study cognitive modeling technology. When the cognitive electronic warfare system works, it is required to be able to perceive the target and the surrounding battlefield environment information in real time. However, in the current high-density complex electromagnetic environment, the number of radiation sources is huge and the signals of different radiation sources are quite different. Therefore, in order to implement cognitive reconnaissance quickly, accurately, and comprehensively, it is necessary to conduct dynamic cognitive modeling of the electromagnetic environment around the system and describe different types of information through a unified model architecture. The information described can be divided into static parameter information and dynamic parameter information. The contents involved include frequency, repetition frequency, direction of arrival, bandwidth, waveform characteristics, protocol, electronic protection mode, and function intention, etc.

(3) Jamming measure synthesis technology
Jamming measure synthesis technology is the key link to implement electronic attack on target object. The key to the breakthrough of this technology also lies in the design of software algorithm. The intelligent idea is introduced to develop an intelligent optimization algorithm for jamming measure synthesis. During the design of the algorithm, it is necessary to consider the relationship between the threat level of the target and the synthesis of countermeasures, the setting of jamming parameters of different types of countermeasures, and the intelligent generation or selection of jamming strategies.

(4) Effectiveness evaluation technology
In the cognitive electronic warfare technology, it is necessary to develop a new effectiveness evaluation method, that is, a technology to evaluate the effectiveness of jamming according to the changes of signal characteristics and working

state of the target before and after jamming. The research of this technology needs to analyze and summarize the various working states of various targets and the characteristic parameters in different working states in detail, so as to form an intelligent reasoning mechanism, so as to infer the current state of the target, even the intention of the target, and then guide the optimization of the synthesis of jamming measures to achieve the best jamming effect.

10.2.2 Distributed Cooperative Jamming

With the continuous development of modern radar technology, the continuous improvement of radar performance and the application of modern radar networking technology, the requirements for jammers are becoming higher and higher. In many cases, it is difficult for a single jammer or multiple jammers that do not work together to effectively jam a single radar or netted radar with excellent performance. For example, a single jammer is limited by the working frequency band, jamming power, and jamming style, which cannot effectively jam the networked radar, nor can it jam a single high-performance radar; under the non-cooperative working mode, multiple jammers will cause insufficient jamming to the radar with high threat level, and there will also be repeated jamming to several radars with low threat level, resulting in a waste of jamming resources. Distributed cooperative jamming technology is an efficient "face-to-face" jamming method that combines multiple jammers organically and realizes a certain working mode. According to the battlefield electromagnetic situation, determine the number of jammers, spatial distribution mode, adopted jamming style, jamming frequency, jamming power, jamming beam direction and other strategies, and conduct cooperative jamming on a key radar and radar network, so as to achieve better jamming effect.

The schematic diagram of distributed cooperative jamming is shown in Fig. 10.5. It is assumed that there are multiple radars and jammers in the scene, and the distributed cooperative jamming is composed of jamming control center, multiple jammers, and communication links. The workflow of the distributed cooperative jamming system is as follows: the reconnaissance equipment of each jammer reconnoiters the electromagnetic signals at the deployment position. After necessary preprocessing, it is uploaded to the control center through the communication link. The control center performs information fusion, comparison and verification of the reported signals, comprehensively studies and judges the battlefield electromagnetic situation, determines the threat level of the target signal, and formulates the jamming scheme according to a certain cooperative jamming strategy, including the target, jamming style and jamming parameters of each jammer responsible for jamming, and distributes information and instructions through the communication link. Each jammer performs cooperative jamming according to the information or instructions sent by the control center.

The cooperation strategy among multiple jammers mainly includes the following aspects: time domain cooperation, multiple jammers cooperate in time domain,

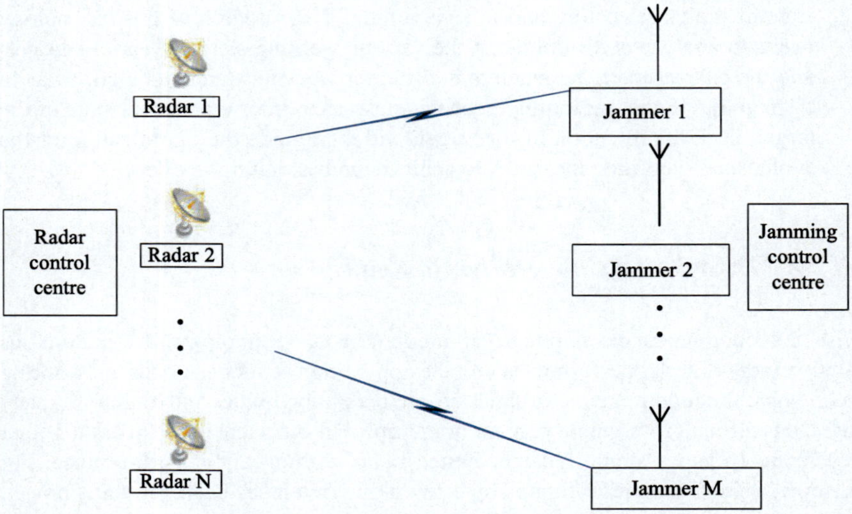

Fig. 10.5 Schematic diagram of distributed cooperative jamming

which can better cover radar signals in time domain; frequency domain cooperation, multiple jammers cooperate in frequency domain, which can cover the working frequency of distributed radar more effectively and flexibly; airspace cooperation, cooperative jamming can timely adjust the direction of jamming beam according to the demand, and effectively cover important radar targets; jamming style cooperation, cooperative jamming can select the most effective jamming style combination according to the demand to maximize the jamming efficiency; energy domain cooperation, cooperative jamming can concentrate jamming resources to focus on high-threat radar targets. Therefore, distributed cooperative jamming is an important development direction of radar jamming in the future.

The key technologies to realize distributed cooperative jamming mainly include [26–29]:

(1) Reconnaissance information fusion technology. Multiple jammers reconnaissance the electromagnetic environment at the deployment location. After necessary pretreatment, they report to the control center, which fuses, compares and verifies the reported reconnaissance information to obtain an accurate and objective description of the battlefield electromagnetic environment.

(2) Comprehensive research and judgment technology of battlefield electromagnetic situation. After the reconnaissance information fusion processing, the control center conducts comprehensive research and judgment on the battlefield electromagnetic situation according to the prior information such as the threat signal database or the operator's manual intervention, selects the jamming target signal, and sorts the threat level.

(3) Cooperative jamming strategy. According to the distribution characteristics and performance parameters of the jammer and the electromagnetic situation of the target signal that needs to be interfered, the cooperative jamming strategy is formulated, including the airspace cooperative strategy, the frequency domain cooperative strategy, the time domain cooperative strategy, the energy domain cooperative strategy and the jamming style cooperative strategy.

(4) Cooperative jamming effect evaluation technology. Jamming effect evaluation is the main basis for judging jamming effect, adjusting jamming strategy and optimizing jamming resource allocation. Distributed cooperative jamming is a new system-to-system jamming mode. The traditional jamming effect evaluation method of single jammer jamming single radar has been difficult to continue to apply. It is necessary to study the jamming effect evaluation technology of this system against the system.

10.2.3 Microminiature

With the further development of semiconductor integrated circuit technology, which is reduced in proportion to the feature size, silicon complementary metal oxide semiconductor (CMOS) technology is limited by a series of basic physical characteristics, physical scale limits, input–output cost performance ratio in terms of speed, integration, power consumption, and investment cost, and Moore's law is challenged. Blindly pursuing the transistor density and chip scale of integrated circuits brings huge technical and economic risks. The actual meaning of More than Moore is to jump out of Moore's law blindly pursue smaller process feature size, constantly challenge process limits and improve circuit integration, and advocate using the existing mature semiconductor technology to pursue chip integration of diversified functional circuits and system-level packaging integration of multiple chips. It is reported that in 2012, a joint research team composed of 9 research institutions and enterprises in the EU developed a miniature high-frequency radar, whose surface is equivalent to the size of the fingernail, about 8 $\times$ 8 mm, the working frequency is 120 GHz, and the detection distance is 3 m. Doppler effect can be used to measure the velocity of moving objects. It has a wide application prospect in automobiles, mobile electronic equipment, robots, and other fields. Radar technology and radar jamming technology are interlinked in many aspects. This kind of high frequency radar with fingernail cover size also provides a good reference for the miniaturization of radar jamming equipment.

Microminiature technology is a comprehensive technology that integrates architecture, algorithms, microelectronics and other elements, adopts new design ideas, design methods and manufacturing processes, and integrates some or all functions of the system. The miniaturization of radar jamming system is not only conducive to reducing the loading space, reducing the pressure on the platform and improving the efficiency by using the quantitative advantage in combat, but also can bring many benefits such as lower price, lighter carrying and more energy saving, and so on. The miniaturization of radar active jamming system has broad application prospects in airborne, spaceborne, missileborne, unmanned aerial vehicle (UAV), separated decoy and other platforms, as well as in operational occasions such as one-time launch of jammers.

The miniaturization development of radar active jamming system mainly includes: using small antennas and batteries to reduce the size and weight of the system; chips are developing toward smaller nanoscale processes; improving the power efficiency of digital signal processing and RF system; multi-functional integration of sensors, RF circuits, passive circuits, high-power microwave circuits and energy electronics. The development of miniaturized systems will involve micro-jammer architecture, high-speed digital and analog processing, three-dimensional high-density integration, electromagnetic compatibility, new materials, and other technologies.

The miniaturization of radar active jamming system has many technical similarities with general integrated circuits. The key technologies involved mainly include the following:

(1) Microelectromechanical system (MEMS) technology
MEMS technology is a typical semiconductor process integration technology developed in accordance with the so-called More than Moore concept, it is also the basic technology to realize microsystems. It has developed rapidly in recent years. RF MEMS is the most widely used MEMS product in radar countermeasure equipment, mainly including MEMS capacitors, inductors, transmission lines, switches, filters, phase shifters, switch filter combinations, switching networks, voltage control oscillators, etc. These devices based on MEMS technology have the characteristics of small size, low loss, and low power consumption and have good application prospects.

(2) System-on-Chip (SoC) technology
System-on-Chip (SoC) refers to the system on chip that uses CMOS technology to process and store information in digital domain such as central processing unit (CPU), memory and logic circuit. With the development of the semiconductor process technology, the system-level integration of RF front-end and digital baseband has been realized on a homogeneous single substrate, namely RF SoC. This new concept product has been able to replace some microwave multi-chip modules (MMCM), which will greatly reduce the number of devices and the area and volume of the whole system. In recent years, 3D RF integrated circuits of More than Moore have gradually matured, using ultra-large-scale integration silicon-based technology and through-silicon-via (TSV) technology to achieve vertical interconnection 3D integration, increase integration density, and obtain more powerful and smaller on-chip systems.

(3) System in package (SiP) system
System in package (SiP) refers to the integration of traditional non-digital and diversified functional circuits such as analog/RF, passive, high voltage/power supply, sensing excitation and biological logic, and multi-functional system-level hybrid packaging based on this. SiP usually refers to a technology that

integrates multiple chips and components of different materials in one package to form a system function module. In radar countermeasure equipment, functional chips and devices, such as antenna reception, ADC/DAC, and microwave signal limiting, filtering, distribution, amplification, control, and quantification are often realized in different substrate materials. Through system simulation design, the above functions are integrated into a physical entity by packaging process technology to realize RF system-level packaging.

10.2.4 Comprehensive Countermeasures

With the continuous progress of radar technology, radar anti-jamming technology has developed in the direction of integration, such as phased array scanning, single pulse tracking, sidelobe cancelation, pulse compression, coherent accumulation, variable carrier frequency, variable repetition frequency, variable pulse width, variable waveform, and so on. These mutually compatible anti-jamming measures can be comprehensively applied in a radar and a working mode, greatly improving the comprehensive anti-jamming ability of radar. In addition, the anti-jamming technology of multiple radars has developed in the direction of networking. The networked radars cross-cover in spatial domain, time domain and frequency domain, and the detection information is interconnected, complementary, and applied to each other, which greatly reduces the jamming efficiency of the traditional "one-to-one" (i.e., one jammer against one radar). Therefore, comprehensive confrontation is an important development direction of radar countermeasures.

Comprehensive confrontation is to reduce or weaken the working efficiency of enemy radar equipment or system, and comprehensively utilize various jamming measures compatible with each other to interfere with enemy radar equipment or system. The connotation of comprehensive confrontation includes the following contents:

(1) Comprehensive use of suppression jamming and deception jamming. When releasing the noise suppression jamming, it also releases the false target deception jamming. The suppression jamming covers and disturbs the echo information of the true target. The deception jamming makes it difficult for radar to distinguish between true and false targets, which further improves the jamming effect.

(2) Comprehensive use of active jamming and passive jamming. Let the radar be under active and passive dual jamming, so that some effective anti-jamming measures against passive jamming of the radar are damaged by active jamming, or effective anti-jamming measures against active jamming are powerless against passive jamming.

(3) Comprehensive use of intra-platform jamming and extra-platform jamming. For example, the integrated application of the gate pull-off jamming inside

the platform and the false target jamming outside the platform is easy to lure the radar terminal guidance and tracking system to the false target outside the platform.

(4) Comprehensive use of electronic jamming and hard destruction. While releasing decoys and false targets to lure radar to power on, they use anti-radiation weapons to destroy the radar.

(5) Comprehensive use of squad escort jamming and platform self-defense jamming. The squad escort jamming reduces the pressure of the platform's self-defense jamming and enables it to concentrate on the main targets that threaten the safety of the platform.

(6) Comprehensive use of high-power centralized jamming and low-power distributed jamming. The high-power centralized jamming is used to cover the launch of the low-power distributed jammers, and then the low-power distributed jammers close to the enemy position are used to cover the penetration of attack aircraft groups.

(7) Comprehensive use of long-range support jamming and target stealth technology. Target stealth reduces the required power of long-range support jamming and increases the effective cover airspace of long-range support jamming. Long-range support jamming reduces the use time of self-defense jamming of stealth targets and reduces the exposure probability of stealth targets.

(8) Comprehensive use of radar, communication, and navigation and other jamming technologies. In the implementation of military tasks, radar, communication, and navigation are all electromagnetic support for the successful completion of the task. Only one aspect of jamming is difficult to work. Comprehensive use of radar, communication and navigation and other jamming technologies can greatly improve the success probability of jamming.

References

1. Li XQ et al (2008) Integrated electronic warfare – killer mace in information war, 2nd edn. [M]. National Defense Industry Press, Beijing
2. Li X, Fan M (2012) Research progress on cognitive radar and its key technologies [J]. J Electron 40(9):1863–1870
3. Bin W, Jinkuan W, Xin S et al (2011) Adaptive waveform selection algorithm based on Q learning in cognitive radar [J]. Syst Eng Electron Technol 33(5):1007–1012
4. Meimei F, Dongping L, Xiaofeng D (2012) Multi-target recognition cognitive radar waveform adaptive method based on WLS-TIR [J]. J Electron 40(1):73–77
5. Jin L (2013) An Overview of Intelligent Cognitive Radar [J]. Mod Radar 35(11):6–11
6. Jun S (2014) Key technologies in intelligent cognitive radar [J]. Mod Radar 36(10):14–19

7. Li T, Wang X, Zhou Y (2015) Time domain beamforming method for large aperture broadband digital array [J]. Mod Radar 37(6):21–25
8. Wu M (2008) Development and conception of digital array radar [J]. Radar Sci Technol 6(6):401–405
9. Romero RA, Goodman NA (2011) Adaptive beamingsteering for search-and-track application with Cognitive Radar Network [C]. In: IEEE Radar conference, pp 1091–1095
10. Wu M (2006) Digital array radar and its progress [J]. J Chin Acad Electron Sci 1(1):11–16
11. Zengping C, Yue Z, Qinglong B et al (2010) Advances in digital array radar and its key technologies [J]. J Nat Univ Defense Technol 32(6):1–7
12. Wu Y, Li W, Zhang Y et al (2015) Design of aroadband digital array radar receiving system based on OpenVPX Platform [J]. Signal Process 31(11):1484–1488
13. Weiqing Z, Jianmei T, Han C (2008) Application of DDS in digital array radar [J]. Radar Sci Technol 6(6):467–471
14. Yanbin S, Zhemin D (2005) Overview of research on software radar technology [J]. Avionics Technol 36(4):50–53
15. Dong X, Hao C, Wei Z et al (2012) Software radar based on software product line technology [J]. Mod Radar 34(4):32–35
16. Tang J, Wu H, Wei K et al (2015) Software radar technology research [J]. J Radar 4(4):481–489
17. Gongyou Z, Shuanghua H, Deqiang W et al (2009) Software radar bus technology [J]. Ship Electron Eng 29(3):86–87
18. Lai J, Li N, Tang J et al (2008) A kind of software architecture of software radar [C]. In: The papers of the tenth national radar academic annual conference, pp 322–326
19. Taylor JD (2001) Ultra-wideband radar technology [M]. CRCPress, London
20. Li H, Yang R (2001) The development, present situation and application of ultra-wideband radar [J]. Remote Sens Technol Appl 16(3):178–183
21. Huang Q, Wu S, Meng S et al (2011) The moving human target tracking imaging algorithm based on UWB radar [J]. J Electron 39(3):531–537
22. Yang L, Xu Z, Su W et al (2010) Ultra-wideband radar clutter modeling and simulation based on subband synthesis [J]. J Electron Inform 32(9):2172–2178
23. Ni S, Long X, Wei X (2011) Analysis of the development of cognitive technology in electronic warfare equipment [J]. Electron Inf Countermeasure Technol 26(6):22–26
24. Jia X, Zhu W, Qu W (2015) Concept and key technology of cognitive electronic warfare [J]. J Equipment Inst 4:96–100
25. Zhou F, Xu T (2010) High solution methods for electronic counter measures environments establishing and side lobe cancellation in cognitive radar [C]. In: IEEE International conference on intelligent computing and intelligent systems, pp 805–808
26. Jun H, Hong D, Yi M et al (2011) Research on cooperative jamming task allocation based on distributed optimization [J]. Comput Eng 37(21):264–266
27. Ge L (2015) Research on array characteristics in distributed regional precise jamming [J]. Wireless Commun Technol 41(1):67–70
28. Jieyu C, Peiyang Y, Jian T et al (2014) Multi-UAV distributed collaborative dynamic target allocation method [J]. J Air Force Eng Univ 6:11–16
29. Sang H, Li Z (2001) Principles of smart antenna, research progress and application of adaptive beamforming algorithm [J]. J Natl Defense Univ 23(6):85–91